FROM MANTLE TO METEORITES
A Garland of Perspectives

A Festschrift for Devendra Lal

FROM MANTLE TO METEORITES
A Garland of Perspectives

A Festschrift for Devendra Lal

Edited by

K GOPALAN
V K GAUR
B L K SOMAYAJULU
J D MACDOUGALL

Indian Academy of Sciences
Bangalore

Published by

Indian Academy of Sciences
P.B. No. 8005
C V Raman Avenue
Bangalore 560 080

Telephone: 34 25 46   Telex: 0845-2178 ACAD IN

Distributed by
Oxford University Press
YMCA Library Building, Jai Singh Road, New Delhi 110 001
Bombay Calcutta Madras
Oxford New York Toronto
Melbourne Tokyo Hong Kong

ISBN 0-19562581-1

Copyright © 1990 by the Indian Academy of Sciences

Typeset at Thomson Press, New Delhi and printed by Macmillan India Press, Madras.

CONTENTS

.... ix

.... P B Price xiii

... forces in nature .... Ramanath Cowsik 1

...anetesimal accretion ....
.... D A Mendis and G Arrhenius 29

...e effects: Applications to the pre-solar nebula,
...terstellar space.......................................
... Mark H Thiemens and S K Bhattacharya 35

...uclides: What can we do with a few atoms?
.... James R Arnold 61

...atory .... Helmut Moritz 67

...orm mass spectroscopy of xenon..........
.... P McConville and J H Reynolds 71

...ns of new collections of Greenland and
...aurette, Claus Hammer and Michel Pourchet 87

...errestrial system .... G Cini Castagnoli 127

...time .... Bruce A Bolt 137

...netary degassing based on ^{40}Ar systematics
.... Karl K Turekian 147

Rare gases and hydrogen in native metals.........................
.... R J Poreda, K Marti and H Craig 153

Emanation of radon from rock minerals .... Rama 173

Several considerations on the early history of the Earth .... E M Galimov 177

Life on the early Earth: Bridgehead from Cosmos or autochthonous phenomenon?..................................... Manfred Schidlowski 189

Continued

**From Mantle to Meteorites:
A Garland of Perspectives**

A Festschrift for Devendra Lal

Edited by **K. Gopalan**, National Geophysical Research Institute, Hyderabad, India, **V. K. Gaur**, Department of Ocean Development, Delhi, **B. L. K. Somayajulu**, Physical Research Laboratory, Ahmedabad, India, and **J. D. MacDougall**, Scripps Institution of Oceanography, La Jolla, USA

Professor Devendra Lal, a distinguished physicist and former Director of the Physical Research Laboratory in Ahmedabad has been widely honoured for his contributions to science by various bodies both within India and abroad. This volume with contributions from his colleagues and admirers is presented as a tribute to this versatile and evergreen scientist on the occasion of his sixtieth birthday.

Contributors: P.B. Price; R. Cowsik; D.A. Mendis; G. Arrhenius; M.K. Thiemens; S.K. Bhattacharya; J.R. Arnold; H. Moritz; P. McConville; J.H. Reynolds; M. Maurette; C. Hammer; M. Pourchet; G.C. Castagnoli; B.A. Bolt; K.K. Turekian; R.J. Poreda; K. Marti; H. Craig; E.M. Galimov; M. Schidlowski; T-H. Peng; T-L. Ku; J. Southon; C. Measures; W.S. Broecker; P.K. Das; A.P. Dube; T.N. Narasimhan; S.K. Banerjee; P.J. Wyllie; J. Akella & S.K. Sharma.

342 pages, halftones, line drawings and tables throughout OUP India/Indian Academy of Sciences March 1991
0-19-562581-1 £12.95 Paper covers

Factors controlling the distribution of ^{10}Be and ^{9}Be in the ocean .... *Tsung-Hung Peng, Teh-Lung Ku, John Southon, C Measures and W S Broecker*   201

Asymptotes of coastal upwelling .... *P K Das and A P Dube*   205

Hydrogeology. A new focus ...., ... *T N Narasimhan*   213

Geomagnetic paleointensity from quaternary sediments: Methodology, results and future prospects .... *Subir K Banerjee*   229

Steps toward understanding the Earth's dynamic interior .... *Peter J Wyllie*   239

Application of diamond-anvil-cell technique to study *f*-element metals and some materials relevant to planetary interiors .... *Jagannadham Akella*   249

Applications of Raman spectroscopy in earth and planetary sciences .... *Shiv K Sharma*   263

Subject index   309

Author index   314

List of Publications of D Lal

Devendra Lal

INTRODUCTION

For a physicist, the best chemist there is

In the above words, did Professor J R Arnold describe Devendra Lal, his friend and colleague of many years, on the occasion of Lal's election as a Foreign Associate of the US National Academy of Sciences. Indeed, there are many qualities responsible for Lal's distinguished scientific career spread over nearly four decades. Even a casual acquaintance will not fail to notice his obvious genetic endowments: a questing mind, tenacity of purpose and infinite energy. But his real strength stems from what has been so incisively summed up by Professor Arnold.

Professor Lal combines a deep understanding of the physics of high energy particle interactions with a rare insight into the evolution of the tenuous trails which they leave behind in naturally occurring chemical systems. Thus, by tracking down the exotic products of these high energy interactions into the specific niches of various low energy chemical reservoirs, he has made pioneering contributions to such diverse fields as astrophysics, planetary evolution, oceanography and archaeology. This volume contributed by his colleagues and admirers is presented as a tribute to this versatile and evergreen scientist on the occasion of his sixtieth birthday.

Devendra Lal was born in the ancient city of Varanasi (Banaras) on 14 February 1929. He was educated in the same city, receiving his Masters Degree in Physics in 1949 from Banaras Hindu University. With the overwhelming appeal of high energy physics to bright young physicists in those days, the young Lal naturally gravitated to the then newly established national centre for such studies, the Tata Institute of Fundamental Research in Bombay. Here, he began his research career under the stimulating guidance of Prof. Bernard Peters, studying high energy cosmic ray interactions through their tracks recorded in nuclear emulsion plates. Even in the very early years of his research in this highly competitive field, Lal made some very important discoveries such as the associated production of K-mesons and negative K-mesons.

Prompted perhaps by the new and important discovery by Prof. W F Libby of cosmic ray-produced C-14 in the atmosphere, Lal, following the suggestion of Professor Peters, began to look for the presence of cosmic ray-produced Be-10 in ocean sediments. Since this isotope has a long life of 1·5 million years, its variation with depth in these sediments could be used to reconstruct the history of cosmic radiation in the near earth environment over this scale of time. Realizing the far-reaching significance of these investigations in tracing the evolution of various earth processes, Lal took perhaps the most crucial decision in his professional life—to switch from what he could do *for* cosmic rays to what he could do *with* them. His marriage at this time with a charming and gracious young lady, Aruna, was a happy coincidence. As Professor Price recalls in the following article, Aruna's emotional and intellectual identification with Lal's scientific interests contributed in no small measure to his scientific achievements.

The chemical isolation of Be-10 from ocean sediments and detection of its extremely weak beta radioactivity (much weaker than that of C-14) in 1957 was an important scientific achievement. In the following years, Lal and Rama together with a team of young dedicated workers toiled night and day to isolate and detect many other isotopes of widely different half-lives and chemical affinities, which are produced in the atmosphere but quickly sequestered in different chemical reservoirs of the earth system. The list includes Be-7, Be-10, P-33, Mg-28, Cl-38, S-38 and Si-32, the last one discovered in collaboration with Prof. E D Goldberg during Lal's first visit to Scripps Institution of Oceanography in 1958. In a classic paper, Lal and Peters worked out the production rates of these isotopes in the atmosphere to serve as a basis for their use as probes of atmospheric, hydrological and oceanic processes. A crucial finding was the usefulness of Si-32 for determining the residence time of water in deep seas. Around this time, Lal also organized at the Tata Institute of Fundamental Research the first Indian laboratories for radiocarbon dating and tritium measurements for application to archaeological and hydrological research.

If cosmic ray interaction with a simple and diffuse gaseous mixture like the terrestrial atmosphere could spawn some dozen radioactive isotopes, its interaction with chemically more complex solid objects should produce a far greater variety of isotopes with a wide range of half-lives. This realization led Lal, Arnold and Honda to develop a new and extremely fruitful line of research on the detection and significance of radioactive isotopes produced by cosmic ray bombardment of atmosphere-free extra-terrestrial objects such as meteorites and the moon.

With his early training in the use of nuclear emulsion plates to record cosmic ray interaction events, Lal was quick to realize the potential of the discovery of microscopic damage trails in rocks and minerals due to passage of high energy uranium fission fragments by R L Fleischer, P B Price and R M Walker. He and his colleagues saw in it an opportunity to unfold the records of ancient cosmic rays preserved in meteorites and lunar surface materials. Through imaginatively designed experiments to study cosmic ray-produced radioactivity, implanted atoms as well as tracks in key meteorites and lunar soils, they illuminated a large number of unexplored vistas. Notable examples are the temporal variation in the flux and energy spectrum of cosmic rays back to the beginning of planetary formation, evolutionary history of meteorites and lunar soil column and solar radiation history of silicate grains before their compaction into what are known as gas-rich meteorites.

While still actively involved in the study of meteorites and lunar samples, Lal became deeply interested in the problems of modern oceanography following his joint faculty appointment in 1967 at the Scripps Institution of Oceanography. Thus began his collaboration with Professors H Craig, H E Suess, G Arrhenius and other leading lights in oceanographic research, and his nurturing of an active group of young scientists in this country to address themselves to significant problems in oceanography. The GEOSECS programme set up to acquire coordinated physico-chemical data from the oceans also appeared at this juncture. Lal's participation in this programme led to the investigation of the large scale circulation of oceans using Si-32 as a tracer, bottom water diffusion employing $Fe(OH)_3$ impregnated acrilon fibres to concentrate key trace elements and isotopes, and the role of micron-size particulate matter in transporting elements from surface to deep waters.

Lal succeeded Vikram Sarabhai as Director of the Physical Research Laboratory (PRL), Ahmedabad a constituent laboratory of the Indian Department of Space, in

Introduction

1972. This placed new responsibilities, not all scientific, on his shoulders. But, whatever he touched bore the imprint of his vigorous mind and unflagging spirit. He restructured the ongoing programmes of this laboratory until then primarily in atmospheric and nuclear physics, and started new programmes in experimental plasma physics, climatology and infra-red astronomy. He also brought his former research group to PRL to address basic questions in the earth and planetary sciences.

Professor Lal relinquished the Directorship of PRL in 1983 to return to full time and 'hands on' research. In the years following, he has resumed his earlier interest in tracing the evolution of cosmic ray-produced isotopes in terrestrial settings with renewed vigour, using the powerful tool of accelerator mass spectrometry. Characteristically, his investigations are addressed to a host of exciting problems such as the links between C-14 and geomagnetic variations, direct measurement of the erosion rates of rocks and nutrient pathways in the oceans. He wears his sixty years of terrestrial residence very lightly indeed, and if anything, has redoubled his activity.

Professor Lal has been widely honoured for his contributions to science by various bodies both within India and elsewhere. He was awarded the M S Krishnan Gold Medal by the Indian Geophysical Union (1965), the S S Bhatnagar Prize by the Council of Scientific and Industrial Research (1967), a medal for outstanding scientist by the Federation of Indian Chambers of Commerce and Industry (1974) and the Jawaharlal Nehru Medal for science and technology by the Government of Madhya Pradesh (1986). In recognition of his outstanding contributions to science, Professor Lal was elected Fellow of the Indian Academy of Sciences (1964), Fellow of the Indian National Science Academy (1971), Foreign Associate of the National Academy of Sciences, USA (1975), Fellow of the Royal Society of London (1979), Founding Member of the Third World Academy of Sciences, Italy (1983), and Fellow of the American Academy of Arts and Sciences (1989). The Government of India conferred on him the National Honour of Padma Shri in 1971.

Professor Lal has served numerous national and international scientific bodies with distinction—as President, International Union of Geodesy and Geophysics (1983–87); President, International Association for Physical Sciences of the Ocean (1979–83); Executive Committee Member, International Council of Scientific Unions; Vice-President, Indian Academy of Sciences; Member, Indo-US Joint Commission on Science and Technology; Member, Indo-Soviet Joint Commission on Earth Sciences and Member, International Association of Geochemistry and Cosmochemistry. He has also served on the editorial board of leading journals: Moon and Planets, Earth and Planetary Science Letters, Space Science Reviews and Proceedings of the Indian Academy of Sciences on Earth and Planetary Sciences.

We have had a most pleasant but difficult task in bringing out a commemoration volume of scientific papers contributed by Prof. Lal's friends and admirers on his sixtieth birthday, which would be a fitting tribute to his vigorous intellect. Although a volume of invited articles on a fairly general theme still remains the best way of honouring a highly regarded fellow-scientist, we are keenly aware of the increasing concern of active scientists in contributing papers to festschrift publications in view of their limited reach and inevitable delays entailed in such publications. We therefore chose to invite authors to contribute papers outlining their futuristic perceptions and visualizations of unchartered areas. We felt that this would have a special appeal to young enquiring scientists who may prefer the glimpse of a half-formed flower to a guided tour of a well-manicured garden. Moreover, restrained speculation on conceptual and technical advances coupled with a vision of new directions would inspire the younger generation to leave the trodden path for the venturesome and the unknown. The response has been truly gratifying and covers a wide range of topics from the earth's **mantle** to meteorites and beyond, many of which Professor Lal himself has done so much to enrich. Most affectionately do we present this garland of perspectives to Professor Lal and join the authors of this volume as well as other friends and colleagues throughout the world in wishing Devendra and Aruna many more years of their truly remarkable and enviable companionship and enduring contributions to science.

National Geophysical Research Institute, Hyderabad   K Gopalan
Department of Ocean Development, New Delhi.   V K Gaur
Physical Research Laboratory, Ahmedabad.   B L K Somayajulu
Scripps Institution of Oceanography, La Jolla.   J D Macdougall

A list of selected publication of D Lal appears at the end of the volume

A friendship with Lal

P B PRICE
Physics Department, University of California, Berkeley, California 94720, USA

In 1964, a couple of years after Bob Walker, Bob Fleischer and I began developing scientific and technological applications of the track-etching technique, we got a letter from Lal asking if one of us would be willing to spend some time at the Tata Institute of Fundamental Research. He wanted to learn first hand how to use fossil tracks in solids to study the history of terrestrial and extraterrestrial rocks. The technique fascinated him; it meshed perfectly with his broad interests in geophysics and radioactive dating. If we hadn't invented it, he might very well have done so himself.

Walker was committed to a sabbatical year in Paris, but I was free and, to tell the truth, I was so excited about the prospect of visiting India that I could hardly sleep at night. My wife and I read all we could about Indian culture, religion, art, and architecture. Meanwhile, Lal was becoming interested in work being done at UCLA by Willard Libby. By the time we arrived, with three small children, in Bombay in September, 1965, we found that India was at war with Pakistan and that Lal and Aruna would be departing in a few weeks for their own sabbatical with Libby in Los Angeles. The good news was that after they left we could live in their lovely penthouse apartment overlooking the Arabian Sea.

While Jo Ann, with Aruna's help, looked for a temporary place to stay and settled two of our three children in elementary schools, I got acquainted with many of the physicists at TIFR. One of them, the Director, Homi Bhabha, was killed in a plane crash on Mont Blanc a couple of months after I arrived, but not before I had acquired from him a taste for modern painting and sculpture. With others, such as Menon, Daniel, Biswas, Cowsik and Sreekantan, I developed lasting friendships. As Lal hoped, I introduced a number of his colleagues, including Rajan, Rama, Biswas, Bhandari, Durgaprasad, Tamhane and Iyengar, to the art of track-etching. Rajan and Tamhane were just beginning research and were assigned to work with me. Lal himself was so busy tying up loose ends before departing for UCLA that we had few uninterrupted conversations until he later visited me in the United States.

While he and Aruna were with Libby in Los Angeles, Rajan, Tamhane and I mapped out the track densities in surface samples from a large stony-iron meteorite, Patwar, which had fallen in India. The work took a very exciting turn. Some of the samples not only had a very high density of tracks of very heavy cosmic rays but also had large gradients of track density on a scale of a millimeter. With the help of calculations done by Rajan and Tamhane, I realized that these gradients were possible only if the samples had come from within ~ 1 cm of the preatmospheric surface and if the rate of erosion in space had been less than $\sim 10^{-7}$ cm/year.

When Lal returned from Los Angeles, he seemed to appreciate the future applicability of our new method for studying erosion even more than we did. He became interested in our modelling of the track density profiles and taught me the Laplace transform method for solving simultaneous differential equations, which we needed in order to correct for fragmentation of the cosmic ray nuclei as they penetrated the meteorite. Possibly his current project to infer terrestrial erosion rates using accelerator mass spectrometry to map gradients of cosmogenic radionuclides may have been stimulated by our work at TIFR in 1965.

Jo Ann and I returned to Schenectady, New York, in the spring of 1966. For about six years Lal and I kept up a lively correspondence averaging about eight letters apiece per year. The dates on those letters, typically with a two-week turnaround, showed that mail delivery was much better in both the U.S. and in India in those days than now. During these six years our interests overlapped on several fronts—studies of heavy and ultraheavy cosmic ray nuclei via tracks in meteorites and in plastic detectors; detection of fossil fission tracks from the extinct nuclide ^{244}Pu; the many applications of tracks in lunar samples; and the analysis of tracks of energetic magnetospheric heavy ions in plastic detectors mounted outside the Skylab. The ideas flowed furiously in those letters, as we exchanged ideas for improving the techniques for etching various minerals and for identifying the nuclei producing tracks in a specific mineral.

We recognized that heavy ion accelerators held the key to putting "trackology" on a quantitative basis. We approached people at Berkeley and at Dubna in the Soviet Union, the only two places where heavy ions could be accelerated to sufficiently high energies to be useful. One of our minor triumphs was to persuade Flerov at Dubna to permit one of his scientists, Perelygin, to expose crystals to energetic Zn ions for us, in exchange for co-authorship on a paper. (In the late 60's the cold war made such collaborations difficult.) Later, of course, the Lawrence Berkeley Laboratory Superhilac was able to produce energetic ions up to uranium, and calibrations were no longer a problem.

During that six-year period before Lal began to assume the heavy administrative responsibility of the Directorship of PRL, his letters were very long and full of detailed ideas, questions, and answers. In those days he would have benefitted greatly if computer mail had existed. Between the time I left India in 1966 and the time I moved to Berkeley in 1969 and Lal began his pilgrimages to La Jolla, we only saw each other once, but we got to know each other well through the many letters. Excerpts from a very long letter dated October 8, 1968, provide the flavour of our correspondence:

"Your letter arrived when I was mailing mine to you yesterday. Thanks for your quick and *kind* reply. Rama and one university student have been working on dating mica and some very exciting results have come out.... Can you please send us a small piece of synthetic mica?... Thank you so much for sending the cellulose triacetate sheets and the Makrofol.... We tried to see if the gas-rich meteorites had unusually high track density in some regions due to early bombardment. If grains are exposed in free space for some time before compaction and are not heated enough subsequently, they should have high density..." [Comment: they eventually did make this exciting discovery. Next Lal put down a long list of scientific ideas that he developed while in bed with a terrible case of Hong Kong flu.] "... What is said above has occurred to everyone and what irritates is that nothing much has been done to settle the scores. Can we jointly break the hell loose? Several experiments are clear:..."

A week later he still has the flu: "... It is a catastrophic virus invasion—clearly indicative of the awful possibilities of any biological warfare. Aruna had it first and

A friendship with Lal

then I. First symptoms are simply hoarseness in throat. Then fever rising.... Then immense weakness follows if one is still living. So our vacation plans ended in togetherness at home.... In the delirium of my fever, I have been thinking of all sorts of experiments and many in fact were a great success in my dreams. So I thought I should write to you about several of these...."

On November 6, 1968: "Thanks again for a quick reply. I am now digesting your intensive comments and criticisms and will write to you again. Thank you for sending Lexan sheets. Do you have thinner Lexan sheets? We would be grateful if you could send more, in sheets about half as thick. I am sending you a parcel post by air of Rosaries. The original colour of the big beads is black—they are painted brown and will turn black slowly. The Saints prefer it that way. Also, you will note that there are always 108 beads in each rosary. That is the Indian magic number for obtaining salvation...."

On December 16, 1968: "I have been thinking of a scheme.... This is so simple and enjoyable experiment that even if it fails, it does not matter...."

On January 3, 1969: "... What was new and exciting at Cameron's Relativistic Astrophysics conference? Incidentally, before going into scientific matters, let me thank Helen Couch for the hybrid tomato seeds. I have a huge big tomato in my terrace garden. Yellow and solid...."

On May 26, 1969: "... I am very happy to learn that you have finally decided for Berkeley and will be moving out there in September. May this be the most gorgeous and fruitful move for you. Amen!..."

Now a passage of a letter from me to Lal on November 24, 1970, indicating a problem: "... I am afraid I simply haven't found time to sit down and give thoughtful replies to your many letters. With teaching plus a growing group I have little free time anymore...."

Another letter from Lal, dated February 17, 1971: "... I would very much like to urge the following. We trackologists get down to business on energy spectrum and erosion on Apollo 14 rocks. Can we get together—joined hands—and sample, say, ten rocks, sit together with the results and go back to selected important slices and then two years later, we put out a Berkeley-St. Louis-Schenectady-Bombay bulletin on the history of cosmic radiation. More specifically, could I invite you for a Berkeley-Bombay collaboration on studies of... I know you are a very busy man now—teaching, research and then wow—the lovely balloon flight you had, probably does not let you find time to walk down to the Hare Krishna Square...." [He refers to Sproul Plaza.]

Part of the difficulty of collaborating closely with Lal's group was simply that I was responsible for training graduate students to think independently, and I felt it was important to let them learn from their own mistakes rather than supervising them as closely as would be needed to keep a collaboration on schedule.

On October 20, 1972, Lal wrote with interesting news: "I have been invited to join PRL in Ahmedabad.... So the above is a new landmark in my life and I hope that my decision is not too wrong a one. My own research work will be affected heavily but having realised that I should begin some time by helping develop science in India, I have emotionally adjusted to this situation...."

A year later Jo Ann and I spent an enjoyable week in Ahmedabad at the Silver Jubilee of PRL. Six months later Lal wrote with bad news: "As I write this letter there is a partial workers' strike in PRL set off by my terminating the services of one of the employees on account of indisciplined misbehaviour. There is a general mood for all organizations in the country to entertain this pleasure, to organize strikes, etc. We have become a part of this whirlpool...."

His next six months' visit to La Jolla was cut short by further labour problems at PRL but while he and Aruna were visiting us in Berkeley we talked for hours about how best to run a laboratory. Lal and I disagreed on procedures for both directing students and directing a laboratory, with me taking perhaps too soft a line and with him taking perhaps too hard a line. Often in those days he would ask me why none of his former students would be likely to do research of Nobel-prize calibre.

As I moved away from lunar sample studies and he moved more into oceanographic studies at La Jolla and deeper into administration at PRL, our letters became infrequent, and his were often handwritten on hotel stationery. From Bern, dated July 8, 1974: "... In any case, the entire Lexan methodology is beautiful [referring to its use on Skylab]. New methods like this need to be developed now and then to make life worth living...."

The most recent letter from Lal, dated a few months ago, is what I call a Generic Lal Letter: "In our oceanographic studies, we need to separate.... It occurred to me that the angstrom size silica which you use for making Cerenkov detectors may well serve our purpose.... Could you please give me the addresses of suppliers of the fine-grained silica.... If you have some quantity of silica to spare... I would very much like to have some for tests rightaway.... I was thinking of passing a colloidal...." By the time I found time to call an expert at LBL about colloidal silica, he told me a man named Lal had just called him to ask him for samples of silica gel. Lal knows how to hedge his bets.

Our two families enjoy each other's company immensely. We have a pleasant tradition of meeting at least once every two years that he and Aruna come for their six month stay in La Jolla. In the good old days they would stay with us in Berkeley and play with our young children, whom they always brought the latest electronic gadgets and puzzles. Our children remember those visits fondly. Lal would try out his latest crop of word-games, riddles and coin and match-stick tricks on them. He, Aruna, Jo Ann and I would stay up late at night discussing Indian and U.S. politics, and Lal and I would sit, completely relaxed, talking about crazy ideas for experiments. Sometimes we would go to a science-fiction movie. The four of us saw 2001: Space Odyssey, Star Wars, and many others over the years. As we were walking out of the theater where we saw 2001, Lal said he had decided he would go into motion picture production once he reached the age of 60. We shall see.

Even today, with our now grown children, whenever we leave the beaten path on a sightseeing trip to some out-of-the-way place, one of them is likely to say with a smile, "I have to obey Lal's Rule." [Lal taught them years ago never to pass up an opportunity to visit a toilet whether it is needed or not, a Rule he probably formulated in India.]

Today Lal is no longer a Laboratory Director, but he is still a driven person. His imagination is still as fertile as ever, and he still has an insatiable curiosity. It is all he can do to resist twirling knobs on pieces of equipment. Secretaries dread his visits, because he gives them a long list of things to do even on a short visit. After he leaves, with their work for him incomplete, I tell them it's all right not to finish the tasks. I have a gauge that tells me whether Lal is starting a new project. If I get a spurt of letters and phone calls asking me for obscure references, for samples of various laboratory items or detectors, for old computer codes, and for information on doing a particular calculation, I know that he is thinking of something new.

Visits to his home in La Jolla are stimulating. Library books, as well as his own books, are piled everywhere, and papers are strewn on all available surfaces. A computer is left permanently on. Breakfast is a multi-media event: Aruna is bustling in the kitchen, the television is on loudly, Lal is on the rug in his pajamas doing his exercises, and I am trying to eat and answer a question he has just put to me.

What brings order into his life and keeps his imagination from running away from him is Aruna, who helps him with his phoning, his letters, his disorderly files, his laboratory sample preparation, his book orders, and his travel arrangements. She is a wonderful diplomat as well. The two of them operate very effectively as a team, and this volume should be dedicated not only to Lal, but also to Aruna, who is largely responsible for Lal having achieved so much.

I hope the next twenty-five years of scientific and social interactions with Lal will be as pleasant as the last twenty-five have been.

Geophysical searches for new forces in nature

RAMANATH COWSIK

Tata Institute of Fundamental Research, Bombay 400 005, India and McDonnell Center for the Space Sciences, Washington University, St. Louis, MO 63130, USA.

Abstract. Various experiments searching for new intermediate range forces weaker than gravity using the earth's material as the source are reviewed against a backdrop of historical developments in the field of gravitation and the theoretical motivations for the possible existence of such a force. The results of these experiments are compared with those obtained from laboratory studies and then a plan for a future geophysical programme for such a search making use of the unique opportunities offered on the Indian subcontinent is briefly outlined.

Keywords. Equivalence principle; new forces; gravity measurements; geophysics.

1. Historical backdrop

Gravitation is the weakest force known to man: the electrostatic force of attraction between an electron and a proton exceeds the gravitational force between them by 33 decades in strength! The very strength of electromagnetism forces most of the objects on this earth and in the heavens to be electrically neutral so that the gravitational force manifests itself more clearly, especially since there exist no known particles with negative gravitational masses, i.e. particles which repel rather than attract other masses. In a 'through-the-looking-glass' world made up of both positive and negative gravitating masses, the effects of gravity would be feeble in the extreme and, perhaps, never would have been discovered!

When we talk of gravitation, the name of Sir Isaac Newton immediately springs to mind. He described gravitation as the force that is exerted by all bodies 'according to the quantity of solid matter which they contain and propagates on all sides to immense distances, decreasing always as the inverse square of the distances'. A hundred years before Newton, Galileo Galilei discovered that all bodies fall at a rate independent of their mass and composition. We are told that he reasoned that if a large mass is thought of as two smaller masses and each of these were to fall slower than their composite then a contradiction would follow. An actual experiment was carried out from the tower in Pisa which confirmed his intuitive reasoning. Newton, who was certainly aware of all these developments, concluded on the basis of his second law that the force exerted by gravitation was proportional to the mass of the body on which it acts. Also his assertion that the gravitational field generated by a body is proportional to its gravitational mass, stated in quotes earlier, follows from the third law. He tried to check, more precisely than what Galileo could do, that indeed all bodies irrespective of their composition or mass had the same acceleration in a given gravitational field. He studied the frequency of oscillation of simple pendula, whose bobs were of different substances, and found they were independent of the composition but decreased as the square root of their length, as had been noticed earlier by Galileo. It is with these aspects of history in our mind that we think often of new forces as possibly dependent upon composition. Newton had realized that the "inertial mass", m_i, that appears in his second law might not precisely be the same as the "gravitational mass", m_g, which appears in his law of gravitation. In standard notation we have

$$\mathbf{a} = \mathbf{F}/m_i; \tag{1}$$

$$\mathbf{F} = \mathbf{g} m_g; \tag{2}$$

$$\mathbf{a} = (m_g/m_i)\mathbf{g} \tag{3}$$

where $\mathbf{g}$ is the given gravitational field. The essential question is, 'does the ratio, (m_g/m_i) depend upon the composition or any other aspect of the body'?

This question of the composition independence of the acceleration suffered by a body was taken up in an important way by Baron Roland von Eötvös. He devised an ingenious scheme that showed that the ratio (m_g/m_i) was independent of the composition to an accuracy of a few parts in 10^9! A photograph of his apparatus is shown in figure 1 and an explanatory sketch in figure 2. Eötvös hung two masses A and B made of different substances, say Cu and Pb, from the ends of a 40 cm long beam which is suspended on a fine wire from its midpoint just like the Cavendish balance. A telescope views a mirror attached to the beam and the whole set-up is mounted on a smooth turnstile.

The sketch also points out the gravitational forces acting towards the centre of the earth, $m_g g_\oplus$ and the horizontal component $m_i a$ due to the centrifugal acceleration by the Earth's rotation. Thus the suspension fibre does not point exactly towards the centre of the earth and a torque T acts on the balance, of magnitude

$$T = \kappa m \{1 + (\Delta m/m)^2\} r_\oplus \omega_\oplus \sin b \cos b \sin \theta. \tag{4}$$

Here, $m = m_A \approx m_B =$ the mass of the suspended objects; $\Delta m = m_A - m_B$; $\kappa = (m_i/m_g)_A - (m_i/m_g)_B$; $r_\oplus =$ radius of the earth, $\omega_\oplus =$ angular frequency of earth's rotation $= 2\pi$ day^{-1}, $b =$ the latitude of the laboratory where the balance is located, and $\theta =$ the angle the beam makes with the North-South line. This torque T can be measured by setting the balance at different values of θ by means of the turnstile and measuring the deflection $\phi = T/k$ of the balance using the telescope. Here k is the torsion constant of the suspension fibre. With this apparatus Eötvös and his collaborators compared (m_i/m_g) ratios of a wide variety of substances; the common metals, and alloys, and also exotic substances like snake wood, radioactive materials, etc. His observations showed that the inertial and gravitational masses of all substances bore the same ratio to each other (chosen conveniently as 1), correct to at least a few part in 10^9. Einstein was much impressed and stimulated by this equivalence between inertial and gravitational masses, and enunciated his principle of the equivalence of gravitational field and acceleration thus: "We arrive at a very satisfactory interpretation of this law of experience, if we assume that the systems K and K' are physically equivalent, that is, if we assume that we may just as well regard the system K as being in a space free of gravitational fields, if we then regard K as uniformly accelerated. This assumption of exact physical equivalence makes it impossible for us to speak of the absolute acceleration of the system of

Figure 1. Photograph of the torsion balance developed by Eötvös, now at the Eötvös Museum, Tihany, Balatonfured, Hungary.

reference, just as the usual theory of relativity forbids us to talk of absolute velocity of a system; and it makes the equal falling of all bodies in a gravitation field seem a matter of course".

1.1 Theoretical motivations

Such experiments set very stringent limits on the strength of any nongravitational forces that might exist in nature. Even before the modern experiments which showed the extreme stability of the proton, baryon number was thought of as being conserved. Consequently Lee and Yang (1955) considered the possibility of an exactly gauged

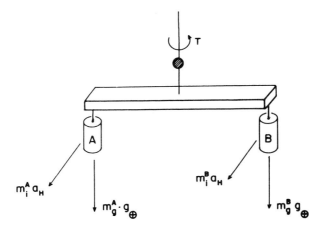

Figure 2. Sketch showing the principle of the Eötvös balance; the equilibrium orientation of the balance is controlled by the gravitational forces $m_g g_\oplus$ and the inertial forces $m_i a_H$ where the horizontal component of earth's centrifugal acceleration, $a_H = r_\oplus \omega_\oplus^2 \sin b \cos b \cos \theta$.

baryon number in analogy with the conserved electric charge and showed that Eötvös's experiments indicated that any new kind of "electrostatic" force in which the number of baryons plays the role of electric charge has to be much weaker than gravitation. This however did not preclude the existence of forces with somewhat greater strength if the range, λ, of the new force was sufficiently short in comparison with the radius of the earth.

In an attempt to preserve the conservation of the product of parity and charge conjugation (PC) Bernstein et al (1964) suggested the possibility that the PC-violating decays of K_L^0 observed by Christenson et al (1964) could be possibly due to the regeneration of K_S^0 in the K_L^0 beam due to a long range vector interaction with material of the earth. However, experiments which were performed soon after this suggestion failed to observe the quadratic increase of the rate of this hypothesized regeneration with the Lorentz factor of the K_L^0 beam. At the same time, in a very important paper Weinberg (1964) pointed out that if such a hyperphoton v were to exist, then the rate for the decay $K_S^0 \to 2\pi + v$ with respect to the standard mode $K_S^0 \to 2\pi$ would increase catastrophically as $f^2 m_v^{-2}$ so that the observational limits on decays involving a hyperphoton in the final state implied a very stringent limit on the strength of the new coupling $f^2 < G$ for $m_v^{-1} \sim 10^3$ cm. These considerations involving the $K_0 - \bar{K}_0$ system as a probe for new forces have been taken up in several recent studies (Aronson et al 1986; Suzuki 1986). Also within the context of modern gauge theories, there have been many suggestions as to how new intermediate range forces could arise, for example through the spontaneous violation of scale invariance, lepton number and other symmetries. Even a millenium before Newton there was an awareness and a concern about the nature of the decrease of the strength of the force with distance and since Newton there have been studies to see how closely the decrease actually follows the inverse square law propounded by him. More recently in an attempt to quantize gravity within the context of extended supergravity theories Deser, Zumino, Scherk and Fayet have suggested that vector bosons of finite rest mass could be present mediating a repulsive force between particles, thereby leading to a violation of the inverse square law. More broadly speaking the new forces could be considered as a 'low energy' vestige of Grand Unification. A slightly more detailed discussion of the theoretical motivations, along with references to a few recent papers is given by Cowsik et al (1988). It is customary to parametrize such a force through a potential having the Yukawa form:

$$V = G_N(\alpha m_1 m_2/r)\exp(-r/\lambda) \qquad (5)$$

with obvious notation.

1.2 Resurgence of interest

Experimental and theoretical investigations related to the possible existence of forces weaker than gravity have become vigorous during the last three years, stimulated by the work of Fischbach et al (1986) who noticed that the residuals in the famous experiments of Eötvös correlate well with the differences in the number of baryons per atomic mass unit, $\Delta B/\mu$ of the two substances placed on either side of the balance, as shown in figure 3. They suggested that the force could indeed have the form generated by a potential given in equation (5) with the value of α chosen as

$$\alpha = \xi q_{\text{source}} \times q_{\text{test}}, \qquad (6)$$

where the new charge q could be adequately parametrized as a linear combination of the baryon number $B = N + Z$ and the isospin $I = N - Z$ of the nuclei involved

$$q_i = \cos\theta(N+Z)_i/\mu_i + \sin\theta(N-Z)_i/\mu_i. \qquad (7)$$

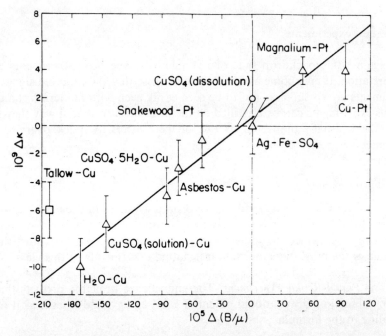

Figure 3. Value of the horizontal acceleration differences $\Delta\kappa$ versus $\Delta(B/\mu)$ measured by Eötvös et al (1922) (after Fischbach et al 1986).

Here μ_i is the mass of the nucleus in atomic mass units. Notice that since ξ and λ are coupled in the expression for the potential it is not possible to determine each of them separately from a single observation of the new force at one particular distance, and Fischbach *et al* (1986) brought Airy's geophysical method of measuring the value of the Newtonian gravitational constant G_N to bear on the problem (Rapp 1974). There was a long standing discrepancy between the value of G measured in the laboratory, for example by Luther and Towler (1982) and that measured by the geophysical method, reviewed by Stacey *et al* (1987) as shown below.

$$G_{\text{lab}} = (6\cdot 6726 \pm 0\cdot 0005) \times 10^{-8} \text{ cm}^3 \text{g}^{-1} \text{s}^{-2}, \tag{8}$$

$$G_{\text{Airy}} = (6\cdot 72 \pm 0\cdot 024_{\text{systematic}} \pm 0\cdot 002_{\text{fitting}}) \times 10^{-8} \text{ cm}^3 \text{g}^{-1} \text{s}^{-2}$$

The geophysical value is about 1% larger and could be explained if there existed a repulsive force coupling to baryons which acted with a range of ~ 20–100 m, corresponding to the typical depths up to which the Airy's method of measurements was carried out. The point is that at these depths the new finite range force acts essentially isotropically and is nullified. On the other hand at the surface of the earth the force repels the body upwards counteracting gravity. Thus for a purely baryonic coupling ($\cos\theta = 1$), they suggested $\xi \approx 0\cdot 01$ and $\lambda \sim 20$–100 m. Since the earth's surface is an equipotential with respect to the sum of gravitation and the new Yukawa form, at least in principle, it was immediately clear that to get the correlation with the baryon number shown in figure 2 one has to evoke large horizontal gradients in the field at the site of Eötvös experiments. Without such geophysical anomalies the Eötvös experiments set upper bounds on ξ for baryonic couplings at the level of $\sim 10^{-5}$ for $\lambda \gtrsim r_\oplus$.

2. The new experiments

It is against this backdrop that one should view the new experiments and their interpretation. It may come as a matter of surprise that the expected signals in these experiments are so small as not to be seen easily even with modern techniques. The various experiments can be classified on the basis of the essential way they study the earth's field, usually pioneered by one of the masters. A quick overview of these experiments is presented below.

2.1 *Galileo's method*

The essence of Galileo's method was to drop test bodies in the earth's gravitational field, say from a tall tower, and study their motion either in comparison with that of another or with respect to some theoretical expectation. We shall start this review with one of the most ingenious and fascinating experiments of this class.

2.1a *The Pound-Rebka experiment*: The equivalence principle predicts that even a photon falling in a gravitational potential gains energy, and loses it as it moves up, according to the formula

$$\Delta v = (-g_\oplus h/c^2)v = -\Delta\phi v \tag{9}$$

or

$$\Delta v/v = -\Delta\phi = \frac{(980\,\text{cm s}^{-2})(2260\,\text{cm})}{(3\times 10^{10}\,\text{cm s}^{-1})^2} = 2\cdot 46\times 10^{-15}, \tag{10}$$

where all the symbols have their usual meaning and the numerical estimate is made for a tower 22·6 m high available to them for the experiment. Notice the minuteness of the expected frequency shift; Pound and Rebka (1960) took up the challenge to measure it! This is how they did it: they chose the 14·4 keV γ-ray line emitted by ^{57}Fe as their photon source, mounted at the top of a vacuum tube which spanned the height of the tower. Resonance absorption of this γ-ray by a detector situated at the bottom of the tower was possible through the Mössbauer effect which required both the source and the detector to be single crystals. If the intrinsic width of the resonance be Γ corresponding to an intrinsic lifetime for the state of $\sim 10^{-7}$ s ($\Gamma/v \sim 1\cdot 13 \times 10^{-12} \approx 460\,\Delta v/v!$), then the frequency shift due to the fall of the photon in the earth's gravitational field will cause a change in the counting rate in the detector, according to the relation

$$C\alpha\frac{\Gamma^2}{\Gamma^2 + \Delta v} \approx 1 - (\Delta v^2/\Gamma^2) \approx 1 - 2\times 10^{-6}. \tag{11}$$

By keeping the source and the detector at the same height and at the same separation and then by repeating with the source at the top, it would be virtually impossible to achieve the requisite accuracy for measuring the counting rate needed to see the miniscule effect. Pound and Rebka subjected the detector to an oscillatory motion in the direction of the source so that the frequency of the γ-ray line was Doppler-shifted also in an oscillatory fashion. The net frequency change then was the sum of the change due to gravitation Δv_g and that due to Doppler effect:

$$\Delta v = \Delta v_g + v_0(v_0/c)\cos\omega t. \tag{12}$$

Substituting this in (11) the counting rate is seen to be modulated as

$$C \approx 1 - (\Delta v_g/\Gamma)^2 + v_0^2/\Gamma^2\left(\frac{v_0}{c}\cos\omega t\right)^2 + \frac{2\Delta v_g}{\Gamma^2}\frac{v_0}{c}\cos\omega t. \tag{13}$$

Thus the modulation factor is enhanced by a factor $2v_0/c\Delta v_g$ which is as large as $2\cdot 6\times 10^4$ even for $v_0 = 1\,\text{cm s}^{-1}$. In the actual experiment the difference in the modulation obtained by placing first the source at the top of the tower and the detector at its base and later reversing their positions yielded the result $\Delta v/v = (2\cdot 57 \pm 0\cdot 26)\times 10^{-15}$ in good agreement with the expected value of $2\cdot 46\times 10^{-15}$. Besides this large enhancement in the modulation rate, their "trick" made the modulation linear, sensitive to the phase of the oscillation and distinctive in the frequency domain so that the signal could be measured with accuracy and confidence. What a wonderful experiment!

2.1b *Experiments of Faller and Kuroda*: These experiments follow very closely the strategy of the original experiments of Galileo in that the objects of differing composition are dropped in vacuum and their relative motion is monitored interferometrically to look for any differential acceleration. The apparatus designed

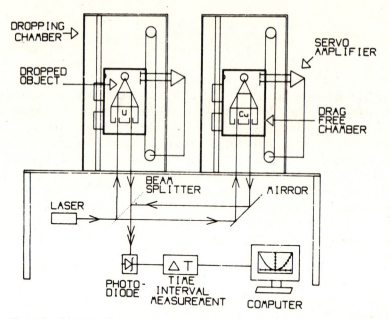

Figure 4. Schematic diagram of the apparatus used in the modern version of the Galilean free-fall experiment (after Niebauer *et al* 1987).

by Faller and his colleagues is sketched in figure 4. Notice that if the hypothetical force couples to baryon number as suggested by Fischbach *et al*, then the expected difference in acceleration between common substances is rather small.

$$\Delta a \approx g_\oplus \xi \Delta(B/\mu) \frac{\lambda}{r_\oplus} \approx 10^{-9} \xi \lambda \, \text{cm s}^{-2}. \tag{14}$$

Here we have taken $\Delta(B/\mu) = (B/\mu)_{\text{Cu}} - (B/\mu)_{\text{Pb}} \approx 1.0011 - 1.0001 \approx 10^{-3}$ and the radius of earth $r_\oplus \approx 6 \times 10^8$ cm. For typical values of $\xi \approx 10^{-3}$ and $\lambda \approx 10^4$ cm, $\Delta a \approx 10^{-8}$ and during a free fall of ~ 500 cm in ~ 0.5 s the path difference that accumulates between the two objects is merely

$$\Delta s \approx \tfrac{1}{2} \Delta a \, t^2 \approx 2.5 \times 10^{-9}. \tag{15}$$

With about ten thousand repetitions of the experiment the best upper bound that Niebauer *et al* (1987) could place on the $\xi\lambda$ product is

$$\xi\lambda < 9 \times 10^2 \, \text{cm}, \tag{16}$$

which corresponds to an uncertainty in the measurement of $\Delta s \approx 2 \times 10^{-7}$ cm. Similar results have also been obtained by Kuroda and Mio (1989). The drag due to residual gas in the vacuum chamber, the gradients in the ambient gravity field and the effects of any unshielded magnetic field on the earth are the main sources of noise in this experiment. In any case the result as it stands is quite interesting as it improves upon the Eötvös test of the equivalence principle in the earth's gravity field. The differential acceleration η in units of the gravitational acceleration on the surface of the earth is

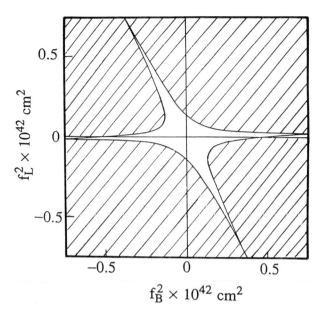

Figure 5. Constraints placed by the studies of equivalence principle in the Earth's gravity field on the coupling constraints in theories of gravity with a non-symmetric metric (after Will 1989).

constrained by these experiments to be

$$\eta < 5 \times 10^{-10}. \tag{17}$$

Will (1989) has shown that any theory of gravitation such as the one due to Moffat (1979, 1987) in which a nonsymmetric metric couples minimally to the electromagnetic field, violates the weak equivalence principle and the result stated in (17) sets very severe bounds on the possible couplings in these theories to baryons and leptons as shown in figure 5.

2.1c *Particles and antiparticles in the Earth's field*: A number of experiments are underway in which elementary particles and antiparticles would be allowed to fall freely under gravity and their rates of acceleration compared. Notice that this class of experiments can probe the existence of new forces which might cancel each other out in a symmetric situation involving particles both as sources and as the test body. Pioneering work in this field was done by Fairbank of the Stanford University; latterly Goldman *et al* (1988) are planning to drop antiprotons and compare their free fall with that of a H^- ion so that the effects of any stray fields may be common both to the particle and the antiparticle. In figure 6 several interesting aspects of the design of their apparatus are brought out. The ingenuity in the experimental design is mainly directed towards cooling the fundamental particles to sufficiently low levels so that the effect of the earth's gravity field on their trajectories become measurable.

2.1d *The KC-135 experiment*: In this experiment Cowsik *et al* (1989) plan to measure the torque on a body of dual composition during its free-fall. A sketch of the test body is shown in figure 7. It consists of two semi-circular cylinders, one of lead and the

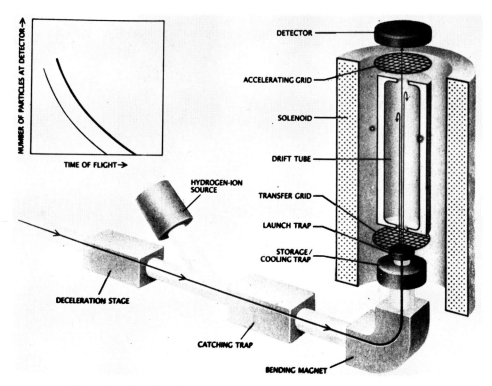

Figure 6. Effect of gravity on antiprotons will be compared with that on H^- ion and will be studied with our apparatus sketched above. (after Goldman et al 1988)

other of copper, encapsulated in an aluminium sheath to form a circular cylinder. Even though the density of lead is about 15% higher than that of copper, suitable grooves cut in the lead half make its mass and first two moments of its mass distribution about the cylinder axis the same as that of the copper half. Further, the height of the cylinder is nearly the same as its diameter so that all the elements of its quadrupole moment tensor are small compared to mr^2. Thus under conditions of free fall the gravitational field and its gradients cannot exert any torques on the body, especially along the axis of azimuthal symmetry. On the other hand if there exists a force which couples differently to copper and lead then there would be a torque

$$T \approx \alpha \, mrg_\oplus (\lambda/r_\oplus). \tag{18}$$

Here, m is the mass of the cylinder, r its radius and the other symbols have their usual meaning. Under the influence of this torque the cylinder will be subjected to an angular acceleration $\theta = T/I$, I being the moment of inertia of the cylinder about its axis

$$\theta \approx \omega_0 t + \tfrac{1}{2}\alpha g_0 \frac{\lambda t^2}{rr_\oplus} \tag{19}$$

$$\approx 100 \alpha t^2 \approx 4 \times 10^{-7} \text{ rad}.$$

Here ω_0 is the initial angular velocity and the numerical estimate assumes $\alpha \approx 10^{-11}$, $r = 5$ cm, $\lambda \approx r_\oplus$ and the time of free fall, $t \approx 20$ s.

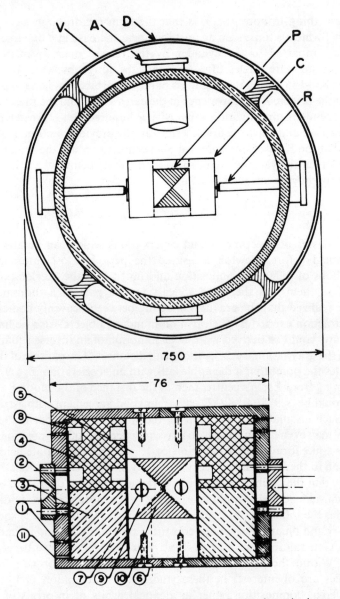

Figure 7. Sketch of the KC-135 apparatus: *D*, aerodynamic dome; *A*, autocollimator; *V*, vacuum chamber; *P*, right angle prism; *C*, cylinder with dual composition and *R*, release mechanism. The lower panel shows the details of the cylinder with dual composition: (1) aluminium sheath, (2) sapphire anvil with conical indentation, (3) semi-circular cylinder of copper, (4) semi-circular cylinder of lead with grooves for matching, mass, moment of inertia, etc with the copper counterpart, (5) aluminium prism holder, (6) window in the sheath for optics, (7) screws holding the various components to the aluminium sheath, (8) step in the sheath to facilitate evacuation of trapped air, (9) aluminium prisms to balance the glass prism reflectors, (10) window in the prism holder for the optics, and (11) vent holes for evacuation of trapped air.

The important thing to note in (19) is that the term quadratic in t contains the interesting signal which is to be seen against the linear increase in θ due to any initial angular velocity imparted to the cylinder by the release mechanism. Since it is relatively easy to measure angles of $\sim 10^{-7}$ rad, the long observation time made available by the KC-135 during its weightless parabolic orbit allows one to probe into the violation of the principle of equivalence in the earth's field and the presence of new composition-dependent forces with a sensitivity a hundred times greater than that possible with earthbound experiments described in the previous section. It should be noted further that the KC-135 environment allows one to reduce the drag forces due to the residual gas in the vacuum chamber to levels much below what one wishes to probe in this experiment, viz. $10^{-11} g_\oplus$.

2.2 Newton's method

There are two classes of experiments and observations which can be attributed to Newton. As noted before, Newton measured the period of oscillation of simple pendula with bobs of different composition and noted that the acceleration due to gravity for all of them was the same at least to one part in a thousand. More importantly he realized that the gravitation acting between heavenly bodies was the same as the attraction exerted by the earth on mundane objects. Also he had shown that Kepler's laws could be derived with the assumption of an inverse square law for the decrease of the gravitational acceleration with distance. "The parable of the apple" merely illustrates the point that if the apple falls with an acceleration of $\sim 978 \text{ cm s}^{-2}$ at a distance of $r_\oplus$ from Earth's centre, then at the distance of the moon $r \approx 60 r_\odot$ its acceleration would be $\sim 978/60^2 \sim 0.27 \text{ cm s}^{-2}$ in good agreement with astronomical observations. Such an analysis sets a limit on the strength of any posited force-field whose coupling may even be universally to mass, independent of composition, but still follow a Yukawa-like form given in equation (5). It is interesting to note here that the first modification to the Newtonian theory of gravity, namely General Relativity, was tested by noting that the perihelion of Mercury was advancing at the rate of $\sim 43''$ per century while no such advance would be expected in the motion of two bodies about each other under the influence of forces which satisfy strictly the inverse square law. Indeed planetary and lunar observations constitute the testing ground for most of the predictions of General Relativity (Will 1981; Ciufolini 1986; Bertotti *et al* 1987; Cowsik 1988). Within the more specific context of the search for new forces the following studies are of interest. In these studies the acceleration of two bodies, possibly of different composition either in identical orbits or in orbits of different apogee are compared, to look for deviations from the predictions of standard dynamics.

2.2a Planetary studies: Clearly, the sensitivity of these studies in limiting the value of α depends on the assumed value of λ, the range of the force: if λ is too large it would constitute merely a renormalization of the value of G and if it is too short it would contribute negligibly at large distances. Modern planetary studies are most useful in setting limits on α for λ in the range 0.1–$10 \text{ AU} \sim 10^{12} - 10^{14}$ cm, corresponding to typical planetary distances; the essential result is $\alpha < 10^{-8}$ (Talmadge and Fischbach 1988).

2.2b *Lunar laser ranging and Eötvös experiments*: In a very thought-provoking paper Nordvett (1988) pointed out that in order to fully utilize the ~ 1 cm accuracy achievable in lunar laser ranging a knowledge of the equality of accelerations in free fall for different substances to an accuracy of $|\delta a/a| \sim 10^{-13}$ is required. During the last two decades nearly continuous measurement of the Earth-Moon distance has been going on, by reflecting laser pulses off the retroreflectors left on the lunar surface by the Appollo astronauts, in an effort to test metric theories of gravity (Williams *et al* 1976; Shapiro *et al* 1976; Will 1981). As Nordvett (1988) pointed out, if these bodies are assumed to have anomalous accelerations towards the Sun at a rate proportional to the fraction of their mass contributed by their gravitational self-energies, according to the formula

$$a_{\text{anomalous}} \approx (Gm/Rc^2)g_\oplus \equiv \Delta g_\oplus, \tag{20}$$

then, since $\Delta_{\text{Earth}} \neq \Delta_{\text{Moon}}$, there would be a polarization of the Earth-Moon distance-vector towards the Sun. Since $|\Delta_E - \Delta_M| \approx 5 \times 10^{-13}$. Should one observe a polarization at this level, it would not be possible to interpret this as due to the "Nordvett-effect" unless we are sure that the acceleration differences are not caused by a violation of the weak equivalence principle at that level as the mass of the Earth gets a substantial contribution from its iron core, in contrast with the Moon which is predominantly made up of silicates. Alternatively the bounds on any such polarization in the Sun's field can be interpreted as a bound on the difference in the coupling of any new force to the materials of the Earth and the Moon acting with a range greater than 1 AU ($\approx 10^{13}$ cm). This foregoing discussion is one of the important motivations for improving the experimental tests of the equivalence principle.

2.2c *Observations with LAGEOS*: Geodynamic satellites were designed and launched specifically to study the gravitational field of the earth by accurately measuring their orbits through the use of lasers and precise clocks. The value of $Gm_\oplus$ obtained from the tracking data and the corresponding Newtonian value of earth's surface gravity estimated assuming a biaxial ellipsoidal model for the earth are

$$Gm_\oplus = 398600{\cdot}4342 \pm 0{\cdot}002 \text{ km}^3 \text{ s}^{-2},$$

$$g_\oplus(0) = 978{\cdot}0324 \pm 0{\cdot}0003 \text{ cm s}^{-2}. \tag{21}$$

This value coincides with remarkable accuracy with the mean value of surface gravity obtained by direct measurements on the earth's surface (Rapp 1974):

$$\bar{g}(0)_{\text{direct}} = 978{\cdot}0326. \tag{22}$$

If a Yukawa term had been present it would have contributed differently to the surface gravity and to the field at the height $z \approx 5900$ km of the satellite an acceleration

$$\delta g(z) = 4\pi G\rho\alpha\lambda \{r_\oplus \cosh(r_\oplus/\lambda) - \lambda \sinh(r_\oplus/\lambda)\}$$
$$\times \frac{\exp[-(r+z)/\lambda]}{r+z} \times \left(1 + \frac{\lambda}{r+z}\right). \tag{23}$$

The bound on δg placed by the above mentioned measurements is less than a part per million and this corresponds to a limit on α of $\sim 10^{-6}$ for $\lambda \sim r_\oplus$ (Rapp 1974; Cohen and Smith 1985; de Rujula 1986; Hughes *et al* 1989). Similar limits are also obtained by comparing the relative accelerations of the moon and the LAGEOS.

2.2d *Laplacian detector*: This detector designed by Paik measures directly the Laplacian of the field in a source-free region using a three-axis gradiometer comprising of superconducting rings which are magnetically levitated. Existence of any Yukawa-type field will manifest itself as a non-zero value for the Laplacian. Even in a ground-based experiment the detector indicated its high sensitivity (Chan *et al* 1982). It is planned to put this detector into an earth-bound orbit where it would function as a sensitive probe of the earth's gravity field, testing the general relativistic effects and searching for Yukawa-like fields (Paik and Will 1989).

2.2e *WEP studies with satellites*: There are essentially two basic designs which are proposed. Worden (1976) propose to levitate two coaxial cylinders of different substances in a superconducting cradle. The position of the two cylinders are monitored continuously as they orbit around the Earth; the differential accelerations could be measured accurately enough to test the weak equivalence principle up to 10^{-17}. The design due to Cowsik *et al* (1989) consists essentially of a torsion balance with the bob having a dual composition not unlike the cylinder described in 2.1c. The torsion fibre is axial, extends symmetrically on both sides of the cylinder and is anchored at both ends. A periodically oscillating torque is expected to act if WEP is violated. It is relatively easy to achieve sensitivities to test the violation up to 10^{-14} level.

2.3 *Airy's method*

This method was originally designed by Airy to measure G using the data on $g(d)$, the acceleration due to Earth's gravity as a function of depth. Extensive discussions of various gravity anomalies and free-air corrections are to be found in standard textbooks so that the Newtonian part of gravity is easily taken care of, once the terrain and density stratification are known. In the presence of a new Yukawa-like force there would be an extra contribution to the variation of g with depth

$$\delta g(h) = 4\pi G \rho \alpha \lambda \frac{(r_\oplus + \lambda)}{(r_\oplus - d)} \exp(-r_\oplus/\lambda) \left\{ \cosh\left(\frac{r_0 - d}{\lambda}\right) - \frac{\lambda}{r_\oplus - d} \sinh\left(\frac{r_\oplus - d}{\lambda}\right) \right\}. \tag{24}$$

Equations (23) and (24) form the basis of the design of several recent experiments looking for new forces feebler than gravity. The purely Newtonian variation of g with depth for a spherically stratified density profile for the Earth can be approximated as

$$\Delta g(d) \approx g(d) - g(0) = d(\gamma - 4\pi G \rho), \tag{25}$$

where the "free-air gradient" γ is the variation in g expected when the effects of finite density ρ can be ignored. At a latitude b for a spherical earth we can approximate the free-air gradient as

$$\gamma \approx (3083 + 5\cos^2 b)\,\text{s}^{-2}. \tag{26}$$

2.3a Measurement in Queensland mines: In §1.2 we already noted how the discrepancy between the value of G estimated from mines and the value from laboratory experiments led to the suggestion of a new force with a strength of about a hundredth of gravity. It is customary to adopt somewhat simpler formulae in fitting the non-Newtonian gravity residuals:

$$\Delta g(h) = \frac{4\pi G \rho \alpha}{1+\alpha}\left\{d - \frac{\lambda}{2}[1 - \exp(-d/\lambda)]\right\}. \tag{27}$$

One should note here that there are several difficulties in the application of this method to determine the Yukawa parameters precisely; (i) the density of the rock in and around the mine has to be sampled extensively with bore holes; (ii) adopting slightly different values for the free-air gradients in subtracting the Newtonian contribution yields different values for the Yukawa parameters; (iii) the parameters α and λ are not each determined uniquely, a whole band in the parameter plane yielding equally good fits to the data (Stacey et al 1987).

2.3b Measurements in bore holes: The main advantage of adopting measurements in bore holes for the search of new forces is that in contrast to mines the location of the bore hole can be chosen to lie in regions where the variations of density are small and the terrain is flat so that the purely Newtonian effects can be estimated accurately and are not confused with a possible contribution from a Yukawa force. With this idea in mind Ander et al (1989) chose to perform measurements of gravity in a hole bored in the Greenland ice-sheet. Their analysis of the gravity variation with depth has suggested the presence of an attractive Yukawa force with $\alpha \sim 0.03$ and $\lambda \sim 225$ m; however they cannot rule out the possibility that the variations with respect to purely Newtonian modelling might have been generated by unusual density distributions in the bed rock below the ice-sheet. One might refer to Rapp (1987) and Hsui (1987) for similar studies with bore-holes on a land mass.

2.3c Dynamical version of Airy's method: In an attempt to avoid the problems related to unknown density variations below the level of observation, Moore et al (1988) compared the weights of two stainless-steel masses suspended in evacuated tubes from the arms of a balance at different depths in a hydroelectric reservoir. The varying level of water provides a well-defined modulation of the signal, thus allowing an accurate measurement. The measurement determines the value of G for an effective separation of ~ 22 m between the gravitating masses; the value obtained by them is $G = 6.689(57) \times 10^{-8}\,\text{cm}^3\text{g}^{-1}\text{s}^{-2}$ which is within 0.6 standard deviation from the laboratory value. The importance of this experiment is that it constrains the possible interpretations of other Airy type experiments as described at the end of §2.3d.

2.3d Tower gravity experiment: The principle of this experiment is complementary to the ones described in the preceding sub-sections in that one measures the variation of gravity above the earth's surface and adopts an equation similar to (23) to describe the non-Newtonian contribution to the gravity (Eckhardt et al 1988). The gravity measurements were performed with a "Lacoste-Romberg" gravimeter (Faller and Rinker 1983) at various levels of a television tower which stands 600 m high in a topographically flat area near Garner, NC, USA. Extensive measurements of surface gravity in an area surrounding the tower allow an accurate upward continuation of the expected Newtonian contribution. The residuals were fit with contributions from two superimposed Yukawa terms;

$$V = -\frac{Gm}{r}[1 - a\exp(-r/v) + b\exp(-r/s)]. \tag{28}$$

The results of their measurement are that the parameters are constrained in a band around $(a - b) \approx 0.007$ and $av - bs = 510$ cm. It is to be noted here that Moore et al (1988) are able to simultaneously fit the data from the Queensland mines (Stacey et al 1987) and that from the TV-tower (Eckhardt et al 1988) within the constraint provided by their dynamical version of the Airy measurements; they favour shorter ranges $v \approx 20$ m and $s \approx 270$ m among the set of values shown in table 1 reproduced here.

2.4 *Eötvös method*

In the introductory section we noted how Eötvös compared the ratio of inertial mass to the gravitational mass of several bodies and found that this ratio was remarkably identical for all bodies. As an off-shoot of his experiment the analysis by Fischbach et al (1986, 1988) and others showed that if there existed a strong horizontal component of gravity and other force-fields owing to a nearby cliff or other such geological feature then one could attribute the residuals seen by Eötvös to a new force which is coupled to baryon number.

This has stimulated the setting up of torsion balances at sites specially chosen for large horizontal components of gravity. For example consider a location as shown in figure 8 by the side of a steep cliff. If there exists a new force coupling to baryon number, say, with a range λ and strength α in units of gravity it would generate a horizontal acceleration of a magnitude.

$$a \approx \pi G\rho\alpha\lambda \sim 6 \times 10^{-5}\alpha\lambda\text{s}^{-2}. \tag{29}$$

Table 1. Simultaneous analysis of gravity measurements in mines and the measurements up a television tower in terms of two Yukawa-like potentials by Moore et al (1988).

a	b	v(m)	s(m)	δg(mGal)	σ(mGal)	$(\Delta G/G)_{\text{lake}}$(%)	$(bs - av)$(m)
0.02	0.01523	0*	618	0.3746	0.04455	1.95	9.4
0.03	0.02387	16.8	270.8	0.4942	0.04173	0.82	6.0
0.05	0.04327	39.3	165.0	0.5523	0.04045	0.54	5.2
0.1	0.09300	59.0	116.0	0.5773	0.04038	0.49	4.9
0.3	0.2929	73.0	91.1	0.5858	0.04072	0.49	4.7
1	0.9928	70.1	75.3	0.5932	0.04379	0.59	4.6
10	9.9927	68.0	68.5	0.5934	0.04631	0.64	4.5

* This solution drives v to zero. A value of 1 m is assumed in fitting $b, s, \delta g$.

Geophysical searches for new forces 17

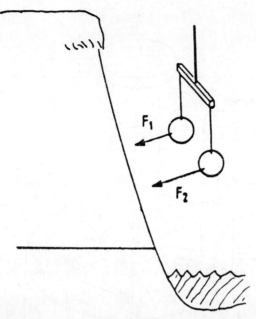

Figure 8. Yukawa forces, F_1 and F_2, acting on two dissimilar substances may not be the same; the couple $(F_1 - F_2) L$ leads to a deflection of the balance.

If there are two substances such as Cu and Pb with $\Delta(B/\mu) \approx 10^{-3}$ then the difference in their accelerations would be

$$\Delta a \approx \pi G \rho \alpha \lambda \Delta B \sim 6 \times 10^{-8} \alpha \lambda \text{s}^{-2} \tag{30}$$

and the observational constraints on Δa could be translated into constraints on the product $\alpha \lambda$.

2.4a *Continuous mode Eötvös experiment*: When Eötvös performed his measurements the torsion balance would be brought to various specific azimuthal orientations and the angular oscillations of the balance were recorded for a long-time interval to locate the "zero" of the balance; the shift in the zero occurs when the balance was rotated through 180° with respect to its earlier east-west orientation. Consider now the possibility that the balance is rotated continuously with an angular frequency ω_D. If there were a cliff nearby, it would exert a periodic torque, T, on the balance

$$T \approx \pi G \rho \alpha \lambda \Delta B m r \sin \omega_D t, \tag{31}$$

where mr is the moment of one of the masses about the suspension axis of the balance. If the damping parameter β of the balance is not too large then this torque would lead to an oscillation of the balance whose amplitude and phase with respect to the driving torque are given by

$$A = \pi G \rho \, \Delta B [r^2 \{(\omega_0^2 - \omega_D^2)^2 + 4\beta^2 \omega_D^2\}]^{-\frac{1}{2}} \alpha \lambda, \tag{32}$$

$$\Delta \phi = \tan^{-1} \frac{2\beta \omega_D}{\omega_0^2 - \omega_D^2}, \tag{33}$$

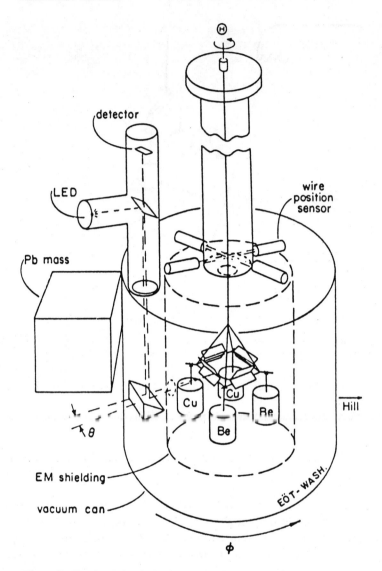

Figure 9. Torsion balance developed for the continuous Eötvös method (after Stubbs et al 1987).

where ω_0 is the natural period of oscillation of the balance. By measuring A the value of $\alpha\lambda$ can be estimated.

This is precisely the design of the instrument built by Stubbs *et al* (1987); a sketch of their apparatus is reproduced in figure 9. With this instrument they constrain any force which couples to baryons by $\alpha\lambda \lesssim 1$ cm.

2.4b *Modulated source experiment*: In certain locations the horizontal component of gravity undergoes repeated changes whose magnitude can be estimated; such a modulation allows a sensitive search for a new force. The main advantage of this method over the previous one is that the disturbances caused due to the rotation of the balance are avoided. Thus Bennett (1989) placed a Cu-Pb torsion balance next to a large lock on the Snake River where 1.7×10^5 tons of water could be 'turned' on or off within 12 minutes. Lateral gravitational accelerations of $\sim 4 \times 10^{-4}\,\mathrm{cm\,s^{-2}}$ could be generated in the absence of any differential acceleration between Cu and Pb at a level of $\sim 3 \times 10^{-8}\,\mathrm{cm\,s^{-2}}$ allowed them to set a limit of $\alpha\lambda < 10\,\mathrm{cm}$ for any isospin coupling.

2.4c *Static balance method*: The standard Cavendish-Eötvös design of the torsion balance is slightly modified to carry an aluminium vane suspended beneath the masses. The vane is constrained to move between capacitor plates at each end and the angle of the balance is sensed capacitively and a feedback circuit applies a d.c. voltage to the plates to keep the balance from deflecting under the influence of any external torque. This d.c. voltage serves as a measure of the torque on the balance. Such balances were developed by Fitch *et al* (1988) and were operated on the steep slopes of Mt. Maurice, Montana, to set limits $\alpha\lambda \lesssim 15\,\mathrm{cm}$ for both baryonic and isospin couplings.

2.4d *Frequency-mode Eötvös experiment*: Instead of attempting to measure the deflection of the torsion balance due to a torque excited by any new force Boynton *et al* (1987) looked for the shifts in the frequency of the balance expected when the composition axis initially normal to cliff face, is rotated through 180° to the antinormal position. Indeed the period of the balance gets modified depending on its angle with respect to the cliff-face according to the formula

$$\frac{\Delta T(\theta)}{T} \frac{G\rho\alpha\theta\Delta BT^2}{8\pi r}\cos\theta. \tag{34}$$

Here r is the radius of the balance made insensitive to gradients in gravity according to the prescription by Cowsik (1981) for example, and a new force coupling to baryon number has been assumed for specificity. The meticulous observations by Boynton *et al* (1987) with the apparatus sketched in figure 10, operated at Mt. Index yield $\alpha\lambda < 10\,\mathrm{cm}$ for the baryonic coupling and $\alpha\lambda < 10\,\mathrm{cm}$ for isospin couplings.

2.5 *Ròzsa-Sélenyl floating accelerometer*

This elegant instrument was first conceived by Ròzsa and Sélenyl (1931); it consists of a sphere which floats submerged in a liquid of a different composition but of identical density. In the presence of a force field which has a composition-dependent coupling the sphere will drift inside the liquid with a speed which is a measure of the difference in the force exerted on the fluid and that on the sphere. Intrinsically this instrument is capable of measuring differential accelerations $\sim 10^{-9}\,\mathrm{Gal} \sim 10^{-12}g_0$. Such an accelerometer was built by Thieberger (1987) and was operated near the top edge of the Palisades cliff, New Jersey, USA. A sketch of the apparatus is shown in figure 11. Initial data from this indicated the possible presence of a composition-dependent force with $\alpha\lambda \sim 100\,\mathrm{cm}$. However the searches by Bizzeti *et al* (1989) with an instrument of improved design near the Vallombrosa ridges, Florence, failed to confirm the earlier result but set the limit $|\alpha\lambda| < 25\,\mathrm{cm}$.

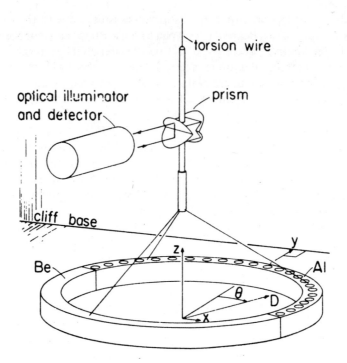

Figure 10. Torsion balance used in the frequency-mode Eötvös experiment (after Boynton et al 1987).

2.6 Dicke method

In the Eötvös experiment the centrifugal acceleration due to the diurnal rotation of the Earth is the basic source of the signal. To measure this the whole apparatus had to be rotated and during this rotation the balance could be disturbed and its sensitivity lowered. To overcome this problem Dicke (1962) noted that the rotation of the Earth had another consequence, namely, that the composition axis of any Eötvös balance with respect to the solar gravity field undergoes a sinusoidal modulation at the diurnal frequency, thus circumventing any need for the rotation of the balance. The magnitude of the solar acceleration is $\sim 0{\cdot}6$ cm s^{-2} and under its influence the 'zero' of any Eötvös balance, with a period P, short compared with 24 hours, will suffer a diurnal modulation with an amplitude given by the formula

$$A \approx a_\odot \frac{\alpha \Delta B \cdot P^2}{4\pi^2 r}. \tag{35}$$

Such experiments were conducted by Roll et al (1964) and by Braginsky and Panov (1972) to show that the weak equivalence principle was valid up to the 10^{-12} level and that long range forces ($\lambda \gtrsim 1$ AU) coupling to baryon number had couplings smaller than 10^{-9} of gravity per atomic mass unit. A floating version of this experiment was conducted by Kayser et al (1981) at a similar level of sensitivity.

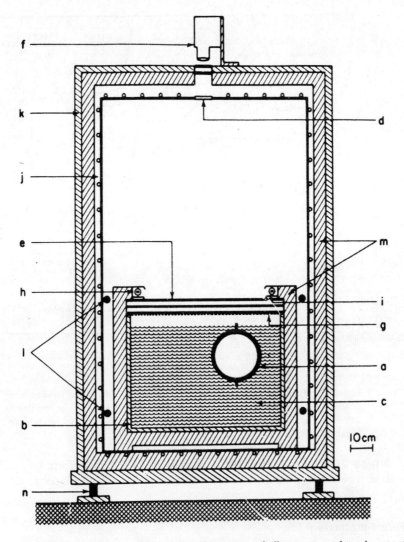

Figure 11. The floating differential accelerometer: *a*, hollow copper sphere; *b*, water tank; *c*, water at 3·9°C; *d* and *e*, observation windows, *f*, television camera; *g*, fine grid of copper mesh; *h*, illumination lamps; *i*, IR fitters; *j*, thermostatically controlled copper shield; *m*, Styrofoam insulation; *k*, magnetic shield; *l*, four coils for positioning the sphere with eddy currents (after Thieberger 1987).

3. Geophysical results and comparison with laboratory experiments

The various results obtained with methods discussed in §2 are summarized in figures 12–15 along with the results obtained with laboratory experiments. The main thing that strikes the eye in these figures is the remarkable variety of experimental techniques which when taken together put rather severe constraints on the strength of any new composition dependent-force, whatever be its range. On the other hand, experimental constraints are less stringent when it comes to forces which couple universally to mass; with a Yukawa-like form, they can indeed have couplings as large

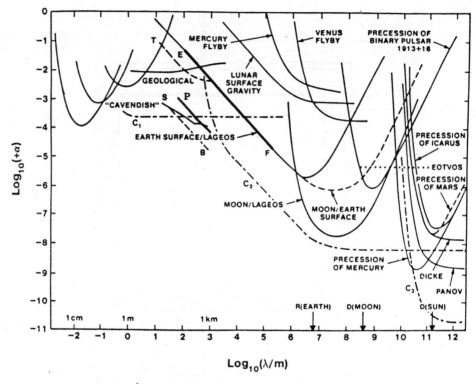

Figure 12. Existing and potential limits on the strength of any new force coupling to baryon number as a function of the assumed range (updated version of the one due to de Rújula 1986). C_1, C_2, C_3 represent capabilities of the instrument developed by the Tata Institute of Fundamental Research.

as ~ 0.01 to 0.001 of gravity if their range is ~ 20–200 m. Efforts to confirm this indication through well-controlled laboratory experiments are underway.

4. Future prospects

In this section we would like to point out the special features available in and around the Indian sub-continent where further studies of the type described in the previous sections can be extended and, in a few instances, improved upon significantly. To be brief, the discussions here would be restricted to three specific points: (1) Eötvös-type studies near mountains, (2) Airy-type studies at the Kolar Gold Fields, and (3) Airy-type studies in the seas.

The main advantage of Himalayan cliffs over the locations of the previous experiments is their height and steepness, which makes them suitable for studying forces with ranges up to 1 km. Secondly, they are situated at latitudes not too much to the south of the 45° parallel where the expected Eötvös signal would be close to its maximum, this being proportional to $\sin b \cos b$. Therefore, a single laboratory set-up there could be used for both the experiments. The selection of a proper site is very important for the success of any of these experiments. The instrument is to be located about half the way up a steep cliff so that the horizontal component of gravity due to

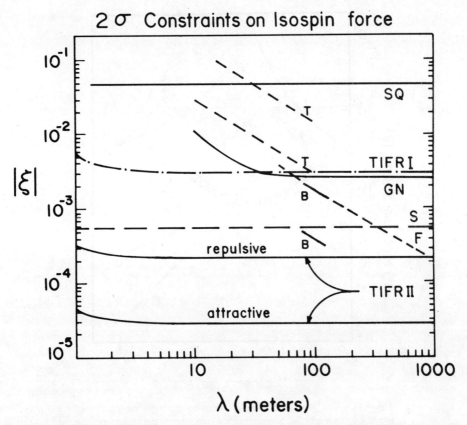

Figure 13. Limits on the strength of intermediate range forces coupling to nuclear isospin (after Cowsik *et al* 1989). The meaning of the symbols are: T, Thieberger (1987); B, Boynton *et al* (1987); GN, Bennett (1989); S, Stubbs *et al* (1989); F, Fitch *et al* (1988) and TIFR I, II, Cowsik *et al* (1988, 1989).

the cliff does not have a large vertical gradient. The cliff should be of uniform density and composition and the terrain both on and beyond the cliff, as also across the ravine or the valley on the other side, should be sufficiently simple and well-mapped so that the expected signals due to any non-Newtonian forces can be estimated accurately. If the cliff forms one face of ravine then the other face has to be removed from this by several kilometers so that the Yukawa fields generated by it are weak at the location of the instrument. If the ravine is considerably deeper than its width we would have the advantage of cancelling out or at least reducing the effects of gravity. The instrument will have to be housed in a cave, either natural or one that is specially excavated to reduce the temperature variations. An active collaboration between physicists and geophysicists would be essential for successfully launching a sensitive search for Yukawa fields at a Himalayan laboratory.

The Kolar Gold Fields, in a like fashion, provide excellent opportunities for Airy-type experiments up to depths of 3000 m! Not merely does this depth exceed that of the previous studies but the occurrence of the ore has been such that a vertical sheet ~ 7000 m long has been excavated up to these depths. This allows a two-dimensional mapping of gravity field which would provide a much more stringent test for the

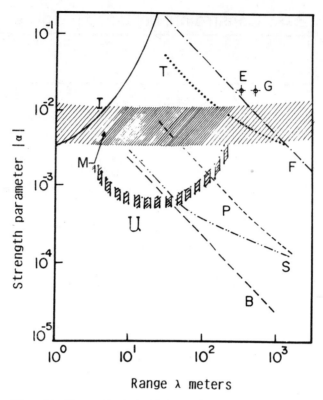

Figure 14. The 'geophysical window' showing the preferred region on the α-λ plane where the new force might lie. Notice that the composition-dependent experiments (broken lines) apparently contradict preferred region of a force of finite strength (solid line, hatched region and points). Existence of universal coupling to mass with a Yukawa like fall off with $10\,\text{m} < \lambda < 100\,\text{m}$ may resolve the contradiction. The references are: I, Panov and Frontov (1984); M, Moore *et al* (1988); T, Thieberger (1987); F, Niebauer *et al* (1987); E, Eckhardt *et al* (1988); S, Stubbs *et al* (1989) and B, Boynton *et al* (1987). The expected sensitivity for a new experiment planned at Uran, Maharashtra is $\alpha \sim 10^{-3}$ in the region $10\,\text{m} < \lambda < 100\,\text{m}$.

presence of any Yukawa field (Gaur *et al* 1988). One cannot omit to mention here that a careful gravitational survey of the mine would be of much interest as it might reveal other ore deposits in the area.

Finally, Airy-type of measurements would be of great interest if performed at sea, as the uniformity of density is guaranteed and one would be able to place much confidence in any results obtained pertaining to the hypothetical Yukawa force. In figure 15 a map of the Bouguer anomalies in the southern part of the Indian sub-continent is shown. This anomaly represents nothing but the difference between the observed value of the acceleration due to earth's gravity and the theoretically calculated one based on a model for the Earth as a whole, without any considerations of local isostacy etc. The measurements are to be carried out up to depths of $\sim 2500\,\text{m}$ at the centre of the extensive abyssal basins found in the Arabian sea, for example at a distance of $\sim 1000\,\text{km}$ off the West Coast of India. Notice that the Bouguer contours are very flat and quite small in this area indicating that there are not any serious density variations below the sedimentary layer. An Indian programme of deep sea studies in this region is bound to be most fruitful (Gaur *et al* 1988).

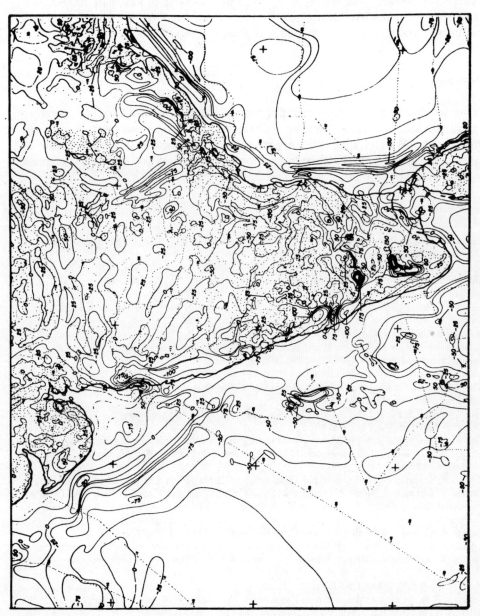

Figure 15. The contour map of Bouguer anomaly of the Indian sub-continent.

5. Summary

During the last few years there has been a tremendous resurgence of interest in geophysical studies of gravitation and feebler forces. Even though there is as yet no difinitive proof for the existence of a new force, there are indications of possible new forces much weaker than gravity and acting with a range of less than 1000 m. It is likely that great discoveries are to be expected in this field and the Indian subcontinent offers unique sites for conducting such studies.

Dedication

Anybody who knows Professor D Lal cannot but be impressed by his enormous enthusiasm for work and his great joive de viure, as I was, time and again over the last three decades. With a feeling of warm personal regard, I dedicate this article to him on his sixtieth birthday and wish him many more decades of joy and scientific enterprise.

References

Adelberger E G, Stubbs C W, Rogers W F, Raab F J, Heckel B R, Gundlach J H, Swanson H E and Watanabe R 1987 *Phys. Rev. Lett.* **59** 849
Ander M E *et al* 1989 *Phys. Rev. Lett.* **62** 985
Aronson S H, Cheng H Y, Fischbach E and Haxton W 1986 *Phys. Rev. Lett.* **56** 346
Bennett Wm R Jr 1989 *Phys. Rev. Lett.* **62** 365
Bizzeti P G, Bizzeti-Sona A M, Fazzini T, Perego A and Iacopetti N 1989 *Phys. Rev. Lett.* **62** 2901
Bertotti B, Ciufolini I and Bender P L 1987 *Phys. Rev. Lett.* **58** 1062
Bernstein J, Cabibbo N and Lee T D 1964 *Phys. Lett.* **12** 146
Boynton P E, Crosby D, Ekstrom P and Szumilo A 1987 *Phys. Rev. Lett.* **59** 1385
Braginsky V B and Panov V I 1972 *Sov. Phys. JETP* **34** 463
Chan H A, Moody M V and Paik H J 1982 *Phys. Rev. Lett.* **49** 1745
Christenson J H, Cronin J W, Fitch V L and Turlay R 1964 *Phys. Rev. Lett.* **13** 138
Ciufolini I 1986 *Phys. Rev. Lett.* **56** 278
Cohen S C and Smith D E 1985 *J. Geophys. Res.* **90B** 9217; 9215–9438
Cowsik R 1981 *Indian J. Phys.* **55B** 497
Cowsik R 1988 in *Highlights in gravitation and cosmology*, Proc. Int. Conf. Gravitation and Cosmology, Goa (eds) B R Iyer, A K Kembhavi, J V Narlikar and C V Vishveshwara (Cambridge: Cambridge University Press) p. 421
Cowsik R, Krishnan N, Tandon S N and Unnikrishnan C S 1988 *Phys. Rev. Lett.* **61** 2179
Cowsik R, Krishnan N, Tandon S N, Unnikrishnan C S and Saraswat P 1988 in Proc. 27th COSPAR meeting, Espoo, Finland in *Advances in Space Research* (eds) R Vessot and R Reasenberg (Oxford, UK: Pergamon Press)
Cowsik R, Krishnan N, Tandon S N and Unnikrishnan C S 1989 *Phys. Rev. Lett.* (in press)
Cowsik R, Krishnan N, Tandon S N, Unnikrishnan C S, Tarle G and Musser J 1989 (in preparation)
Cowsik R, Krishnan N, Unnikrishnan C S, Tandon S N and Saraswat P 1989 *Pramana – J. Phys.* **32** L303
Dicke R H 1962 *Phys. Rev.* **125** 2163
Eckhardt D H, Jekeli C, Lazarewicz A R, Romaides A J and Sands R W 1988 *Phys. Rev. Lett.* **60** 2567
Eötvös R V, Pekar D and Fekete E 1922 *Ann. Phys.* **68** 11 (Leipzig)
Fairbanks W M 1987 in Rencontre de Moriond, Neutrinos and exotic phenomena, editiones frontiers (ed.) Tran Thauh Van
Faller J E 1986 (private communication)
Fischbach E, Sudarsky D, Szafer A, Talmadge C and Aronson S H 1986 *Phys. Rev. Lett.* **56** 3
Fischbach E *et al* 1988 *Ann. Phys.* **182** 1

Fitch V L, Isaila M V and Palmer M A 1988 *Phys. Rev. Lett.* **60** 1801
Gaur V K, Chalam V and Cowsik R 1988 *NGRI-TIFR report*
Goldman T, Hughes R J and Nieto M 1988 *Sci. Am.* **258** 32
Hughes R, Goldman T and Neito M M 1989 Los Alamos-preprint (LA-Ur-89-1011)
Hsui A T 1987 *Science* **237** 881
Keyser P T, Faller J E and McLagan K H 1981 in *Precision measurement and fundamental constants, II*, (eds) B N Taylor and W D Phillips (Gaithersberg, MD, US: National Bureau of Standards) p. 639
Kuroda K and Mio N 1989 *Phys. Rev. Lett.* **62** 1941
Lee T D and Yang C N 1955 *Phys. Rev.* **98** 1501
Luther G G and Towler W R 1982 *Phys. Rev. Lett.* **48** 121
Moffat J W 1979 *Phys. Rev.* **D19** 3554
Moffat J W 1987 *Phys. Rev.* **D35** 3733
Moore G I, Stacey F D, Tuck G J, Goodwin B D, Linthorne N P, Barton M A, Reid D M and Agnew G D 1988 *Phys. Rev.* **38** 1023
Niebaur T M, McHugh M P and Faller J E 1987 *Phys. Rev. Lett.* **59** 609
Nordvett K 1988 *Phys. Rev.* **D37** 1070
Paik H J and Will C M 1989 Washington University Preprint
Panov V I and Frontov V N 1984 *Sov. Phys. JETP* **50** 852
Pound R V and Rebka G A 1959 *Phys. Rev. Lett.* **3** 439; 1960 **4** 337
Rapp R H 1974 *Geophys. Res. Lett.* **1** 35
Rapp R H 1986 (private communication)
Rapp R H 1987 *Geophys. Res. Lett.* **14** 730
Roll P G, Krotkov R and Dicke R H 1964 *Ann. Phys. (N.Y.)* **26** 442
Rózsa M and Sélenyl P 1931 *Z. Phys.* **71** 814
Rujula de A 1986 *Phys. Lett.* **180B** 213
Shapiro I I, Counselman C C and King R W 1976 *Phys. Rev. Lett.* **36** 555
Speake C C and Quinn T J 1988 *Phys. Rev. Lett.* **61** 1340
Stacey F D, Tuck G J, Moore G I, Holding S C, Goodwin B D and Zhan R 1987 *Rev. Mod. Phys.* **59** 157
Stubbs C W, Adelberger E G, Heckel B R, Rogers W F, Swanson H E, Watanabe R, Gundlach J H and Raab F J 1987 *Phys. Rev. Lett.* **58** 1070; 1989 **62** 609
Suzuki M 1986 *Phys. Rev. Lett.* **56** 1339
Talmadge C and Fischbach E 1988 in *Searches for new and exotic phenomena*, Proc. VIIth Moriond Workshop, Les Arcs, France
Thieberger P 1987 *Phys. Rev. Lett.* **58** 1066
Weinberg S 1964 *Phys. Rev. Lett.* **13** 495
Will C M 1981 *Theory and experiment in gravitational physics*, (Cambridge: Cambridge University Press)
Will C M 1989 *Phys. Rev. Lett.* **62** 369
Williams J G, Dicke R H, Bender P L, Alley C O, Carter W E, Currie D G, Eckhardt D H, Faller J E, Kaula W M, Mulholland J D, Plotkin H H, Poultrey S K, Shelus P J, Silverberg E C, Sunclair W S, Slade M A and Wilkinson D T 1976 *Phys. Rev. Lett.* **36** 551
Worden P W Jr and Everitt C W F 1974 in *Experimental gravitation*, (ed.) B Bertotti (New York: Academic Press) 381–401
Worden P W Jr 1976 *Cryogenic test of the equivalence principle*, Ph.D. thesis, Stanford University, USA

Electrodynamic control in planetesimal accretion

D A MENDIS and G ARRHENIUS
Department of Electrical and Computer Engineering, and Scripps Institution of Oceanography, University of California at San Diego, La Jolla, CA 92093, USA

Abstract. The observed orbital motion of dust particles in the Saturnian and Jovian magnetospheres is governed by electrodynamic and gravitational forces. Orbital evolution under their action leads to radial concentration into ringlets; gravito-electrodynamic focusing tends to agglomerate the dust in the ringlets to form larger bodies. Outside the Roche limit such longitudinal focusing has presumably been responsible for the formation of secondary bodies around their magnetized primaries; satellites around planets, and planets around the Sun.

In general, the electrodynamic forces, observed to pervade the space medium, profoundly influence the dynamics of charged dust particles and their motion relative to larger, gravitationally controlled bodies.

Keywords. Solar system; accretion; planetesimal; gravito-electrodynamics; magnetosphere; plasma.

1. Remodelling of the solar nebula

Attempts to modelling of the early phases in the evolution of the solar system have become increasingly realistic as space exploration has provided relevant observational material in the last two decades. The first set of experiments of importance in this context concerned the dynamic behaviour of interstellar and circumplanetary dust and plasma, particularly in the Jovian and Saturnian environment. Observation and clarification of the combined effects of electromagnetic and gravitational forces on dust particles immersed in the space medium has provided new insights into the motion of such particles in bound orbits around the Sun and the planets, and, thereby into the process of accretion of celestial bodies.

The second set of guiding information comes from the Earth-orbit based astronomical observations of protostars and young stars with masses similar to our Sun's, and with properties suggesting the presence of an accretion disk. From the emission and absorption characteristics of such objects, particularly T-Tauri stars, inferences can be made about the accretionary environment around the early Sun, including the controlling electrodynamic parameters, and the state of excitation of the nebular plasma.

The purpose of the present paper is to draw the attention to the observed behaviour of orbiting dust particles in the circumplanetary plasma of Jupiter and Saturn, the physical explanation of this behaviour, and the fundamental differences in the evolution of these observed systems compared to the accretional development commonly assumed in modelling of the formation of planets and satellites.

2. The formation of secondary bodies around primaries

The dynamic similarity of satellite systems around the outer planets to the planetary system itself leads one to the view that the same basic processes would have governed the formation of all these systems. This so-called heterogonic principle has been emphasized in the discussion of solar system evolution by Alfvén and Arrhenius (1976).

The new observations of protostellar regions (Dupree and Lago 1988) suggest, in agreement with earlier theoretical modelling (see Wood and Morfill 1988 for a review), that stars in the mass range of the Sun in their early luminous stages are surrounded by a thin accretion disc of "gas" and dust, in which planets and satellites are likely to form. The physical and chemical state of the disc medium inferred from observation differs drastically from the electrically neutral dust and gas in its ground state explicitly assumed or implied in most current models. The actual state instead corresponds to the expectations extrapolated from modern space probe measurements of the interplanetary and circumplanetary medium.

The validity of the assumption of a disk of dispersed matter as the penultimate state in the formation of the regular satellite systems is supported by the presently existing ring systems around the outer planets, particularly the highly structured system around Saturn. These systems may be considered to be in a state of arrested development in this embryonic stage by virtue of their being within the Roche limits of the respective planets, wherein the disruptive effects of the gravitational shear of the planet overcomes cohesive forces of agglomeration to larger bodies in the disc. Indeed as first pointed out by Alfvén (1954), the Saturnian ring system may be regarded as a time capsule showing us an earlier stage of satellite and planetary formation. As pointed out by Northrop and Connerney (1987), the material in the rings is continually being renewed by micrometeorite impact, but the distribution of matter, and the processes causing it, are likely to remain the same as in the early stages of satellite formation.

3. Electrodynamic model

In the present discussion we will start with the assumption of an existing disc of dispersed matter around the central body (the Sun or a planet) and discuss the subsequent evolution. We will further assume that the central gravitating body is also magnetized and that the medium in the disc is significantly ionized. Consequently the dust particles in the ring, as today, would be electrically charged by virtue of being immersed in the plasma. Furthermore these charged grains are now subjected to electrodynamic forces, besides the gravitational force of the central body, as they move around it. The third important factor is the intermittent force associated with interparticle collisions, which drives the accretion.

The main difference between this model and the currently most extensively developed planetary accretion models, particularly that of Safronov (e.g. 1969), is the non-inclusion of electrodynamic effects in the latter type. However, the combined importance of gravitational and electrodynamic forces in determining the motion of dust in the Saturnian ring system, manifest in radial spokes, wavy rings, etc. was demonstrated by the recent Voyager 1 and 2 observations (Mendis et al 1984). This underscores the importance of electrodynamic forces in circumsolar and circumplanetary ring systems also at the time of their original formation.

That sufficiently inelastic collision between particles in such rings would lead to radial focusing rather than radial dispersion was emphasized by Alfvén (1969) and analysed in detail by Baxter and Thompson (1973). Indeed the striking ringlet structure first observed around Saturn and subsequently around Jupiter and Uranus (and no doubt to be observed around Neptune during this year's encounter by Voyager 2) was anticipated by Baxter and Thompson (1973) on the basis of what they termed the 'negative diffusion' associated with inelastic collisions (see also Lin and Bodenheimer 1981).

While the ringlet structure can be maintained by the focusing effect of inelastic collisions it may be initiated by another gravito-electrodynamic effect as suggested by Houpis and Mendis (1983). The charged dust grains in the ring under the combined effect of planetary gravitation and Lorentz forces are moving with a speed intermediate between Kepler and co-rotation, while the ambient plasma moves with the co-rotation speed. The dusty plasma disc under these conditions constitutes a novel type of dust-ring current. The dusty plasma analog of the well-known finite resistivity 'tearing' instability of such a current sheet is shown to tear up the ring into ringlets with the typical widths and separations of the order observed.

The main outstanding problem of this model is how the ringlets would give rise to single large bodies (satellites) should they lie outside the Roche limit of this central body. This is a serious problem for a purely gravitational model, for the ultimate state, in the absence of perturbations, are particles of all sizes moving in unison like beads on a string, since the Kepler speed at any given distance is independent of the mass of the individual bodies.

Longitudinal focusing, necessary for the formation of a single body in such a stream, could be produced by the gravitational perturbation by a neighbouring large gravitating body as shown by Trulsen (1972). If the perturbation is too large, scattering rather than focusing, takes place. If the perturbation is sufficiently small, a propagating density wave is produced by the velocity modulation of the perturbation. While longitudinal focusing can take place within such regions of density enhancement, leading to one or several large bodies, the development must proceed rapidly if such bodies are to form. This model also has the unattractive aspect that it requires the pre-existence of large, accretionally unexplained, gravitating bodies in the vicinity of a stream in order to initiate the formation of planetesimals within it.

This difficulty is immediately removed when one recognizes that the charged dust grains within such a ring move in circular orbits with speeds that depend on the particle size. While the larger bodies, which are largely controlled by gravity, will move with velocities close to the Kepler velocity, the smallest grains, which are controlled mainly by electric field, will move with speeds that are close to the co-rotation velocity. The gravito-electrodynamic theory developed to describe this motion of charged grains around a gravitating magnetized planet (Mendis *et al* 1982), shows that the angular velocity Ω_g of a grain in an equilibrium circular orbit is given by

$$\Omega_g = \frac{\omega_0}{2}\left[1 \pm \sqrt{1-4\left(\frac{\Omega_c}{\omega_0}-\frac{\Omega_k^2}{\omega_0^2}\right)}\right], \tag{1}$$

where Ω_k is the local Kepler (angular) velocity, Ω_c is the spin rate of the central body and $\omega_0 = qB/mc$, q, being the electric charge, and m being the mass of the grain, while B is the local magnetic field strength. The $\pm$ signs in front of the square root term corresponds respectively to prograde and retrograde grain orbits.

All grains, large or small acquire the same electric potential ϕ with respect to the plasma. Assuming that they are spherical, with radius $= a$, $q = a\phi$, and $m = \frac{4}{3}\pi a^3 \rho$, ρ being the density. Consequently $\omega_0 \propto 1/a^2$, and so it is clear from (1) that Ω_g depends strongly on the grain radius. Consequently grains of different size within the ring should collide, provided they can overcome the Coulomb barrier, since they have the same electric polarity. This is possible if the energy, E_T, in the center of mass frame of the colliding bodies > the Coulomb potential energy of repulsion, E_c.

Since

$$E_c = q_1 q_2/(a_1 + a_2) = a_1 a_2 \phi^2/(a_1 + a_2) = a_2 \phi^2 (a_1 \gg a_2),$$

and

$$E_T = \tfrac{1}{2}\mu v_{\rm rel}^2 = \tfrac{1}{2}\frac{m_1 m_2}{m_1 + m_2} v_{\rm rel}^2 = \tfrac{1}{2} m_2 v_{\rm rel}^2 (m_1 \gg m_2),$$

$E_T > E_c$ leads to the result

$$v_{\rm rel}({\rm m/s}) > v_{\rm crit} = 0.23 |\phi|({\rm volts})/a_2(\mu{\rm m}). \tag{2}$$

The value to be assigned to ϕ is of course uncertain but if we take $\phi \approx -5\,{\rm V}$ corresponding to a 'warm' ($\sim 2\,{\rm eV}$) plasma, and $a_2 \approx 0.1\,\mu{\rm m}$ corresponding to the assumed smallest particles in the ring, $v_{\rm rel} \geqslant 10\,{\rm m/s}$.

Once again it is not possible to calculate the relative velocities between (say) a $0.1\,\mu{\rm m}$ grain and a larger $1\,\mu{\rm m}$ grain at a given distance from the Sun or any planet using equation (1), without some knowledge of the primordial magnetic field at that distance. But with the present-day values for the outer planets or with values inferred from chemical remanent magnetization in primitive meteorites (Banerjee and Hargraves 1971, 1972; Brecher and Arrhenius 1974; Suguira and Strangway 1988) the relative speeds in question $\gg v_{\rm crit} (\approx 10\,{\rm m/s})$, for the typical range of satellite distances.

We thus expect that this process of agglomeration is active, starting at the lowest end of the dust mass spectrum, and proceeding toward larger bodies. Once a large embryo with radius $\geqslant 100\,{\rm km}$ has been formed, its own gravitation would enhance the accretion rate. The present process is however what is needed to initiate and maintain the early stages of agglomeration in a thin ring. Gravitational instability in the ring is commonly invoked (Goldreich and Ward 1973) on the assumption of a high instantaneous dust mass density in the terrestrial planet region. However, this mechanism fails in the outer planet region because of insufficient density even in an "instantaneous" dust disc; the formation time for Neptune on the basis of solely mechanically controlled accretion would exceed the age of the solar system.

We thus see that the gravito-electrodynamic processes observed in the present-day dusty plasma discs around the outer planets can lead naturally not only to radial focusing and the formation of ringlets but also to longitudinal focusing resulting eventually in a single large body—a satellite or planet.

Acknowledgement

This work was supported by NASA grants NSG 7623 and NAGW 1031. Lively discussion with our great friend and guru Lal has provided a continuous source of inspiration over many years, in this and other fields of science.

References

Alfvén H 1954 *On the origin of the solar system* (Oxford: Clarendon Press)
Alfvén H 1969 Asteroidal jet streams; *Astrophys. Space Sci.* **4** 94
Alfvén H and Arrhenius G 1976 *Evolution of the solar system* (Washington D.C.: U.S. Govt. Printing Office) Special publication 345
Banerjee S K and Hargraves R B 1972 Natural remanent magnetization of carbonaceous chondrites and the magnetic field in the early solar system; *Earth Planet. Sci. Lett.* **17** 110
Baxter D and Thomson W E 1973 Elastic and inelastic scattering in orbital clustering; *Astrophys. J.* **183** 323
Brecher A and Arrhenius G 1974 The palaeomagnetic record in carbonaceous chondrites: natural remanence and magnetic properties; *J. Geophys. Res.* **79** 2081–2106
Dupree A K and Lago M T (eds) 1988 *Formation and evolution of low mass stars* (Dordrecht, Boston, London: Kluwer Academic Publishers) p. 462
Goldreich P and Ward W R 1973 The formation of planetesimals; *Astrophys. J.* **183** 1051
Houpis H L F and Mendis D A 1983 On the fine structure of the Saturnian ring system; *Moon and Planets* **29** 39
Lin D N C and Bodenheimer P 1981 On the stability of Saturn's rings; *Astrophys. J.* **248** L83
Mendis D A, Hill J R, Ip W -H, Goertz C K and Grün E 1984 Electrodynamic processes in the Saturnian ring system; in *Saturn* (eds) T Gehrels and M S Matthews (Arizona: University of Arizona Press)
Mendis D A, Houpis H L F and Hill J R 1982 The gravito-electrodynamics of charged dust in planetary magnetospheres; *J. Geophys. Res.* **87** 3449
Northrop T G and Connerney J E T 1987 Age of the Saturnian ring system; *Icarus* **87** 124
Safronov V S 1969 Evolution of the protoplanetary cloud and the formation of the earth and planets (Nauka, Moscow: Israel program for scientific translations)
Sugiura N and Strangway D W 1988 Magnetic studies of meteorites; in *Meteorites and the early solar system* (eds) J Kerridge and M Matthews (Tucson AZ: The University of Arizona Press)
Trulsen J 1972 Theory of jet streams; in *From plasma to planet* (ed.) A Elvius (New York: Wiley) p. 179
Wood J A and Morfill G E 1988 A review of solar nebula models; in *Meteorites and the solar system* (eds) J Kerridge and M Matthews (Tucson AZ: The University of Arizona Press)

New quantum chemical isotope effects: Applications to the pre-solar nebula, planetary atmospheres and interstellar space

MARK H THIEMENS and S K BHATTACHARYA*

Department of Chemistry, University of California at San Diego, La Jolla, CA 92093, USA
*Present address: Physical Research Laboratory, Ahmedabad 380 009, India

Abstract. A new variety of isotope effects has been discovered in recent years which has the unique feature that their fractionations are independent of mass. Some of these processes are dependent upon molecular symmetry and, hence, are rather general. For oxygen, the anomalous isotopic fractionation pattern duplicates that observed in the stratosphere and carbonaceous chondritic components, strongly suggesting the involvement of this unique quantum mechanical process in the early solar system and the earth's atmosphere. The features of the process enable a unique physical description of a quantum process, inverse predissociation, which previously was unrecognized. Other mass-independent fractionations involving gas phase decomposition, energy exchange and nuclear spin have been measured, and their possible roles in the cosmochemical environment are discussed.

Keywords. Isotope effects; pre-solar nebula; planetary atmospheres; interstellar space; gas phase decomposition; isotopic fractionation; hyperfine nuclear spin effect.

1. Introduction

The measurement of stable isotope ratio variations in planetary atmospheres, meteorites, terrestrial materials and in the world's ocean has been instrumental in detailing an enormous range of physical-chemical processes, ranging from hydrothermal circulation to the evolution of the solar system. These isotopic variations have classically been attributed to well-known, quantitatively characterized processes such as the position of chemical equilibrium in isotopic exchange reactions, phase change phenomena, diffusion and kinetics. The temperature dependency for isotopic exchange between coexisting minerals or mineral-liquids (cf. Urey 1947; Clayton 1981), and the ability to measure, e.g. the $^{18}O/^{16}O$ ratio to high precision ($\pm 0.1‰$), led to the development of the flourishing fields of palaeo-oceanography and palaeo-climatology.

In the field of meteoritics, the measurement of stable isotope ratios has been invaluable in providing details of the earliest history of the solar system. In practice, some of these measurements serve as chronometers (e.g. $^{147}Sm \rightarrow ^{143}Nd$ and $^{87}Rb \rightarrow ^{87}Sr$) where radiogenic decay results in alteration of nuclidic abundances of the daughter species or, as a measure of cosmic-ray exposure, such as in the case of ^{15}N production from spallation of ^{16}O on the lunar surface and in selected meteoritic phases. In the case of nitrogen, determination of the $^{15}N/^{14}N$ ($\delta^{15}N$) ratio in different components (differential thermal release fractions in the lunar case) may provide a direct measure of cosmic-ray exposure history.

Kinetic isotopic fractionations from processes such as diffusion have been instrumental in elucidating the dynamical history of the Martian atmosphere. The kinetic energy (Σ_i) of a particle (atom or molecule) is simply given by:

$$\Sigma_i = \tfrac{1}{2} m_i v_i^2, \tag{1}$$

where the subscript denotes for species i the mass (m_i) and velocity (v_i). The velocity ratio of two isotopically substituted molecules, i and j, is:

$$v_i/v_j = (m_j/m_i)^{1/2}. \tag{2}$$

As a consequence, the lighter isotope has a slightly greater velocity. In the case of the Martian atmosphere, photochemical-electrodissociative reactions occurring in the Martian exosphere produce high velocity nitrogen atoms following N_2 dissociation. Over a geologic time, preferential ^{14}N (with respect to ^{15}N) exospheric escape, due to its 3·5% greater velocity, has produced the observed (relative to terrestrial) 75‰ ^{15}N enrichment. Details of this model and the evolving Martian atmosphere and its isotopic composition are provided by McElroy (1972), McElroy et al (1976), Fox and Dalgarno (1983) and Yanagita and Imamura (1978).

In the field of meteoritics, multi-isotopic ratio mesurements have been invaluable in delineating the extent of admixture of different nucleosynthetic components. The first use of multi-isotopic measurements as a technique to distinguish nucleosynthetic (or in general, nuclear) from physical-chemical processes was by Hulston and Thode (1965a). Such distinction is based upon the premise that all physical and chemical fractionation processes are dependent upon mass. In the case of sulphur, with stable isotopes ^{32}S, ^{33}S, ^{34}S, the fractionation ratio for $\delta^{34}S$ ($^{34}S/^{32}S$) and $\delta^{33}S$ ($^{33}S/^{32}S$) is given by:

$$\delta^{33}S/\delta^{34}S \cong \frac{(1/m_{32}) - (1/m_{33})}{(1/m_{32}) - (1/m_{34})}, \tag{3}$$

or $\delta^{33}S = 0·515 \, \delta^{34}S$. In general then, a process producing a 2‰ fractionation in $\delta^{34}S$ correspondingly produces approximately half that in $\delta^{33}S$. Likewise for oxygen, the relation $\delta^{17}O \cong 0·5 \, \delta^{18}O$ is obeyed with the value of the $\delta^{18}O$ coefficient dependent upon the molecular mass of the oxygen species (cf. Matsuhisa et al 1978; Thiemens 1988). Figure 1 is a global representation of this phenomena, where it may be seen that lunar and terrestrial materials possess differing $\delta^{18}O$ values, but all are colinear along a fractionation line defined by $\delta^{17}O/\delta^{18}O \cong 0·5$. This derives from the common ancestry of the oxygen which has subsequently been submitted to a different physical-chemical, mass-dependent fractionation process.

Clayton et al (1973) were the first to observe a mass-independent fractionation in meteoritic material. As shown in figure 2, mineral separates from the Allende meteorite define a line with slope $\sim 1·0$, rather than 0·5. As discussed by Clayton et al (1973), it is assumed that a chemical production of mass-independent isotopic fractionations is impossible, therefore deviations, such as those observed in figure 2, must be the result of nuclear processes. In this specific case, the component is suggested as deriving from explosive carbon or helium burning, since essentially pure ^{16}O is produced in these processes and the observed meteoritic anomalies then derive from admixtures of relict grains from an astrophysical environment isotopically distinct from the earth and moon. As figure 3 shows, different nebular oxygen isotopic reservoirs are apparently defined by the characteristic bulk meteoritic $\delta^{17}O$, $\delta^{18}O$

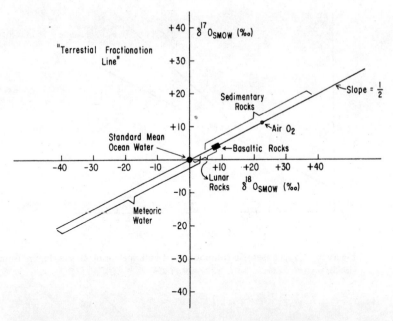

Figure 1. Oxygen isotopic composition of various lunar-terrestrial samples which define a mass-dependent fractionation line.

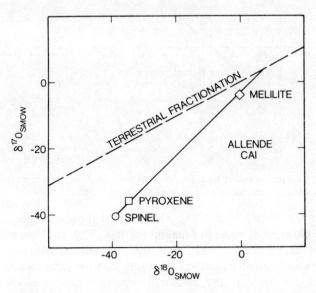

Figure 2. Oxygen isotopic composition of different mineral components of the Allende meteorite. Data are from Clayton and the figure is from Thiemens (1988).

signatures which are both enriched and depleted with respect to the terrestrial fractionation line. A recent review of meteoritic oxygen isotopes and the nebular history is given by Thiemens (1988). A striking feature of the measurements is the magnitude of the effect. The anomalies exist on a bulk sample scale in the most abundant element, yet no other element exhibits such behaviour. Large ^{50}Ti anomalies

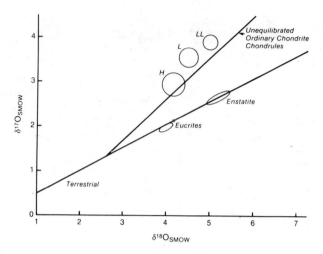

Figure 3. Oxygen isotopic composition of bulk meteoritic classes. Figure from Thiemens (1988).

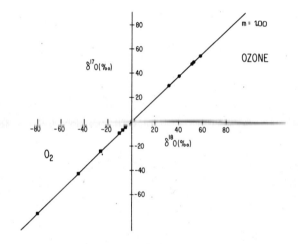

Figure 4. Oxygen isotopic composition of ozone from the dissociation of molecular oxygen at liquid nitrogen temperatures. Figure from Thiemens (1988).

have been detected (e.g. Fahey et al 1985), which exceed those of oxygen; however, the abundance of titanium is orders-of-magnitude less. The Ti/O cosmic ratio is 8.3×10^{-4} (Anders and Ebihara 1982), thus the anomaly in oxygen represents processing of significantly greater numbers of atoms than for any other element.

The fundamental premise that any measured departure from a mass-dependent isotopic fractionation in a multi-isotopic system (e.g. ^{16}O, ^{17}O, ^{18}O or ^{32}S, ^{33}S, ^{34}S, ^{36}S) must reflect some nuclear process such as radiogenic decay, spallation or nucleosynthesis is now known to be invalid. Thiemens and Heidenreich (1983) and Heidenreich and Thiemens (1983) have shown that the gas phase formation of ozone from $O + O_2$ produces a mass-independent isotopic fractionation with equal ^{17}O, ^{18}O enrichment of the O_3, as shown in figure 4. The slope of the O_2–O_3 line is, in fact, identical to that observed for mineral separates from the Allende meteorite (figure 2). The striking similarity raises the distinct possibility that the meteoritic oxygen isotopic composition may reflect a chemical process rather than nuclear; at the very least, it is no longer possible to simply presume that a mass-independent fractionation must be due to a nuclear process, particularly for oxygen.

Heidenreich and Thiemens (1986a) have suggested that ozone isotopic fractionation results from isotopic symmetry factors in the excited state O_3 stabilization process after experimentally ruling out the dissociation process (Heidenreich and Thiemens 1985). Thiemens (1988), based on the symmetry reaction mechanism, has suggested that reactions in an energetic environment such as $O + SiO$, $O + AlO$, $O + CO$, $O + FeO$ and $O + MgO$ would not only produce the $\delta^{17}O = \delta^{18}O$ fractionation, but reactions are also likely to occur in the very first stages of solar system formation. Further attention has been focussed on the symmetry isotopic fractionation process with the observations that stratospheric O_3 is enriched in ^{18}O by some 40% (Mauersberger 1981), and in a mass-independent fashion (Mauersberger 1987).

It is apparent that this new variety of isotopic fractionation is a relevant process in not only the early solar system, but planetary atmospheres as well. In addition, other non-Boltzmann types of fractionation processes may have application in geo-cosmochemical situations. This paper presents an overview of such mechanisms.

2. Isotopic fractionation processes and applications

This section will treat the conventional isotopic fractionation processes and, particularly, the new variety of mass-independent fractionation which may be relevant to geo- and cosmochemistry.

2.1 *Conventional equilibrium processes*

The process of isotopic fractionation, in a general sense, may be considered as a function of the differential partitioning of energy for isotopically substituted molecules. The total energy (E_T) for an ensemble of molecules in thermal equilibrium is describable in terms of the electronic, vibrational, rotational and translational energies. The Born-Oppenheimer approximation, which requires that a molecular potential energy surface be unaltered by isotopic substitution, leads to the result that, in thermal equilibrium, electronic effects do not give rise to isotopic fractionations. An excellent and thorough review of equilibrium isotope effects has been given by Kaye (1987). It should be noted, however, that the crossing between potential energy surfaces, such as that occurring in nonadiabatic molecular collisions, may, in fact, produce anomalous isotopic fractionations, as demonstrated by Valentini *et al* (1987) and Valentini (1987).

The distribution of energies of translation, rotation and vibration is then the source of equilibrium isotopic fractionations. However, as will be discussed, rotational effects (e.g. in symmetry processes), the coupling of vibration and translation (Treanor pumping) and nuclear spin may produce large, nonequilibrium fractionations.

The effect of isotopic substitution upon molecular vibrations may be treated by the quantum theory. A potential function, which describes the diatomic molecular vibration (cf. Herzberg 1950, 1966), has generally a form given by:

$$v(r) = D_e\{1 - \exp[\alpha(r - r_e)]\}^2, \tag{4}$$

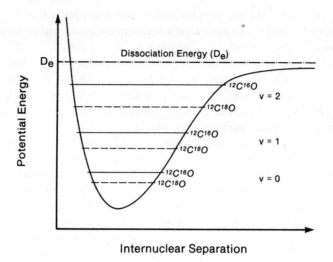

Figure 5. A potential energy diagram for the isotopically substituted CO molecule depicting the different vibrational lines.

with D_e equal to the depth of the potential well (see figure 5), r the internuclear separation, r_e the equilibrium separation distance and α a parameter which describes the steepness of the potential well. Often for simplicity, one may assume a harmonic oscillator potential function. In such case, the precise energy values (E) for the vibrational lines arise from the solution of the harmonic oscillator wave equation of the form:

$$(d^2\psi/dx^2) + (8\pi^2\mu/h^2)(E - \tfrac{1}{2}kx^2)\psi = 0, \tag{5}$$

where x is the internuclear displacement coordinate of the molecule, h the Planck's constant, k the force constant and μ the reduced mass. The solution of (5) yields the single-valued, finite energy wavefunction for only specified vibrational energy levels given by:

$$E(v) = \frac{h}{2\pi}(k/\mu)^{1/2}(v + \tfrac{1}{2}) \tag{6}$$

with the normal restriction that the vibrational quantum number v may only possess single and positive integers (0, 1, 2, 3...). The vibrational energy levels are quantized and discrete, with a minimum or zero point energy equal to $\tfrac{1}{2}hv$, where v is the fundamental vibrational frequency $v = (1/2\pi)(k/\mu)^{1/2}$. It is (6) which gives rise to the source of isotopic fractionations as a result of molecular vibrations. For example, the isotopically-substituted molecules $^{12}C^{16}O$, $^{12}C^{18}O$ have the reduced masses:

$$\mu_{16} = \frac{m_1 m_2}{m_1 + m_2} = \frac{12 \cdot 000 \times 15 \cdot 9949}{27 \cdot 9949} = 6 \cdot 8562,$$

and

$$\mu_{18} = \frac{m_1 m_2}{m_1 + m_2} = \frac{12 \cdot 000 \times 17 \cdot 9991}{29 \cdot 9991} = 7 \cdot 1999$$

and the vibrational frequencies will then differ by a factor

$$v_{18}/v_{16} = (\mu_{16}/\mu_{18})^{1/2} = 0.9758 = \rho.$$

As shown in figure 5, this accounts for the lower vibrational frequency of the heavier isotopic species, $^{12}C^{18}O$.

For the case of a polyatomic molecule, the difference in vibrational frequency may not be explicitly calculated, since the reduced masses are not a simple function and depend upon the vibrational mode (geometry), as well as the masses. As discussed in detail by Herzberg (1966), the Teller-Redlich approximation requires that, for a molecule which possesses multiple vibrations (zero order) $\omega_1, \omega_2, \omega_3, \ldots, \omega_f$ of a given symmetry type, the following relation exists:

$$\omega_1^* \omega_2^* \omega_3^* \cdots \omega_f^* = \rho \omega_1 \omega_2 \cdots \omega_f, \tag{7}$$

where the asterisk refers to the isotopically substituted species. The generalized formula (Herzberg 1945) for ρ is given by:

$$\rho = \frac{\omega_1^*}{\omega_1} \frac{\omega_2^*}{\omega_2} \cdots \frac{\omega_f^*}{\omega_f}$$

$$= \left[\left(\frac{m_1}{m_1^*}\right)^\alpha \left(\frac{m_2}{m_2^*}\right)^\beta \cdots \left(\frac{M^*}{M}\right)^t \left(\frac{I_x^*}{I_x}\right)^{\delta_x} \left(\frac{I_y^*}{I_y}\right)^{\delta_y} \left(\frac{I_z^*}{I_z}\right)^{\delta_z} \right]^{1/2}, \tag{8}$$

where $m_1, m_2 \ldots$ are the masses of the individual atoms, M the total molecular mass, t the number of translations for the specific symmetry type, I_x, I_y, I_z the moments of inertia and $\delta_x, \delta_y, \delta_z$ are either 1 or 0, depending upon whether or not the rotation about a specified axis represents a vibration which is not genuine for the symmetry type considered. It should be specifically noted that, if a molecule possesses multiple isotopic symmetries, such as $^{16}O^{18}O^{16}O$ and $^{16}O^{16}O^{18}O$, different vibrational frequencies will be obtained for a molecule with the same molecular mass but different symmetry. Approximations for the vibrational frequency shifts for isotopic molecules were given by Herzberg (1945). In practice, one determines the specific vibrational energy shift $\Delta\omega_1, \Delta\omega_2, \ldots \Delta\omega_f$ for an isotopically-substituted species for a specific symmetry and, if a molecule possesses more than one symmetry, all symmetries are accounted for.

As a direct consequence of the derived relations for vibrational frequencies for isotopically-substituted molecules, a precise equilibrium constant may be obtained (Urey 1947). As will be discussed later, components of this derivation are important in detailing the mechanisms in non-Boltzmann, mass-independent isotopic fractionation processes. Since the process of isotopic exchange is well known, only the key aspects will be touched upon; the reader is referred to reviews by Clayton (1981), Kaye (1987) and Thiemens (1988).

For an isotopic exchange which may be considered as a reaction, the equilibrium constant is expressed as the ratio of the molecular partition functions (Q) as:

$$K = \frac{\prod Q_{\text{products}}}{\prod Q_{\text{reactants}}}. \tag{9}$$

The total partition function of a molecule consists of components of translation, vibration and rotation. For the exchange of a common atom between molecules 1 and 2 (in diatomics), the ratio of the partition functions can be reduced to (Urey 1947)

$$Q_2/Q_1 = \frac{\sigma_1 u_2}{\sigma_2 u_1} \frac{\exp(-u_2/2)}{1-\exp(-u_2)} \frac{1-\exp(-u_1)}{\exp(-u_1/2)} \qquad (10)$$

and, for polyatomics it is:

$$Q_2/Q_1 = \frac{\sigma_1}{\sigma_2} \Pi_i \frac{u_{2i}}{u_{1i}} \frac{\exp(-u_{2i}/2)}{1-\exp(-u_{2i})} \frac{1-\exp(-u_{1i})}{\exp(-u_{1i}/2)}. \qquad (11)$$

These reduced partition functions by themselves then, in fact, represent the equilibrium constant for exchange reactions between the molecule and the separated atom. For example, in table X of Urey (1947), it may be seen, in $^{18}CO/^{16}CO$, the value for Q_2/Q_1 at 298·1°K is 1·1053 which, in the δ notation, denotes that, in exchange with the oxygen atom, CO is enriched in $^{18}O/^{16}O$ by 105·3‰, compared to the O-atom (the heavier isotope is always sequestered in the molecule).

For isotopic exchange between two different molecules, e.g. $H_2^{18}O$ and $^{16}O_2$, the equilibrium constant for exchange is simply the ratio of the reduced partition functions for the molecular species or:

$$K = (Q_2/Q_1)_{O_2}/(Q_2/Q_1)_{H_2O} \qquad (12)$$

which at 298·1°K, using the data in table X of Urey's paper, is

$$K = 1\cdot 0818/1\cdot 0667 = 1\cdot 0142, \qquad (13)$$

which, in δ notation, means that the $\delta^{18}O$ of O_2 will be 14·2‰ heavier than H_2O after equilibrium isotopic exchange at 298·1°K (assuming equal molar reservoir sizes). The absolute value for the difference in δ between the molecules is a function of temperature and increases with decreasing temperature. The temperature dependency of the equilibrium constant forms the cornerstone of paleo and geo-thermometry and need not be addressed here (cf. Clayton 1981). In meteoritic oxygen isotopic measurements, the principle of isotopic exchange may be used to place temperature constraints on events such as dust-gas exchange and reaction. For example, figure 6 is a schematic representation of the isotopic exchange of different meteoritic components. Chondrules separated from unequilibrated ordinary chondrites display isotopic variability and define a trend between 2 and 3 in figure 6, which has been suggested as deriving from incomplete isotopic exchange between a solid (or liquid) (square 3) and gas (circle 2) (Clayton et al 1983). Allende chondrules also appear to exhibit a similar feature with exchange occurring between a solid at square 4' and the same gas at circle 2 (Clayton et al 1983). The oxygen isotopic reservoirs in the early solar system were discussed by Thiemens (1988). An important feature of these exchange processes is the requirement of temperature. In the process of isotopic exchange between two reservoirs which do not lie along a slope $\frac{1}{2}$ line initially, there are constraints on the final isotopic composition. When two such isotopic reservoirs undergo exchange, a new line will be established which has a slope $\sim \frac{1}{2}$ (Matsuhisa et al 1978). Figure 7 illustrates the effect of isotope exchange between two initial

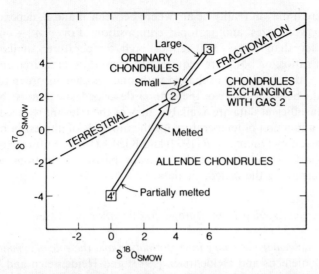

Figure 6. Process of oxygen isotopic exchange between meteoritic chondrules and a nebular gas (from Thiemens 1988).

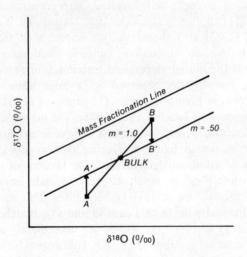

Figure 7. Effect of isotopic exchange between two components A and B with bulk isotopic composition at the star (*) after equilibration, line segment A'-B', passing through the bulk, is defined.

reservoirs A and B of equal size, with a beginning composition of slope = 1·00 for $A - B$ and a bulk isotopic composition, which lies mid-segment $A - B$. As exchange proceeds, the isotopic composition of each reservoir, A and B, moves towards a secondary mass fractionation line $A' - B'$ of slope = 0·50. For a "mixing line" to occur colinearly between A and B, the value of the isotopic difference between A and B $\delta(A) - \delta(B)$ must be significantly larger than $\delta(A') - \delta(B')$, otherwise a secondary $A' - B'$, as shown in figure 7, will be defined with the length and direction determined by the temperature and values of the molecular reduced partition function ratios for A and B. Note also that the bulk isotopic composition was arbitrarily chosen as mid-segment, where it may in reality be anywhere between A and B, depending upon their relative reservoir sizes and isotopic composition. For $\delta(A') - \delta(B')$ to be significantly smaller than $\delta(A) - \delta(B)$, either high temperature, similar reduced partition function ratios, or both, are required. In meteoritic measurements then, if two reservoirs (e.g. dust and gas) exchange with one another in order to define a colinear set of data between the two, the criteria described above must be met. At the present time, insufficient data are available, such as the relevant reduced partition function ratios as a function of temperature for the minerals of interest in meteorites, other than those given by Onuma et al (1972) and Kieffer (1982). Future experimental and theoretical work in determining the relevant partition functions will be of importance in interpreting the meteoritic data.

2.2 Non-Boltzmann mass-independent isotopic fractionation processes

2.2a The effect of molecular symmetry in O_3 formation and the role of symmetry in the pre-solar nebula: Thiemens and Heidenreich (1983) and Heidenreich and Thiemens (1983) clearly demonstrated that chemically produced mass-independent isotopic fractionation processes are possible. Although it was first suggested that the anomalous fractionation process was due to isotopic self-shielding, later experimental data (Heidenreich and Thiemens 1985) definitively ruled this out. Navon and Wasserburg (1985) and Kaye and Strobel (1983) also theoretically demonstrated that a self-shielding process does not produce an anomalous isotopic fractionation in O_3 formation. It is now known that the mass-independent fractionation process is not restricted to tesla discharges, but is also observed in a microwave discharge (Bains-Sahota and Thiemens 1989), at high-frequencies (Thiemens et al 1983) and by UV photolysis of oxygen (Thiemens and Jackson 1987, 1988). The mass-independent effect has been suggested as occurring during the process of inverse predissociation, which has been schematically shown in figure 8. During the process of inverse predissociation, a specific type of combination reaction, the lifetime of the metastable collisional complex, is quite long, as much as 10^{-8} seconds, compared to 10^{-13} seconds for a simple collisional time scale (Herzberg 1966). This stabilization occurs if the energy of the collisional pair, in this case O and O_2, matches that of the narrow range of diffuse energy levels of O_3 which lies immediately above the dissociation energy (1·0 eV). The actual number of $O - O_2$ collisions which result in the formation of a stable O_3 molecule will be determined by the overlap of the energy distribution of $O - O_2$ collision-pair, with the levels contained within the diffuse system. This is indicated in figure 8 as the peaks under the assumed $O - O_2$ energy distribution curve. The larger the number of levels within the diffuse system, the greater the probability of stabilization, since there is an increased number of stabilization modes. In the case of O_3, heteronuclear (C_s type) O_3 ($^{16}O^{16}O^{17}O$, $^{16}O^{16}O^{18}O$) has twice as many levels as homonuclear $^{16}O^{16}O^{16}O$ (C_{2v}), due to the appearance of alternate rotational lines in the diffuse band system for the heteronuclear, and which are absent for $^{48}O_3$. This results in an enhanced probability of stabilization and, as a consequence, O_3 is equally enriched in ^{17}O, ^{18}O on a symmetry, rather than mass-dependent basis. This model predicts that the fractionation process should be a function of isotopic abundance (Heidenreich and Thiemens 1986b) since, at higher abundances of ^{17}O, ^{18}O, the species $^{17}O^{17}O^{17}O$, $^{17}O^{16}O^{17}O$, $^{17}O^{18}O^{17}O$, $^{18}O^{18}O^{18}O$, $^{18}O^{17}O^{18}O$, $^{18}O^{16}O^{18}O$ may become signi-

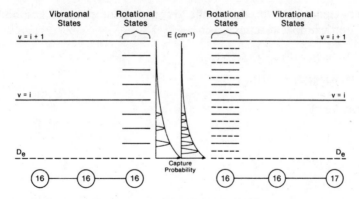

Figure 8. Schematic diagram of the process of inverse predissociation for $O + O_2$ reaction. Note that the asymmetric isotopic species has double the number of rotational states and an increased probability of stabilization.

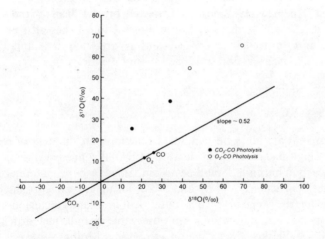

Figure 9. Oxygen isotopic composition of CO_2 produced from $O_2 + CO$ and $CO_2 + CO$ photolysis (starting compositions labelled).

ficant and which are C_{2v} type and, thus, have a decreased probability of stabilization. Yang and Epstein (1987a, b) and Morton *et al* (1988) have experimentally shown that the process is, in fact, dependent upon isotopic abundance.

An alternative mechanism has been proposed by Valentini (1987), resulting from nonadiabatic collision-induced isotopic fractionations between ground-state O_2 and O_2 ($^1\Delta g$) and due to selection rules in the process of potential surface crossing. More recently, however, (Barnes *et al* 1988) visible light O_3 dissociation-reformation experiments, where O_2 ($^1\Delta g$) is completely absent, demonstrate that the same mass-independent fractionation, as reported by Thiemens and co-workers, is observed, which rules out the Valentini mechanism. At present, many details remain to be worked out, but it appears certain that the fractionation process (a) is dependent upon isotopic symmetry and (b) involves the metastable transition state-O_3^*. These two criteria are, in fact, rather general and non-specific, and a wide-range of reactions which may have occurred in the earliest history of the solar system could have produced the oxygen isotopic anomalies observed in meteorites. For example, reactions such as

$$^{16}O + Si^{18}O \rightarrow {}^{16}O\,Si^{18}O, \tag{14}$$

$$^{16}O + Si^{17}O \rightarrow {}^{16}O\,Si^{17}O, \tag{15}$$

and similarly, for,

$$O + FeO \rightarrow O\,FeO, \tag{16}$$

$$O + MgO \rightarrow O\,MgO, \tag{17}$$

$$O + AlO \rightarrow O\,AlO, \tag{18}$$

are all plausible reactions, which fulfil the requirements for production of a mass-independent fractionation with $\delta^{17}O = \delta^{18}O$. As further evidence, Thiemens and Meagher (1987) experimentally observed a mass-independent fractionation in CO reactions, which has later been shown to be due to the $O + CO$ reaction (Bhattacharya and Thiemens 1988a, b), as expected on the basis of the symmetry mechanism. The details will be discussed in a later section. It has also been shown that the effect is not restricted to oxygen and is observed in sulphur as well (Bains-Sahota and Thiemens 1987a, b; 1988a, b), and reactions such as

$$S + FeS \rightarrow S\,FeS, \tag{19}$$

$$S + CS \rightarrow S\,C\,S, \tag{20}$$

plus reactions (14)–(18) are all good candidate reactions. In the case of oxygen, as first discussed by Clayton et al 1973, production of ^{16}O by a nucleosynthetic process, such as explosive carbon burning, should be accompanied by coproduction of integral α-particle products, particularly ^{24}Mg and ^{28}Si. There are no correlated silicon anomalies and, in fact, the range in $\delta^{30}Si$ is quite small in Allende inclusions (5–10‰), and no mass-independent fractionations are observable in bulk meteorites as there are in oxygen (Molini-Velsko 1983; Clayton et al 1985; Molini-Velsko et al 1986), with the bulk $\delta^{30}Si$ essentially identical for all meteoritic classes. If a reaction such as $O + SiO$ were responsible for chemically producing the observed meteoritic oxygen isotopic composition of, e.g. the H, L and LL chondrites (figure 3) from a single reservoir with an isotopic composition on the terrestrial fractionation line, no effect is expected in silicon since it is the rotational character provided by the terminal atoms which gives rise to the effect. In addition, the effect is not restricted to triatomics since it is known that larger numbers of atoms give rise to more vibrational degrees of freedom and a greater number density of levels. A further advantage of the chemical mechanism is that only one oxygen reservoir is required, whereas the nuclear model requires several isotopically distinct reservoirs, yet, as the chondrule data in figure 6 demonstrate, they have a common character in that they appear to have undergone exchange with a reservoir on the terrestrial fractionation line. In addition, the matrices of the same meteorites are different in a $\delta^{17}O$–$\delta^{18}O$ plot from the chondrules, as seen in figure 3, thus, the nuclear model requires, for the two meteorite classes alone (H and CV), at least 5 distinct reservoirs (Allende matrix and chondrules; H-chondrite matrix and chondrules and a gas at circle 2 in figure 6). At the same time that oxygen, the most abundant element in these stones, exhibits this behaviour, no other element demonstrates these characteristics of unique isotopic signatures on a bulk specimen scale. From the standpoint of the chemical model, future research focussed on two primary areas will aid in resolving this question as to the source of the anomalies. First, the demonstration that the anomaly is produced in reactions such as O + SiO will provide further evidence that this process may account for many of the anomalies, since it provides a plausible reaction which has a high probability of occurrence in the early solar system. This would begin to provide a framework for the construction of a kinetic mechanistic model for the synthesis of pre-solar minerals. A second area which may provide relevant information requires further meteoritic measurements. Of all the elements in the periodic table, other than oxygen, sulphur is the only one which (a) is multi-isotopic, (b) is abundant and (c) may participate in reactions which would be subject to the symmetry mechanism. As will be discussed, chemically-produced sulphur isotopic fractionations have now been established, and this premise is not strictly theoretical (Bains-Sahota and Thiemens 1987a, b; 1988a, b).

2.2b *Sulphur isotopic fractionations: Non-Boltzmann and meteoritic*: There are only a few meteoritic sulphur measurements for both $\delta^{34}S$ and $\delta^{33}S$ (plus $\delta^{36}S$) (Hulston and Thode 1965a, b; Rees and Thode 1977; Thode and Rees 1979), all of which demonstrated no isotopic anomaly at the $\sim \pm 0.3$‰ level. A $\delta^{33}S$ anomaly in an acid extract from Allende indicated an anomaly (Rees and Thode 1977), but a repeat of the experiment failed to reproduce the anomaly, and it is unclear whether the effect was real or not. As Pillinger (1984) discussed, there is a strong need for meteoritic sulphur isotopic measurements. In addition, with increased precision and the ability to measure mineral separates, the problems of aqueous alteration and secondary exchange may be avoided by careful choice of the proper refractory and unaltered phases. Such selection is difficult, but not impossible, and the observation of a $\delta^{33}S = \delta^{34}S$ fractionation would be strong evidence for the occurrence of the chemical mechanism in the early solar nebula.

Bains-Sahota and Thiemens (1987a, b; 1988a, b) have shown that a mass-independent fractionation is present in the formation of the symmetric molecule S_2F_{10}. Isotopic exchange between SF_5 radicals and SF_4 adds a second component to the observed fractionation, although calculation of the reduced partition functions allows the magnitude of the mass-dependent component to be estimated. After correction for the exchange, it is found that the magnitude of the mass-independent $\delta^{33}S = \delta^{34}S$ fractionation factor is ~ 5‰, compared to ~ 90‰ for ozone formation (Thiemens and Jackson 1988). The difference in magnitude has been attributed to the difference in the density of the vibrational levels for S_2F_{10}. It is known that the density of vibrational levels rapidly increases as the number of atoms in the molecule increases (Herzberg 1966). If anharmonicity is neglected, the density of levels at a particular energy $E_0(v_1, v_2, \ldots)$ above the lowest vibrational level is approximated as

$$D = \frac{E_0^{f-1}}{(f-1)!\Pi_i v_i}, \tag{21}$$

where D is the density of levels (per cm^{-1}), f the vibrational degrees of freedom and v_i are the fundamental vibrational frequencies (Herzberg 1966). The greater level density for S_2F_{10} results in line overlap; in fact, a limit is eventually approached where the level density exceeds the linewidth. Increased overlap then results in a decreased isotopic selectivity during the lifetimes of the transition state. The true quantum-mechanical basis for the precise role of the level density is not known at the present time. Future research, particularly on molecules of differing degrees of vibrational freedom, is needed to resolve the involvement of the diffuse levels in the isotopic fractionation process. This will be particularly important for future modelling of the early pre-nebular history where a wide variety of reactions and molecular constituents are possible. At present, however, it is certain that a chemically produced mass-independent fractionation does occur in sulphur, and it almost certainly involves symmetry.

2.2c *Treanor pumping and related isotopic fractionation*: It was shown by Treanor et al (1968) that, for a set of anharmonic oscillators, significant deviation from a vibrational Boltzmann distribution may take place, provided vibrational energy exchange is the dominant energy transfer mechanism operative after initial excitation of the system. The basic physical concept was extended to isotopic diatomic molecules by Belenov et al (1973), and can be explained as follows. Let us take a diatomic gas having two types of isotopic molecules, heavy (H) and light (L), whose fundamental vibrational frequencies are given by v_H and v_L obtained from the relation,

$$v_i = \frac{1}{2\pi}(k/\mu_i)^{1/2} \quad i = H, L, \tag{22}$$

where μ_H and μ_L denote the respective reduced masses of the molecules. Since $\mu_H > \mu_L$, we have $v_H < v_L$. The vibrational energy levels are given by

$$E_i = (V + \tfrac{1}{2})hv_i, \tag{23}$$

where V is the vibrational quantum number. The successive energy levels are separated by $\Delta E_i = hv_i$ and, therefore, the spacings are smaller for the heavier molecules, e.g. for carbon monoxide species $^{13}C^{16}O$ has spacing of $2122\,cm^{-1}$, compared to $2170\,cm^{-1}$ for $^{12}C^{16}O$. When such a gas is vibrationally excited (e.g. by glow discharge using microwave, RF, tesla coil or UV light) but kept cool rotationally or translationally, the energy redistribution among the isotopic molecules by collisions can only take place by $V - V$ exchange. For any given vibrational level V, $E_L > E_H$ and so by collisional transfer the light molecule can excite a ground-state heavy molecule to the level V, but a heavy molecule can only excite a ground-state light molecule to level $(V - 1)$. Analysis of such exchange, using the principle of detailed balance, shows that the closer spacings for H molecule result in relative overpopulation of the vibrational states of the H molecule as the whole system relaxes in climbing down the vibrational ladder. Clearly, such energy distributions for H and L molecules deviate significantly from equilibrium Boltzmann distribution during the relaxation period, as shown in figure 10. The non-equilibrium vibrational distribution is thus characterized by higher vibrational temperature T_H for the heavier molecules relative to T_L for L molecules. If during the relaxation time a chemical reaction becomes operative, the rate constants for the isotopic species are given by

$$k_i = A_i \exp[-E^*/kT_{\text{vibration}}], \tag{24}$$

where E^* is the activation energy, assumed equal for both H and L. Since $T_H > T_L$,

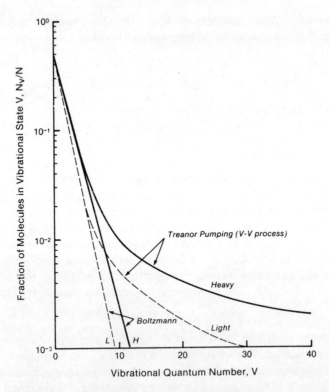

Figure 10. Vibrational excitation of isotopically substituted heavy and light molecules by Treanor pumping mechanism. The normal Boltzmann vibrational distribution shape is shown.

we have $k_H > k_L$, and the products of the reaction will be enriched in heavier isotopes. Such an enrichment has been observed spectroscopically in CO gas excited by infrared laser where the proposed reaction is

$$CO^* + CO \rightarrow CO_2 + C, \tag{25}$$

where CO* denotes the vibrationally excited species. Rich and Bergmann (1979) have observed that C_2, formed by gas phase $C + C$ reaction following (25), has a ^{13}C enrichment of several hundred per cent at room temperature. Excess ^{13}C ($\delta^{13}C = +1100‰$) has been observed in Murchison and Allende (Swart *et al* 1983) which has been attributed to a nucleosynthetic process and interstellar grain transport. The vibrational non-equilibrium isotopic fractionation process is ideally suited for the interstellar environment with many available sources of excitation, low pressure and a predominance of CO. Low temperature is not required either and, therefore, the possibility that the large observed meteoritic D, ^{15}N and ^{13}C enhancements may be derived from non-equilibrium isotope effects must be explored.

Recently, Thiemens and Meagher (1987) performed a discharge (0·5 MHz RF) experiment on pure CO and analysed the product CO_2 for oxygen isotopes, after converting it to O_2 by fluorination technique. They showed that the CO_2 is enriched in ^{17}O and ^{18}O, relative to CO (by about (10–30)‰ in $\delta^{18}O$), which can be explained by a minor involvement of Treanor pumping. In addition, they also noted the presence of a smaller mass-independent component. The ratio of the reaction rate constants for H and L type molecules is given by (Basov et al 1976),

$$k_H/k_L = \exp\left[\left(\frac{1}{T} - \frac{1}{T_H}\right)\frac{v_L - v_H}{v_L} \cdot \frac{E^*}{k}\right], \tag{26}$$

where T is the translational temperature and T_H the vibrational temperature for the H molecule. Under thermodynamically non-equilibrium conditions $T_H \gg T$, thus

$$k_H/k_L = \exp[(E^*/kT)(\Delta v/v_L)], \tag{27}$$

where $\Delta v = v_L - v_H$. For a typical example, assume $T = 300°K$, $E^* = 2.5\,eV$, then $k_{17}/k_{16} = 3.5$ and $k_{18}/k_{16} = 10.5$. For tesla discharge, experimental $k_{18}/k_{16} \sim 1.030$. By using this value, we can get an equivalent ε^*/T value for this reaction and obtain an estimate of k_{17}/k_{16} of about 1.016, which gives a slope of 0.52 in three-isotope plots. Since the data points lie above the line denoting the mass-dependent slope, presence of a second minor process involving mass-independent reaction is required. Based on the results presented in the next section, we postulate that this could be due to $O + CO \rightarrow OCO$ contribution which involves symmetry and mass-independence.

2.2d *Oxygen isotope studies in $O + CO \rightarrow CO_2$ reaction: Energy constraints in symmetry selective fractionation:* As we explained in the section on non-classical isotope effects, if a molecule having rotational asymmetry is formed via inverse predissociation, it is expected to show a mass-independent isotope effect in a three-isotope diagram. Carbon dioxide is a linear symmetric molecule having structure O C O in the ground-state. Consequently, $^{16}O-^{12}C-^{16}O$ is rotationally symmetric, whereas $^{17}O-^{12}C-^{16}O$ and $^{18}O-^{12}C-^{16}O$ molecules (the heavy isotopic species) are not. Thus, if CO_2 can be formed by an inverse predissociation mechanism through the diffuse band system, e.g. by a reaction of the type $O^* + CO \rightarrow CO_2$, the product CO_2 should show a mass-independent effect, i.e. $\delta^{17}O = \delta^{18}O$. However, if oxygen atoms do not have the correct energy to match the diffuse band levels, the symmetry selective fractionation may not be operative and the effect will disappear.

These ideas are being investigated, and preliminary results are available (Bhattacharya and Thiemens 1988a, b). The experiments have been designed to provide sources of oxygen atoms which may be energetically and eventually isotopically characterized, and which react with ground-state CO of known isotopic composition.

The first source of oxygen atoms was obtained from decomposition of ozone in a chamber filled with CO by a Hg UV continuum lamp (180–260 nm) which produces electronically excited $O(^1D)$ atoms. The product CO_2 was separated and analysed for oxygen isotopes ($\delta^{17}O$ and $\delta^{18}O$) after converting to O_2 by fluorination with BrF_5. Preliminary results indicate that the isotopic composition of the product CO_2 lies approximately midway between the CO and a mass fractionation line which passes through the O_3 isotopic composition, suggesting no symmetry selection in the process. Assuming that the dominant photodissociation is by the relatively more intense band around 2500 Å for Hg UV lamp, the total photon energy is about 5 eV. In the dissociation of O_3, where the sum of bond energy ($\sim 1.0\,eV$), the electronic energy of $O(^1D)$ ($\sim 2.0\,eV$) and $O_2(^1\Delta)$ ($\sim 1.0\,eV$) is 4 eV, the total available kinetic energy of the O atoms, and O_2 molecule is then only 1 eV, or about $\sim 0.5\,eV$ for the O atom.

In a second experiment, mixture of O_2 ($\sim$(25–50) μmol) and CO ($\sim$ 30 cm pressure in 5L vessel) was irradiated by a Kr UV continuum lamp (130–150 nm). O_2 is dissociated by $O_2 \rightarrow O(^3p) + O(^1D)$ and, in this case, the released kinetic energy of O-atoms has a distribution with a maximum at about 1·3 eV (Okabe 1978). This value is obtained from an energy balance: The peak energy of the incident photon is $\sim$ 9·7 eV (at 1400 Å peak region) minus the bond dissociation energy for O_2 ($\sim$ 5·1 eV) and the energy of excitation of the $O(^1D)$ ($\sim$ 2·0 eV), which results in a net remaining energy of 2·6 eV distributed between two O atoms. The product CO_2 in these experiments exhibits a large mass-independent enrichment of ^{17}O and ^{18}O (figure 9). These results suggest that the higher energy of the $O(^1D)$ atoms (1·3 eV relative to 0·5 eV) in O_2 dissociation initiates in the $O + CO \rightarrow CO_2$ reaction a symmetry selective effect via the inverse predissociation mechanism, since, in this specific case, the reaction pairs energy overlaps the diffuse band system, whereas, in the O_3 photolysis experiments, the O–CO reaction pair had insufficient energy to reach the diffuse band system.

The above results are preliminary and need to be further verified; they do seem to be consistent with the symmetry selective case of ozone formation and its modelling (Heidenreich and Thiemens 1986a). Since CO is an astrophysically important molecule and a dominant O-bearing species, the mass-independent effect observed in its reactions would be of direct application to the formation of isotopically fractionated reservoirs in the presolar nebula. Since CO is a major precursor oxygen source for planetary objects, which are likewise predominantly oxygen, studies of fractionation processes in CO gas phase chemistry are of importance for understanding presolar and, in general, interstellar chemistry.

2.2e *Hyperfine nuclear spin effects*:   A novel class of non-equilibrium isotope effects, which have been suggested to be potentially relevant for organic phases, are hyperfine nuclear spin effects, which are a specific manifestation of chemically-induced dynamic nuclear polarizations (CIDNP). The unique feature of these reactions is their kinetic mediation by nuclear spin, rather than activation energy. The spin of a nucleus is a quantum property which arises, due to the angular momentum resulting from rotation about its own axis. The spin quantum number may be positive or half-integral numbers, depending upon the specific nucleus. Nuclei such as ^{12}C or ^{16}O do not possess a nuclear spin, whereas ^{13}C and ^{17}O do ($+\frac{1}{2}$ and $\frac{5}{2}$, respectively). A reaction which proceeds by some mechanism mediated by nuclear spin would then selectively fractionate that specific isotope. Furthermore, since nuclei which possess a nuclear spin generate a magnetic moment, such a reaction would then be a magnetic isotope effect. One of the first discussions of this effect was by Buchachenko (1976) which was later experimentally demonstrated by Turro and Kraeutler (1978). Wigner's rule of total electron spin conservation in a reaction is well known not always to be obeyed. In particular, in a longlived state where a transition between states of non-identical spin and multiplicity may occur, magnetic effects may become important. The most prominent class of reactions in which such effects may occur are those involving unpaired electrons or radicals. If one considers, in general, a molecule $R_1 - R_2$, the electron spins of molecules $R_1 - R_2$ form spin pairs which cancel out; that is, the total electronic spin is zero (or a singlet, S). If the $R_1 - R_2$ molecule dissociates, the spins of the R_1, R_2 fragments may be either singlet or triplet (T). In order for $R_1 + R_2$ recombination to occur, the pair must have their spins antiparallel (S), which then requires that those fragments in the parallel, or triplet state, must "flip" from the T_+, T_0, T_- to the singlet state. The probability of such a transition is partly controlled by the interaction of the magnetic moment generated by the unpaired electron in the radical with the magnetic moment of the nucleus (hyperfine coupling). The magnetic torque exerted by hyperfine coupling ultimately results in a spin flip, with concomitant $T \to S$ conversion (intersystem crossing). Complete details of this process are given in Turro (1978). The most important aspect is that the rate of the recombination reaction ultimately depends upon the molecules' hyperfine coupling, and only a nucleus with a nuclear spin ($I \neq 0$) possesses a hyperfine component. For example, ^{17}O, with $I = 5/2$, would have a faster reaction rate than ^{16}O or ^{18}O which do not have a hyperfine moment and, therefore, no available mechanism for intersystem crossing. The $I = 0$ species do react, however, since in the process of dissociation approximately 25% of the radicals R_1, R_2 are in the singlet state and do not require intersystem crossing for reaction. The work of Turro and Kraeutler (1978) first demonstrated the magnetic isotope effect for ^{13}C, although, as discussed by Lawler (1980), a minimum of three isotopes is required in order to clearly demonstrate the existence of a nuclear spin isotope effect. Galimov (1978) reported that, in the auto-oxidation of ethylbenzene, a non-mass-dependent ^{17}O enrichment of $+130‰$ was observed due to the nuclear spin effect. Later duplication of these experiments by Thiemens and Clayton (1979) found strictly mass-dependent fractionation and concluded that the reported Galimov ^{17}O enhancement was due to sample contamination from OOH, produced in the mass spectrometer from the product which is a carboxylic acid. The OOH radical contributes to the mass 33 used for the $\delta^{17}O$ measurement of O_2 and, as a result of the low ^{17}O abundance, the presence of minor contamination produces major isobaric interference in $^{17}O^{16}O$. It now appears that the ^{17}O effect has been unambiguously demonstrated in the thermolysis of endoperoxides (Turro and Chow 1979). Regarding meteoritic and astronomical applications, magnetic isotope effects will not be of importance for gas phase reactions, since the diffusion of the radical pair is too fast to maintain the required cogency of the radical pair. However, since it is now well documented that an ordered surface dramatically enhances the magnetic isotope fractionation (Turro 1978; Epling and Florio 1981), the potential for occurrence in the early solar system (or interstellar space) on grain surfaces must be considered. Haberkorn et al (1977) presented a schematic model for ^{13}C, ^{15}N, ^{17}O enrichments of interstellar molecules on grain surfaces by chemical hyperfine interactions. Regarding heavier elements, there is experimental evidence that ^{117}Sn is enhanced in reactions involving $HSn(CH_3)_3$ (Podoplelov et al 1979), although there have been no reported duplications of this work. Since the general role and mechanisms of heterogeneous organic grain chemistry in the presolar nebula are poorly understood, it is impossible to evaluate the possible role of magnetic isotopic fractionations. However, the mechanism by which the effect occurs, and the large number of experimental observations demonstrating the ability of ordered surfaces to enhance the effect render it as at least a possible source of some of the measured anomalies in D, ^{13}C, ^{15}N in meteoritic material. Future work on the general aspects of organic reaction mechanisms on grain surfaces will be important for determining whether such isotopic fractionations occurred in the presolar nebula

2.2f *Isotopic fractionation in ozone decomposition*: As discussed in §2.2a, ozone formation in both the laboratory and atmosphere is accompanied by equal ^{18}O, ^{17}O enrichments. A quantitative model for the effect is still not available, owing mainly to the limited knowledge of ozone formation mechanism and the disparity between the large enhancement of ^{18}O in the stratosphere ($\sim 40\%$), and the $\sim 10\%$ maximum enhancement in any of the laboratory experiments remains a puzzle.

Dynamically, an ozone reservoir is subjected to both dissociation and production. Since it is known that ozone production is associated with a mass-independent fractionation, it is conceivable that the reverse process of ozone dissociation may not be mass-dependent. To test this premise, ozone dissociation studies were done recently by Bhattacharya and Thiemens (1988c), and the results are discussed below.

Dissociation of ozone is carried out by UV photolysis or heat, and the δ-value of the initial ozone and product oxygen are determined. The product O_2 is always (15–20)‰ lighter in ^{18}O than the progenitor ozone, and the fractionation of ^{17}O and ^{18}O is not strictly mass-dependent. The slope between ozone and oxygen, $\Delta(\delta^{17}O)/\Delta(\delta^{18}O)$ is ~ 0.59, compared to 0.51 expected for a mass-dependent process.

It was shown (Bhattacharya and Thiemens 1988a, b, c) that the mass independent component in ozone decomposition is intrinsic to the process and is not due to any secondary effect. The relevant reactions are:

$$O_3 + h\nu (\text{or } \Delta) \rightarrow O_2 + O, \tag{28}$$

$$O\cdot + O_3 \rightarrow 2O_2. \tag{29}$$

It is not clear, at present, whether reaction (28) or (29), or both, contribute to a component having mass-independence. The theory of thermal unimolecular dissociation (Troe 1977) applied to this case predicts a mass-dependent-fractionation, and reaction (28) is probably not responsible for the anomalous effect.

Dissociation of ground electronic state molecules occurs when the collisional energy transfer mechanism concentrates more than the dissociation energy (DE) in some fraction of the total molecules. Since the probability of such occurrence depends upon the magnitude of DE, which in turn depends upon the mass via the zero-point energy, any model of unimolecular ground-state decomposition is expected to produce a mass-dependent isotopic fractionation. On the contrary, if decomposition occurs in the excited electronic state through predissociation or other such processes, the transition selection rules for the isotopically substituted molecules do not strictly depend upon mass. For example, Valentini (1987) has shown that photolysis of ozone in the excited $^1B_2$ state by (230–310) nm UV absorption results in two products: $O_2(^1\Delta g) + O(^1D)$ and $O_2(^3\Sigma g) + O(^3P)$. These two exit channels result from two potential energy surface crossings at a particular configuration. Valentini (1987) demonstrated both experimentally and theoretically that such surface crossing is subject to symmetry and parity constraints, and the net effect is equal ^{17}O and ^{18}O isotopic enrichment of the $O_2(^3\Sigma g)$ and concomitant ^{17}O and ^{18}O isotopic depletion of the $O_2(^1\Delta g)$ state. If there is a chemical reaction subsequent to the curve-crossing selection which results due to the quantum-mechanical selectivity, a mass-independent fractionation will be sequestered.

Since the ozone molecule possesses a large number of low-lying excited state, it is likely that some of them are thermally populated. Consequently a mass-independent component may be present even in thermal decomposition, since different crossings may occur. Isotopic studies of ozone decomposition and similar symmetric molecules will be of great importance in clarifying mechanisms of dissociation in such cases, and will provide a more highly resolved picture of transition state and intermediate processes in unimolecular decomposition. Such information may eventually be important for the high temperature reactions in the early solar system.

2.2g *Self-shielding and isotopic fractionation*: Different isotopic molecules in a gas absorb at different wavelengths. Since these molecules also have different abundances, the incident photon flux is differentially modified after traversal inside the gas due to self-absorption. This may, in certain cases, give rise to local isotopic heterogeneity, as can be illustrated by the example of the oxygen-ozone system. Oxygen has three major isotopic species with abundances: $^{16}O^{16}O \sim 99.52\%$, $^{16}O^{17}O \sim 0.08\%$ and $^{16}O^{18}O \sim 0.40\%$. The major source of oxygen atoms in the atmosphere is photodissociation of O_2 in the wavelength region of 1759–1950 Å (Schumann-Runge band) where, due to symmetry considerations, $^{16}O^{16}O$ has half as many absorption lines as due to $^{16}O^{18}O$ and $^{16}O^{17}O$, and the various absorption lines, in general, do not overlap with each other. Even though $^{16}O^{16}O$ has less lines, each line has an absorption cross-section σ_{16} which is about twice the value of σ_{17} or σ_{18} ($^{16}O^{18}O$ or $^{16}O^{17}O$ lines). Since the spacing between these lines exceed the line width, most of the $^{16}O^{17}O$ and $^{16}O^{18}O$ lines fall in the windows of the $^{16}O^{16}O$ absorption spectrum; thence, the photon flux in wavelengths absorbed by $^{16}O^{16}O$ (most abundant) is much faster attenuated, with the result that the available photon fluxes corresponding to $^{16}O^{17}O$ and $^{16}O^{18}O$ lines remain relatively unchanged (very slow decrease with penetration depth). Consequently, as a function of distance from the source, photolysis of $^{16}O^{16}O$ decreases faster than the other two species. Hence, the $^{17}O^{16}O$ and $^{18}O^{16}O$ in product atoms increases equally with distance (see figure 11). Consequently, ozone produced from these atoms will carry the effect. Precise calculation of this effect, first proposed by Cicerone and McCrumb (1980), was done by Navon and Wasserburg (1985) and show that large mass-independent fractionations can be produced. However, in order to preserve the fractionation effect, two conditions must be fulfilled. First, the product atoms must retain the isotopic characteristics. Second, the product ozone must be trapped and removed. Kaye and Strobel (1983) pointed out that, for ozone formation, the rate of isotopic exchange between ground-state oxygen atoms ($O\,^3P$) and molecules ($O_2\,^3\Sigma$) is significantly faster than the rate of O_3 formation from $O + O_2$, thus, any self-shielding effect resulting during the isotopically selective O_2 photolysis is immediately lost when the atom exchanges with the essentially infinite O_2 reservoir. Navon and Wasserburg (1985)

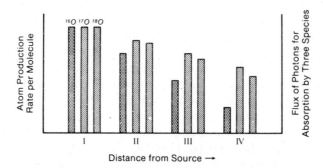

Figure 11. Schematic representation of isotopic self-shielding by $^{16}O_2$, $^{16}O\ ^{17}O$ and $^{16}O\ ^{18}O$. Note that the abundant $^{16}O\ ^{16}O$ species preferentially becomes opaque due to optical attenuation.

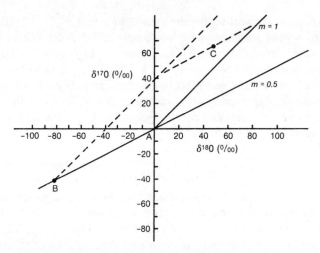

Figure 12. Representation of exchange of oxygen atom with its source oxygen molecule reservoir at A, acquiring an isotopic composition at B. Reaction with O_2 at A, and with a single stage $\delta^{17}O = \delta^{18}O \cong 100‰$ formation, produces O_3 at C.

theoretically demonstrated that, if a rapid trapping of the O_3 occurs, such as in the liquid nitrogen temperature experiments of Thiemens and Heidenreich (1983), the self-shielding effect may be preserved. Recent experiments by Thiemens and Jackson (1988) suggest that the premise may be significantly more complex. It is observed that O_3 produced by room temperature photolysis is essentially isotopically identical to that produced at liquid nitrogen temperature, except for a modest amount of fractionation resulting from secondary O_3 photolysis. This observation is somewhat paradoxical, since the rate constants for exchange are well established and the atom molecule reaction should occur. Yet, if one considers figure 12, an atom produced by O_2 photolysis, regardless of its isotopic composition, if it undergoes exchange with the dominant O_2 reservoir (at A), it will acquire an isotopic composition of $\delta^{18}O = -81.8‰$, $\delta^{17}O = -42.1‰$ (point B) according to the rules of isotopic exchange and the value of the reduced partition function for O_2, as described in §2.1. If an atom at B reacts with a molecule at A, and a mass-independent fractionation with $\delta^{17}O = \delta^{18}O \cong 100‰$ then occurs, O_3 at point C should be produced. If complete exchange does not occur, then the O_3 would be somewhere within the rhombus in figure 12. In addition, exchange is extremely temperature-dependent, e.g. for $O + O_2$ exchange, the $\delta^{18}O$ for the atom ranges from -81.8 to $-37.4‰$ between 300 and 500 degrees. For the photolysis experiment, a similar temperature difference is employed, yet the same isotopic results are obtained; thus, it is difficult to imagine how isotopic exchange could occur, at least in the conventional sense. At present, it is not known if this may be a result of the fact that isotopic exchange rules are derived for ground-state species and in this situation excited states are involved, or possibly a different transition state process is involved. Obviously, a great deal of work is needed to understand this paradox, and the fundamental nature of the transition state needs higher resolution.

Regarding applications of the general process of isotopic self-shielding, Bally and Langer (1982) have measured the ^{12}CO, ^{13}CO, $C^{18}O$ ratios in the molecular cloud boundary layer near the HII region S68 and suggested that an isotopic self-shielding photodestruction occurs, accounting for the measured larger $^{13}C/^{12}C$ and $^{18}O/^{16}O$ variations through ~ 9 arc minutes of the molecular cloud. Such effects have also been observed in the molecular clouds Taurus and ρ Ophiuchi by Langer and co-workers. Observations by Penzias (1983) of the same isotopic ratio in the galactic giant molecular clouds NGC 2264 and W3(OH) also exhibit radial isotopic abundance fractionations; however, these particular variations have been attributed to an association between velocity wings and cloud edges, whereby the fractionation derives from large scale radial motion.

Self-shielding has also been observed to occur in other non-molecular cloud astronomical environments. Morris and Jura (1983) have observed isotopic self-shielding by CO in molecular outflow of IRC 10216. Chu and Watson (1983) have theoretically modelled the effects of isotopic self-shielding in the general interstellar environment and have shown that the process may be significant for $^{18}O/^{18}O$, but less so for the $^{13}C/^{12}C$, primarily due to the abundances. Their model has also built in the mediating factor of equilibrium isotope exchange, unlike other models. As essentially all the workers conclude, the process of isotopic self-shielding needs experimental studies to fortify the astronomical observation measurements. At the present time, there are *no* relevant experimental observations.

3. Conclusions

It is now well known that mass-independent isotopic fractionations may be produced by chemical processes. These unique isotope effects appear to be rather general, and a wide extent of natural environments, ranging from stratospheric ozone, the pre-planetary nebula to stellar outflows and interstellar molecular clouds, may produce such fractionations. In addition to the information these particular processes provide, laboratory isotopic analysis of these unique fractionations detail physical-chemical and quantum effects which are essentially unobservable by any other spectroscopic technique. The role of symmetry during inverse predissociation, transition state effects during unimolecular decomposition, quantum selection rules during potential surface crossing, nuclear spin mediation of reaction rates and vibrational energy transfer are all specific examples of processes which have been elucidated, using isotopic analysis. These specific examples are also isotopic fractionations which may be relevant to different natural situations and eventually may provide more highly resolved details of different astrophysical processes.

Although major advances have been achieved in experimental studies of mass-independent isotopic fractionation processes, a prodigious amount of research remains to be done. As discussed in this text, symmetry-dependent reactions need to be further investigated, particularly for molecules of differing composition, atom number and bond strength. Such investigations are especially important for studies of the early solar system where a wide variety of reaction types, which are symmetry-dependent, may occur and subsequently account for the observed meteoritic oxygen isotopic compositions. At present, insufficient physical-chemical details of the fractionation process are available to construct a solar-system wide model.

Simple processes, such as thermal decomposition appear to produce mass-independent fractionations and, of course, occur in a wide extent of cosmochemical-atmospheric situations. At present, the mechanism for this observed anomalous behaviour is completely unknown and warrants further investigation.

The process of isotopic self-shielding has been theoretically explored and has been attributed as the source of observed carbon and oxygen isotopic compositions in several interstellar clouds and molecular outflow regions. There are no relevant laboratory studies of such processes, though such research would be of value to astrophysicists.

Acknowledgements

M H Thiemens wishes to acknowledge support from NASA NAG 9-83 and NSF ATM 85-15648.

References

Anders E and Ebihara M 1982 Solar system abundances of the elements; *Geochim. Cosmochim. Acta* **46** 2363-2380
Bains-Sahota S K and Thiemens M H 1986 Mass-independent oxygen isotopic fractionation in the microwave region; *Lunar and Planet. Science* XVII, (Houston, Texas: Lunar and Planet. Inst.) 20-21
Bains-Sahota S K and Thiemens M H 1987a Mass-independent oxygen isotopic fractionation in a microwave plasma; *J. Phys. Chem.* **91** 4370-4374
Bains-Sahota S K and Thiemens M H 1987b A chemically-produced non-mass-dependent sulphur isotope effect; *Meteoritics* **22** 320-321
Bains-Sahota S K and Thiemens M H 1988a A mass-independent sulphur isotope effect: relation to ozone; *Eos* **69** 308
Bains-Sahota S K and Thiemens M H 1988b A mass-independent sulphur isotope effect in the non-thermal formation of S_2F_{10}; *J. Chem. Phys.* (submitted)
Bally J and Langer W D 1982 Isotope selective photodestruction of carbon monoxide; *Astrophys. J.* **255** 143-148
Barnes J, Morton J and Mauersberger K 1988 Laboratory studies of ozone isotopic enrichment: ozone re-formation after visible light dissociation; *Eos* **69** 308
Basov N G, Belenov E M, Isakov V A, Markin E P, Oraevskii A N, Romanenko V I and Ferapontov N B 1976 Kinetics of nonequilibrium chemical reactions and separation of isotopes; *Sov. Phys.-JETP* **43** 1017-1019
Belenov E M, Markin E P, Oraevskii A N and Romanenko V I 1973 Isotope separation by infrared radiation; *JETP Lett.* **18** 116-117
Bhattacharya S K and Thiemens M H 1988a Oxygen isotope fractionation in O + CO reaction; *Eos* **69** 308
Bhattacharya S K and Thiemens M H 1988b Oxygen isotope studies in O + CO reaction: energy constraints in symmetry selective fractionation. *Lunar and Planet. Science XIX* (Houston, Texas: Lunar and Planet Inst.) 71-73
Bhattacharya S K and Thiemens M H 1988c Isotopic fractionation in ozone decomposition; *Geophys. Res. Lett.* **15** 9-12
Buchachenko A L 1976 Magnetic effects in chemical reactions; *Russ. Chem. Rev.* **45** 375-390
Chu Y -H and Watson W D 1983 Further analysis of the possible effects of isotope-selective photodissociation on interstellar carbon monoxide; *Astrophys. J.* **267** 151-155
Cicerone R J and McCrumb J L 1980 Photodissociation of isotopically heavy O_2 as a source of atmospheric O_3; *Geophys. Res. Lett.* **7** 251-254
Clayton R N 1981 Isotopic thermometry; in *Advances in physical geochemistry, volume 1, thermodynamics of minerals and melts* (eds) R C Newton, A Navrotsky and B J Wood (New York: Springer-Verlag)
Clayton R N, Grossman L and Mayeda T K 1973 A component of primitive nuclear composition in carbonaceous meteorites; *Science* **182** 485-488
Clayton R N, Mayeda T K and Molini-Velsko C A 1985 Isotopic variations in solar system material: evaporation and condensation of silicates; in *Protostars and Planets II* (eds) D C Black and M S Matthews (Tucson, Arizona: Univ. of Arizona Press)

Clayton R N, Onuma N, Ikeda Y, Mayeda T K, Hutcheon I D, Olsen E and Molini-Velsko C 1983 Oxygen isotopic compositions of chondrules in Allende and ordinary chondrites; in *Chondrules and their origins* (ed.) E A King (Houston, Texas: Lunar and Planet Inst.)

Epling G A and Florio E 1981 Isotope enrichment by photolysis on ordered surfaces; *J. Am. Chem. Soc.* **103** 1237–1238

Fahey A, Goswami J N, Mckeegan K O and Zinner E 1985 Evidence for extreme ^{50}Ti enrichments in primitive meteorites; *Astrophys. J.* **296** L17–20

Fox J L and Dalgarno A 1983 Nitrogen escape from Mars; *J. Geophys. Res.* **88** 9027–9032

Galimov E M 1978 Isotopic nuclear spin effect: a new type of isotope effect; *Fourth Int. Conf. on geology, geochronology and isotope geology* (Colo: Snowmass)

Haberkorn R, Michel-Beyerle M E and Michel K W 1977 Isotope effects in interstellar molecules by chemical fine interaction; *Astron. Astrophys.* **55** 315–318

Heidenreich J E and Thiemens M H 1983 A non-mass-dependent isotope effect in the production of ozone from molecular oxygen; *J. Chem. Phys.* **78** 892–895

Heidenreich J E and Thiemens M H 1985 The non-mass-dependent oxygen isotope effect in the electro-dissociation of carbon-dioxide: A step towards understanding NoMad chemistry; *Geochim. Cosmochim. Acta* **49** 1303–1306

Heidenreich J E and Thiemens M H 1986 A non-mass-dependent oxygen isotope effect in the production of ozone from molecular oxygen: The role of molecular symmetry in isotope chemistry; *J. Chem. Phys.* **84** 2129–2136

Heidenreich J E and Thiemens M H 1986b The effect of isotope abundance of Nomadic chemistry; *Lunar and Planet. Sci. XVII* (Houston, Texas: Lunar and Planet. Inst.) pp. 329–330

Herzberg G 1945 *Molecular spectra and molecular structure. II. Infrared and Raman spectra of polyatomic molecules* (New York: Van Nostrand Reinhold)

Herzberg G 1950 *Molecular spectra and molecular structure. I Spectra of diatomic molecules* (New York: Van Nostrand Reinhold)

Herzberg G 1966 *Electronic spectra of polyatomic molecules* (New York: Van Nostrand Reinhold)

Hulston J R and Thonde H G 1965a Variations in the ^{33}S, ^{34}S and ^{36}S contents of meteorites and their relation to chemical and nuclear effects; *J. Geophys. Res.* **70** 3475–3484

Hulston J R and Thode H G 1965b Cosmic ray produced ^{36}S and ^{33}S in the metal phase of iron meteorites; *J. Geophys. Res.* **70** 4435–4442

Kaye J A 1987 Mechanisms and observations for isotope fractionation of molecular species in planetary atmosphere; *Rev. Geophys.* **25** 1609–1658

Kaye J A and Strobel D F 1983 Enhancement of heavy ozone in the earth's atmosphere?; *J. Geophys. Res.* **88** 8447–8452

Kieffer S W 1982 Thermodynamics and lattice vibrations of minerals: 5. Applications to phase equilibria, isotopic fractionation and high-pressure thermodynamic properties; *Rev. Geophys. Space Phys.* **20** 827–849

Lawler R G 1980 Criteria for establishing the existence of nuclear spin isotope effects; *J. Am. Chem. Soc.* **102** 430–431

Matsuhisa Y, Goldsmith J R and Clayton R N 1978 Mechanisms of hydrothermal crystallization of quartz at 250°C and 15 kilobars; *Geochim. Cosmochim. Acta* **42** 173–180

Mauersberger K 1981 Measurement of heavy ozone in the stratosphere; *Geophys. Res. Lett.* **8** 935–937

Mauersberger K 1987 Ozone measurements in the stratosphere; *Geophys. Res. Lett.* **14** 80–83

McElroy M B 1972 Mars: an evolving atmosphere; *Science* **175** 443–445

McElroy M B, Yung Y L and Nier A O 1976 Isotopic composition of nitrogen: implications for the past history of Mars' atmosphere; *Science* **194** 70–72

Molini-Velsko C A 1983 *Isotopic composition of silicon in meteorites* Doctoral Dissertation, University of Chicago, p. 174

Molini-Velsko C A, Mayeda T K and Clayton R N 1986 Isotopic composition of silicon in meteorites; *Geochim. Cosmochim. Acta* **50** 2719–2726

Morris M and Jura M 1983 Molecular self-shielding in the outflows from late-type stars; *Astrophys. J.* **264** 546–553

Morton J, Schueler B and Mauersberger K 1988 Ozone formation using isotopically enriched molecular oxygen; *Eos* **69** 308

Navon O and Wasserburg G J 1985 Self-shielding in O_2 – a possible explanation for oxygen isotopic anomalies in meteorites?; *Earth Planet Sci. Lett.* **73** 1–16

Onuma N, Clayton R N and Mayeda T K 1972 Oxygen isotope cosmothermometer; *Geochim. Cosmochim. Acta* **36** 169–188

Okabe H 1978 *Photochemistry of small molecules* (New York: Wiley-Interscience) pp. 431

Penzias A A 1983 Isotopic fractionation and mass motion in giant molecular clouds; *Astrophys. J.* **273** 195–201

Pillinger C T 1984 Light element stable isotopes in meteorites-from grams to picograms; *Geochim. Cosmochim. Acta* **48** 2739–2766

Podoplelov A V, Leshina T V, Sagdeev R Z, Molin Y N and Goldanskii V I 1979 Application of the magnetic isotope effect to the separation of heavy isotopes with Sn as an example; *JETP Lett.* **29** 419–421

Rees C E and Thode H G 1977 A ^{33}S anomaly in the Allende meteorite; *Geochim. Cosmochim. Acta* **41** 1679–1687

Rich J W and Bergman R C 1979 C_2 and CN formation by optical pumping of CO/Ar and CO/N_2/Ar mixture at room temperature; *Chem. Phys.* **44** 53–64

Swart P K, Grady M M, Pillinger C T, Lewis R S and Anders E 1983 Interstellar carbon in meteorites; *Science* **220** 406–410

Thiemens M H and Clayton R N 1979 A search for isotopic nuclear spin effects; *Meteoritics* **14** 545

Thiemens M H and Heidenreich J E 1983 The mass-independent fractionation of oxygen: a novel isotope effect and its possible cosmochemical implications; *Science* **219** 1073–1075

Thiemens M H, Gupta S and Chang S 1983 The observation of mass-independent fractionation of oxygen in RF discharge; *Meteoritics* **18** 408

Thiemens M H and Meagher D 1987 Demonstration of a mass-independent isotopic fractionation in CO reaction; *Lunar and Planet. Sciences XVIII* (Houston, Texas: Lunar and Planet Inst.) 1006–1007

Thiemens M H and Jackson T 1987 Production of isotopically heavy ozone by ultraviolet light photolysis of O_2; *Geophys. Res. Lett.* **14** 624–627

Thiemens M H and Jackson T 1988 New experimental evidence for the mechanism for production of isotopically heavy O_3; *Geophys. Res. Lett.* **15** 639–647

Thiemens M H 1988 Heterogeneity in the Nebula: Evidence from stable isotopes; in *Meteorites and the early solar system* (eds) J F Kerridge and M S Matthews (Tucson, Arizona: Univ. Arizona Press)

Thode H G and Rees C E 1979 Sulphur isotopes in lunar and meteorite samples; *Proc. Lunar and Planet. Sci. Conf. X.* (Houston, Texas: Lunar and Planet. Inst.) 1629–1636

Treanor C E, Rich J W and Rehm R G 1968 Vibrational relaxation of anharmonic oscillators with exchange dominated collisions; *J. Chem. Phys.* **48** 1798–1807

Troe J 1977 Theory of thermal unimolecular reactions at low pressures, II, Strong collision rate constants: Applications; *J. Chem. Phys.* **66** 4758–4775

Turro N J and Kraeutler B 1978 Magnetic isotope and magnetic field effects on chemical reactions, sunlight and soap for the efficient separation of ^{13}C and ^{12}C isotopes; *J. Am. Chem. Soc.* **100** 7432–7434

Turro N J 1978 *Modern molecular photochemistry* (California: Benjamin-Cummings) p. 628

Turro N J and Chow M F 1979 Magnetic field effects on the thermolysis of aromatic compounds. Correlations with singlet oxygen yield and activation entropies; *J. Am. Chem. Soc.* **101** 3701–3703

Urey H C 1947 The thermodynamic properties of isotopic substances; *J. Chem. Soc.* (*London*) 562–581

Valentini J J 1987 Mass-independent isotopic fractionation in nonadiabatic molecular collisions; *J. Chem. Phys.* **86** 6757–6765

Valentini J J, Gerrity D P, Phillips D L, Nieh J C and Tabor K D 1987 CARS spectroscopy of $O_2(^1\Delta g)$ from the Hartley band photodissociation of O_3 dynamics of the dissociation; *J. Chem. Phys.* **86** 6745–6756

Yanagita S and Imamura M 1978 Excess ^{15}N in the Martian atmosphere and cosmic rays in the early solar system; *Nature* (*London*) **274** 234–235

Yang J and Epstein S 1987a The effect of the isotopic composition of oxygen on the non-mass-dependent isotopic fractionation in the formation of ozone by discharge of O_2; *Geochim. Cosmochim. Acta* **51** 2011–2018

Yang J and Epstein S 1987b The effect of pressure and excitation energy on the isotopic fractionation in the formation of ozone by discharge of O_2; *Geochim. Cosmochim. Acta* **51** 2019–2024

Cosmogenic and radiogenic nuclides: What can we do with a few atoms?

JAMES R ARNOLD

Department of Chemistry, University of California at San Diego La Jolla, CA 92093, USA

Abstract. At present, using accelerator mass spectrometry and other methods, it is practical to measure the abundance of rare nuclides at levels near 10^6 atoms. Many things are being learned at this level, by Lal and coworkers and others. This paper considers what samples, and what nuclides, are likely to be interesting at the level of 10^3 atoms, when measurements become possible in that range. Some applications are simply extensions of present techniques and studies. Others, such as natural ^{244}Pu, and most secondary radiogenic nuclides in crustal rocks, have not been reached at all at present levels.

Keywords. Rare nuclides; cosmogenic; radiogenic; crustal processes.

1. Introduction

It is now about forty years since W F Libby developed the art of low level counting and used it in a spectacular way to develop ^{14}C dating. In later years a number of workers, Prof Lal and his coworkers prominently among them, have adapted and extended this technique to many other radionuclides. Their researches have dealt with a wide range of terrestrial and extraterrestrial problems.

A decade ago a second revolution occurred, when accelerator mass spectrometry (AMS) was introduced, so that longlived nuclides could be measured by counting atoms rather than decays. The result was the reduction of the necessary sample size for such measurements typically by four orders of magnitude (for million-year nuclides such as ^{10}Be, ^{26}Al, and ^{36}Cl). The number of atoms of a rare nuclide required for a measurement is now of the order of 10^6. Isotope ratios of 10^{-14} or even less are now sufficient for measurement (Gove *et al* 1987).

Meanwhile mass spectrometric measurements of rare gases have reached a comparable level of sensitivity, of the order of 10^5–10^6 atoms.

Since this sensitivity level has not changed much in recent years, the opportunities for using the cosmogenic radionuclides and stable rare gas nuclides for cosmochemical and geochemical studies are being vigorously pursued in laboratories around the world. Probably not all the niches are filled, because of the limitations of our imagination and intelligence, but there cannot be many gaps.

The purpose of this paper is to ask the question: what about the next revolution? For nuclides produced by cosmic rays (cosmogenic), or secondary radioactive processes (radiogenic), 10^3 atoms counted with 100% efficiency give a precision of 3%, enough for many scientific studies.

There have already been indications that such sensitivities may be attainable. In the AMS field, improvements in ion sources, and discriminators such as gas-filled magnets, promise further gains. The use of two- or three-photon laser techniques to detect individual atoms of a rare nuclide has been reported in laboratory trials, and this method is beginning to find uses in geoscience, though at present only for elements, not isotopes (Letokhov 1987).

2. Distribution of rare nuclides

The first thing we must understand is that the production rate of radioactive and rare gas nuclides is such that they are almost everywhere we look if methods are sensitive enough. Consider the case of ^{36}Cl ($t_{1/2}$ 3×10^5 years). It is made in the atmosphere from ^{40}Ar, and hence is found in rain, snow, ice, and in ground water. In addition, as shown first by Davis and Schaeffer (1955) using β-counting and more recently by Phillips *et al* (1986) by AMS, it is produced *in situ* in the surface rocks of the earth by ^{35}Cl(n, γ) and spallation reactions. The ^{36}Cl/^{35}Cl ratios are on the order of 10^{-12}. At deeper levels production by μ^- reactions becomes important.

In still deeper layers where the α-emitting members of Th and U decay chains are present at typical crustal levels, the alpha particles produced can react with Li and Be by (α, n) to produce neutrons. Lal (1988) calculates an expected ^{36}Cl/^{35}Cl steady state on the order of 10^{-14}, still (just) in the present detectable range. It is only in the earth's mantle, where all these trace elements are less abundant, that the steady state isotope ratios must fall below present limits, probably in the range 10^{-17}–10^{-16}. But as stated above, even this range may be attainable before long.

Below some tens of meters depth in rock, this radiogenic mechanism must dominate. Using Lal's (1988) estimates, I have listed in table 1 the expected longlived radionuclides, along with isotope abundance and thermal neutron cross sections. Most of the target elements are trace elements in typical rocks. Hence the predicted steady state concentrations of the radionuclides in atoms/gram of rock are

Table 1. Interesting long-lived radiogenic radionuclides.

Nuclide	Half-life	Target isotope abundance	Thermal neutron cross-section	Notes
^{14}C	5730 y	^{14}N, 99.6%	1.8 b	(n, p) reaction
^{36}Cl	3×10^5	^{35}Cl, 75.8%	43	
^{41}Ca	10^5	^{40}Ca, 96.9%	0.4	
^{59}Ni	7.5×10^4	^{58}Ni, 68.3%	4.6	
^{79}Se	6×10^4	^{78}Se, 23.5%	0.4	
^{93}Zr	1.5×10^6	^{92}Zr, 17.1%	0.2	
^{94}Nb	2×10^4	^{93}Nb, 100%	1.1	
^{97}Tc	2.6×10^6	^{96}Ru, 5.5%	0.25	Non-isotopic, decay of ^{97}Ru
^{99}Tc	2.1×10^5	^{98}Mo, 24.1%	0.13	Non-isotopic, decay of ^{99}Mo
		^{235}U, 0.71%	580×0.06	Fission product
^{107}Pd	6.5×10^6	^{106}Pd, 27.3%	0.28	
^{129}I	1.6×10^7	^{128}Te, 31.7%	0.22	Non-isotopic, decay of ^{129}Te
		^{235}U, 0.71%	580×0.06	Fission product
^{137}La	6×10^4	^{136}Ce, 0.19%	60	Non-isotopic, decay of ^{137}Ce
^{186m}Re	2×10^5	^{185}Re, 37.4%	0.3	
^{205}Pb	1.4×10^7	^{204}Pb, variable	0.66	
^{210m}Bi	3×10^6	^{209}Bi, 100%	0.0146	

comparatively low. However, at the level of 10^3 atoms total this is not a major problem. For example, let us take a target nuclide with $A = 100$, an abundance in rock of 10 ppm by weight and a 1 barn production cross-section. Assume the radionuclide's half-life is 3×10^5 years. Then the number of target atoms per gram rock is 6×10^{15}. The calculated production rate is given by

$$\frac{dN \text{ production}}{dt} = r \times f = \frac{dN \text{ decay}}{dt} \text{ (steady state)}, \quad (1)$$

where r is the neutron production per gram and f the fraction captured by this target nucleus. The steady state concentration of the product is then the production rate times the mean life. In the unfavourable example given, this is about 2 atoms/gram using Lal's (1988) estimates of neutron flux and total cross-section. Thus a 500 gram rock sample would contain 10^3 atoms of the product mixed with 3×10^{18} atoms of target, for an isotope ratio of $\sim 3 \times 10^{-16}$ (less if there are other stable nuclides). Similar results would be obtained for most radionuclides in table 1.

It is not trivial to extract a trace element fraction from a sample of this size, but Lal, Honda, and I were doing similar things already decades ago. For trace target elements, mineral separation will often be an effective first step, making the chemist's task much simpler.

A few nuclides in table 1 are especially attractive, because the products of neutron reactions are chemically different from the targets. Especially notable are the technetium isotopes, ^{97}Tc and ^{99}Tc, and ^{129}I. In these cases the isotope ratio will not normally present problems. Since Tc has no stable or very longlived isotopes, it must always be rare. Its chemical behaviour in natural settings is unknown, but it may well follow Mn in most rocks. Anyway we can find out. For practical extraction a carrier such as ^{98}Tc, which can form in natural settings only as a rare, shielded product of ^{235}U fission, may have to be added in processing.

The isotope ratio ^{129}I/^{127}I will depend on the relative abundances of I and the target elements Te and U (again nuclear fission). It should be favourable in many cases, especially after separation of minerals.

The absence of any abundant "natural" isotopes of Tc is in fact not an unmixed advantage to the experimenter. It is well known that "carrier-free" radioactive nuclides, at molar concentrations below 10^{-7} M or so in aqueous solution, can behave quite differently than they do in the more familiar world of greater abundances. Adsorption on solids and walls is one common mechanism. I am not aware of any studies of this phenomenon in silicate melts, hydrothermal systems, or the complex environments in which sedimentary rocks form and evolve. Researchers should be alert to possible segregation and concentration mechanisms.

Table 2 lists other significant sources of rare nuclides at the 10^3 atom level or above. One is a special case: ^{244}Pu. If the literature half-life of $8 \cdot 05 \times 10^7$ years is correct, one gram of this nuclide incorporated in the earth $4 \cdot 55 \times 10^9$ years ago would leave 10^{-17} grams or $1 \cdot 4 \times 10^4$ atoms surviving today. Detection and measurement of terrestrial ^{244}Pu would give important information on the last stage of nucleosynthesis, and on the early history of crust-mantle fractionation. Earlier claims at comparable concentrations (Hoffman et al 1971) for natural ^{244}Pu have apparently not been confirmed. However, much more sensitive measurement techniques will make a new search worthwhile. We must remember that (i) a 1% error in the half-life

Table 2. Terrestrial long-lived radionuclides produced by other mechanisms.

Primordial nucleosynthesis: ^{244}Pu (half-life 8.05×10^7 y).

Spallation in the atmosphere: ^{10}Be (N, O targets), ^{36}Cl, ^{26}Al (half-life 7×10^5 years, Ar target), ^{81}Kr (half-life 2.1×10^5 years, Kr target), ^{129}I (Xe target).

Spallation in surface rocks: ^{10}Be (O target). ^{26}Al (Al, Si targets), ^{41}Ca (Ca, Ti, Fe targets), ^{53}Mn (3.7×10^6 years, Fe, Mn targets). Heavier radionuclides in table 1 can also be produced, in varying yield, by this mechanism.

Extraterrestrial influx: This mechanism exists for all long-lived radionuclides. It is relatively important for ^{53}Mn and ^{59}Ni.

---

changes this figure by 30%, (ii) other Pu isotopes, especially ^{239}Pu, now are found everywhere on the earth's surface in larger amounts, and (iii) the geochemistry of Pu is unknown, though reasonable inferences are possible. Again mineral separation may be the key to success.

The production rates and concentrations of other entries in table 2 vary widely. By far the highest production rate and concentration on the earth's surface is that of ^{10}Be in the atmosphere. The production rate is on the order of 10^6 atoms/cm^2 surface year, and ^{9}Be is a rare nuclide. The steady state concentration is more than 10^{12} atoms/cm^2. Still when this nuclide has been incorporated in sediments, when the sediments have been subducted, and this material incorporated in lavas from island arc volcanos (see below) the full resources of AMS are required for measurement. At the other extreme of the abundance scale are ^{81}Kr and ^{129}I, produced in the atmosphere, because of the low abundance of the targets.

Extraterrestrial influx is of minor importance for most nuclides produced in other ways, even though the specific activity of the incoming material is high. The mass influx to the earth is only on the order of 10^7 kg/year. For ^{53}Mn, the most favourable case, the steady state surface concentration is estimated to be $\sim 10^3$ atoms/cm^2.

3. Applications

What can we hope to learn from measurements of rare nuclides at the 10^3 atom level? Since, as shown above, virtually all rocks and sediments—indeed virtually all natural materials—in the earth's crust contain at least some radionuclides with half-lives from 5×10^3 to 2×10^7 years at the level of 10^3 atoms/gram or greater, the problem is one of choice. One can use these nuclides as natural tracers to follow material flows and chemical transformations, and especially to derive time scales of a great variety of processes. In fact, given this wealth of opportunity, the problem is best put the other way around. What are the most interesting questions in earth science? For a given problem, which nuclides and which samples are most suitable to answer them?

In this paper I can only give a few examples, probably not the best ones. The first is plate tectonics. Brown and coworkers (Brown 1984) have published extensively on the use of ^{10}Be produced in the atmosphere and deposited in deep-sea sediments ("garden variety" is Lal's term for this component) in tracing subducted sediments through various stages to their re-emergence in lavas from volcanoes associated with plate boundaries. With higher sensitivity other tracers with different half-lives and chemical properties, such as ^{36}Cl, ^{26}Al, and ^{41}Ca, could be used, especially to clarify the chronology of these stages.

Rubin and MacDougall (1988) have used radioactive disequilibrium in the U and Th decay chains to trace processes leading to MORB basalt formation, on time scales as short as 10^3 years. Measurements by the newer, more sensitive techniques may have wide applicability in these decay chains also, especially for shorter-lived daughters such as ^{226}Ra.

More broadly a great deal of effort is being devoted to the study of "pristine" mantle rocks, to derive information on the chemical composition and evolution of the mantle. These tracer nuclides could be valuable in understanding the level of crustal contamination in these materials, which must in general be in contact with crustal rock on the way to the surface. An improvement in our ability to estimate, and to reduce, the crustal component in mantle materials will be well worth pursuing.

An important phenomenon for study is that of persistent undersea hot spots, especially the "black smokers," which seem to play a significant role in forming ores of many less common elements. Concentrations of tracer nuclides could permit quantitative understanding of the processes taking place in these systems. Since the whole ocean cycles through these systems over long periods, the consequences are of major importance.

The trace elements in the ocean, present at the parts per billion level or below, can also be studied directly with these nuclides. This requires preconcentration, a technique already well developed by Lal, Somayajulu, and coworkers (Krishnaswami et al 1972). Sources and sinks can be better understood.

There are fields now studied by AMS where a further reduction of $\sim 10^3$ in required sample size would be of great value. For example, in our joint effort on ^{10}Be and ^{26}Al in quartz (Lal and Arnold 1985; Nishiizumi et al 1986), typical sample sizes would shrink from grams to milligrams. Thus often single grains could be studied. The most attractive aspect of this is that rare minerals would become candidates for analysis. Diamond, garnet, and other gemstones, magnetite, chromite, and other spinels are all likely to have been closed systems since formation. In some cases studies of these systems have already begun, but they are very much handicapped at present by limitations of available sample size.

One can even imagine applications in the crime laboratory, where these nuclides could provide a signature of the source.

There is enough excitement here to keep Prof Lal and his friends busy for decades.

References

Brown L 1984 Applications of accelerator mass spectrometry; *Ann. Rev. Earth Planet. Sci.* **12** 39–59
Davis R Jr and Schaeffer O 1955 Chlorine-36 in nature; *Ann. NY Acad. Sci.* **62** 105–122
Gove H E, Litherland A E and Elmore D (eds) 1987 Accelerator mass spectrometry; *Nucl. Instrum. Methods* **B29** (Amsterdam: Holland Phys. Pub.) p. 454
Hoffman D C, Lawrence F O, Newherter J L and Rourke F M 1971 Detection of plutonium-244 in nature; *Nature (London)* **234** 132–134
Krishnaswami S, Lal D, Somayajulu B L K, Dixon F, Stonecipher S A and Craig H 1972 Silicon, radium, thorium and lead in seawater—in situ extraction by synthetic fibre; *Earth Planet. Sci. Lett.* **16** 84–90
Lal D 1988 In situ-produced cosmogenic isotopes in terrestrial rocks; *Ann. Rev. Earth Planet. Sci.* **16** 355–388

Lal D and Arnold J 1985 Tracing quartz through the environment; *Proc. Indian Acad. Sci. (Earth Planet. Sci.)* **94** 1–5

Letokhov V 1987 Laser photoionization spectroscopy (New York: Academic Press)

Nishiizumi K, Lal D, Klein J, Middleton R and Arnold J 1986 Production of ^{10}Be and ^{26}Al by cosmic rays in situ and implications for erosion rates; *Nature (London)* **319** 134–136

Phillips F, Leavy B, Jannik N, Elmore D and Kubik P 1986 The accumulation of cosmogenic Cl-36 in rocks: A method for surface exposure dating; *Science* **231** 41–43

Rubin K and Macdougall J D 1988 Radium-226 excesses in mid-ocean-ridge basalts and mantle melting; *Nature (London)* **335** 158–161

The Earth as a physical laboratory

HELMUT MORITZ
Technical University, Graz, Austria A-8010

Abstract. The paper briefly reviews some actual or possible uses of the Earth as a laboratory for fundamental physics, such as the confirmation of Newtonian mechanics by grade measurements in Peru and Lapland in the 18th century, the Michelson-Morley experiment which provided the experimental base for Special Relativity, the refutation of Whitehead's theory of gravitation by measurements of tidal gravity, and other possible applications to gravitation and particle physics.

Keywords. Geodesy; relativity; gravitation; particle physics.

1. Introduction

In geodesy and geophysics we are accustomed to take fundamental geometry and physics for granted when studying the geometry and physics of the Earth. In the present article, on the other hand, we shall consider cases in which the Earth as a whole has been, or may be, used *to test* fundamental geometry and physics. What we mean will become clearer on considering some cases selected from fields with which the author is more or less familiar.

2. Flattening of the Earth and Newtonian Mechanics

We know that Newton derived his laws of mechanics by a rigorous deduction from Kepler's laws of planetary motion, which belong to astronomy. As all new theories, Newtonian mechanics was rather controversial at the beginning.

Since the Earth is rotating, Newton concluded that it must be flattened at the poles, much in the same way as a sphere of water, originally at rest but beginning to rotate, must take on the form of a flattened ellipsoid.

Now, on the basis of geodetic triangulations and arc measurements in France, J Cassini around 1720 announced that, rather than being flattened, the Earth had the form of an oblong (egg-shaped) ellipsoid. This gave rise to a considerable controversy, in which even national elements were present. Voltaire sarcastically wrote: "In Paris you consider the Earth like a melon, in London it is flattened at both sides".

The French Academy decided to resolve the dispute between the "Newtonians" and the "Cassinians" by making new geodetic arc measurements which were sufficiently accurate to provide a definitive answer. This gave rise to the famous arc measurements performed by Godin, Bouguer and La Condamine in Peru (1736–1743) and by Maupertuis and Clairaut in Lapland (1736–1737). These measurements definitively confirmed the flattening of the Earth and thus Newton's theory.

3. The Michelson-Morley experiment and the special theory of relativity

The Earth moves with a velocity v of about 30 km per second around the Sun, the period being of course a year. Considering the light as vibrations of a space-filling "ether", actual measurements of the light velocity c performed on the earth should range between $c + v$ and $c - v$, the extremes corresponding to diametrically opposed positions of the Earth with respect to the Sun. Measurements should thus range between 300 030 and 299 970 km/s.

Now A A Michelson and E W Morley in 1881–87 performed highly precise interferometric measurements of the light velocity c, which conclusively proved that the measured c does not vary at all! This fact provided a heavy blow to classical mechanics and electrodynamics.

The further story is well known: by a stroke of genius, Albert Einstein made the very constancy of the speed of light, under all circumstances, the cornerstone of his special theory of relativity, which corresponds to a flat (uncurved) spacetime (1905).

Here the Earth has indeed served as a "physical laboratory": how could otherwise a velocity difference of $30 + 30 = 60$ km/s have been achieved at that time?

4. General relativity and tides

By admitting spacetime to be curved, Einstein in 1916 was able to take into account gravitation, thus creating the general theory of relativity.

Many alternative theories of gravitation have been proposed, but none as elegant as Einstein's theory, with the possible exception of a theory published in 1922 by the great British logician and philosopher Alfred North Whitehead. This theory is based on a flat spacetime, in which Whitehead's gravitational theory is the exact tensor analogue of Maxwell's vector equations for electromagnetism.

Both theories have passed all four standard experimental tests of general relativity: gravitational redshift, perihelion shift of Mercury, light deflection by the sun, and time delay of radar measurements of bodies within the solar system. (By the way, in these tests, the solar system, that is, the "environment" of the Earth, is used as a laboratory!) These tests are extensively discussed in Misner et al (1973). Whitehead's theory even gives the same theory of black holes as Einstein's (Schild 1962, p. 74)!

The gravitational theory of Whitehead is contradicted, however, believe it or not, by plain terrestrial gravity measurements: it predicts tidal variations of gravity of order $\Delta g/g \doteq 2 \times 10^{-7}$ with a 12-hour period which, needless to say, have never been observed, as pointed out in 1971 by C M Will (Misner et al 1973, pp. 1067 and 1124).

Here the Earth, through its tides, has proved to be a laboratory of great simplicity, but of unique value.

5. Is the gravitational constant constant?

Some alternative gravitational theories predict a very small temporal change of the gravitational constant G, which, according to Einstein, is a true constant. Some astronomical observations seem to be compatible with a decrease of G by a few parts in 10^{11} per year. The evidence is not clear, however; it seems to be difficult to separate a true change of G from other systematic influences. Dicke (1964, pp. 272–281) discussed a possible connection of a decrease of G with a slow expansion of the Earth. Again, the evidence is inconclusive, but this is another possible use of the Earth as a testing laboratory.

The problem of testing gravitational theories by gravity and other geophysical measurements continues to be of interest, as recent work by H J Treder, F D Stacey, and K Runcorn shows; see also a brief note in Eos, vol. 68, No. 35 (1987), p. 725.

Most theoreticians (including the present author) believe that the general theory of relativity is of such incomparable elegance and perfection that it will remain *the* theory of gravitation.

However, as the famous physicist Wolfgang Pauli once remarked "Elegance is for the taylor", and Einstein's theory might have to be modified in connection with other physical theories, especially elementary particle and quantum theory, to provide the intensively sought, but by no means yet achieved, unification of the gravitational force with the electromagnetic, the strong and the weak force of elementary particle physics, and perhaps with another, still unknown, "fifth force".

This brings us to particle physics, and the reader will see that our arguments become more and more speculative, trying to throw some weak and uncertain flash of light into a dark and unknown future.

6. The Earth as a laboratory for particle physics?

Experiments concerning the physics of elementary particles are nowadays mostly performed by giant particle accelerators. One of the most ambitious enterprises in this field will be the large electron-positron storage ring (LEP) at CERN, Geneva, an underground circular accelerator with a length of 26·7 km, passing underneath the frontier between Switzerland and France. In spite of the large underground extension of the accelerator, we cannot properly speak here of the earth as a laboratory, but what is interesting is the extremely high accuracy specification of the relative alignment of the steering quadrupoles: a root mean square error of 0·1 mm! So may it be permitted to the author, who is a geodesist, to mention this example after all with some pride (Turner 1987).

Other examples are much more speculative. One of the great unsolved problems in modern physics is the existence of magnetic monopoles. The discovery of a magnetic monopole would rank as one of the most fundamental physical achievements in this century, comparable perhaps to the discovery of the positron (both have been predicted by Dirac!). Here again the Earth could possibly serve as a laboratory: monopoles trapped in the Earth would tend to accumulate at two places in the earth's core. Following a reversal of the magnetic field, the two segregated populations of monopoles would migrate through each other. Some pairs of north and south monopoles would meet and annihilate each other. In this process, mass would be converted into energy, which could be detected by heat-flow measurements (Carrigan and Trower 1982).

Until some time ago, protons were believed to be stable particles of infinite lifetime. The theory that unifies electromagnetic and weak interaction which was independently developed by Steven Weinberg and Abdus Salam in 1967–68 and which earned their authors the Nobel Prize in 1979, predicts, however, a finite though very long lifetime (on the order of 10^{30} years) of the proton. To test this, experiments are being undertaken in various mines or tunnels, usually in a depth of several thousand meters to screen off cosmic radiation. For instance, several thousand tons of water, located at such depth, are being watched for Čerenkov radiation produced by the decay of a proton (so far, without success if I am not mistaken); cf. (Weinberg 1981).

Here again, only part of the Earth, so to speak, acts as a physical laboratory, but in honour of Professor Lal, who is a nuclear physicist, I could not resist mentioning this example.

Neutrinos are among the most elusive physical particles. They have an enormous penetrating power because, so to speak, they very successfully avoid interaction with other particles. According to (Miakišev 1979), the Soviet Academician M A Markov has proposed a very interesting idea to send a neutrino beam through the earth, e.g. emitted at the Brookhaven accelerator in USA and received in the Soviet neutrino research centre of Baksan (I do not know how realistic this idea is). A proposal to use the Earth itself as a detector for solar and galactic neutrinos recently appeared in Eos vol. 69, No. 24 (1988), p. 649.

In the past, the Earth and its surrounding space have proven an ideal laboratory for fundamental physical experiments. Given the enormously increasing accuracy of measurements, and the certainly not decreasing inventiveness of theoretical and experimental physicists, it can be expected that this will continue also in the future.

References

Carrigan R A and Trower W P 1982 Superheavy magnetic monopoles; *Sci. Am.* **246** 91–99
Dicke R H 1964 Experimental relativity; in *Relativity, groups and topology* (eds) C DeWitt and B DeWitt (New York: Gordon and Breach) pp. 163–313
Miakišev G J 1979 *Elementary particles* (in Russian) (Moscow: Nauka)
Misner C W, Thorne K S and Wheeler J A 1973 *Gravitation* (San Francisco: W H Freeman)
Schild A 1962 Gravitational theories of the Whitehead type and the principle of equivalence; in *Evidence for gravitational theories* (ed.) C Møller (New York: Academic Press) pp. 69–115
Turner S (ed.) 1987 CERN Accelerator School: Applied geodesy for particle accelerators; CERN 87-01, Geneva
Weinberg S 1981 The decay of the proton; *Sci. Am.* **244** 52–63

A trial of static Fourier transform mass spectroscopy of xenon

P McCONVILLE and J H REYNOLDS

Department of Physics, University of California, Berkeley, CA 94720, USA

Abstract. The field of isotopic studies of noble gases in natural samples continues to be robust but needs a static instrument which is both sensitive and rapid, unlike the most sensitive magnetic sector spectrometers. In principle FTMS, which in dynamic runs can detect 100 ions and which can produce a spectrum of the 7 most abundant xenon isotopes in a sample containing 50,000 ions within a few seconds, meets these requirements if a ~ 5 ma electron current can be used to produce the trapped ions in a small static cell. We have not yet determined whether or not the trapping will survive the transient resulting when the emission is interrupted. Brief tests at the University of California, Irvine of a static system, fabricated at Berkeley, in a borrowed magnet and with borrowed electronics were inconclusive because of (i) insufficient electron current to ionize the static sample rapidly and (ii) contamination of the test xenon sample with lighter background molecules. Static analysis of a large xenon sample (which simulated a clean sample) was successfully carried out, nevertheless, by sweeping out light ions with RF excitation during the ionizing pulse. Errors in the relative abundances (those for ^{128}Xe and ^{130}Xe were markedly low) were attributed to excessive space charge and/or non-uniform excitation across the mass range. We suggest how to proceed further in evaluating the method. As a NASA study group has recently concluded, the ultimate use of the technique may be in conjunction with resonance ionization of Ar, Kr and Xe, although the complex instrumentation required for that would probably not be within the means of an individual investigator to develop and maintain.

Keywords. Fourier transform mass spectroscopy; xenon isotopes; static xenon analysis; FTMS; mass spectrometry; fourier transform ion cyclotron resonance.

Dedication

It is a great pleasure to be able to contribute this paper to a volume in honour of Professor Lal. We have not yet achieved what we believe is possible with static fourier transform mass spectrometry (FTMS) of the noble gases, a success which would have been fittingly described in this book. Nevertheless we feel that our paper responds to the solicitation by this book's sponsors of papers concerned with likely future developments in a field where Professor Lal has been, and continues to be, active and eminent. We wish him many more happy and productive years.

1. Introduction

The field of isotopic studies of the noble gases in natural samples remains robust. Among very productive lines of current research we can list, by no means exhaustively: (i) the detection in separable carbon-bearing phases in the acid-resistant residues from meteorites of unusual isotopic signatures for neon, xenon, and krypton. Increasingly the evidence points to preserved debris from stellar nucleosynthetic processes being the source of these isotopic anomalies (e.g. Tang *et al* 1988a, b, c).

Indeed the focus has shifted from whether or not presolar carbonaceous grains have been preserved in the meteorites to questions such as whether one or more nucleosynthetic settings have contributed and if so to what extent nucleosynthetic parameters have varied among the contributing settings (Ott et al 1988). (ii) Increased use of the ^{39}Ar–^{40}Ar dating method, enhanced in speed, precision, and sensitivity by highly automated laser heating (Wijbrans et al 1987; Drake et al 1988). (iii) Exploitation of the noble gases as tracers in studying the mantle of the earth. While studies of helium isotopes have led the way, isotopic patterns in argon, neon, and xenon have enriched this field (Staudacher and Allègre 1982; Rison and Craig 1983; Mamyrin and Tolstikhin 1984; Kennedy et al 1985; Ozima et al 1985). (iv) Laser decrepitation of microscopic fluid inclusions in ores, providing miniscule samples of noble gases for high sensitivity mass analysis and a means, if the samples have been preirradiated with neutrons, of characterizing the fluids for a number of elements in addition to the noble gases (Kelley et al 1986; Turner 1988; Kirschbaum et al 1987). In all these applications progress would be furthered by more sensitive and more rapid mass analysis than what conventional magnetic instruments provide.

From work in this laboratory (Kirschbaum 1988) we know that conventional magnetic instruments operated statically can detect as few as a thousand xenon atoms in a microsample of the gas, but at this level of sensitivity obtaining only two isotopic ratios requires hours of running time. Conventionally the sample is dispersed in a $\sim$ liter-sized spectrometer volume, V, while an effective volume $\sigma Li/e$ is ionized per second by the electron beam in the ion source, where σ is the ionizing cross section, L is the length of the electron paths, i is the electron current and e is the electron charge. Values for these quantities in the most sensitive magnetic instrument in our laboratory (Kirschbaum 1988) give a volume rate ≈ 0.32 cc/s corresponding in a 2-litre instrument to a value for the mean-life of a noble gas atom before ionization of 6,000 seconds. This figure is a lower limit for the electron ionizing current used (200 μA) because the ion extraction efficiency for the source and the transmission of the mass analyser are both <1. The measured mean-life is about 12,000 seconds so that the system is about 50% efficient. Statistical error alone in counting the small number of ions requires that a substantial fraction of the sample be ionized and brought to the collector. Thus long running times are unavoidable in this technique. The sluggish data rate and inevitable contamination of the sample by memory during these long runs inhibit research progress. There is a need for a technique which can provide rapid analysis and simultaneous collection of all the isotopes of interest.

2. Possible solutions

Two techniques suggest themselves for rapid analysis of the noble gases, resonance ionization mass spectrometry [RIMS] (Hurst and Payne 1988) and Fourier transform mass spectrometry [FTMS] (McIver 1970; Gross and Rempel 1984; Comisarow 1985; Marshall 1985). In RIMS a volume in the ion source is illuminated by pulsed ionizing laser radiation of such high intensity that a large fraction of the sample atoms in that volume is ionized. The resulting ion pulse can then be mass analysed with high efficiency by time-of-flight methods (Thonnard et al 1984). The spectrometer volume can be small and the laser pulse rate reasonably high, leading to rapid mass analysis of xenon, say, with data for all isotopes. The fraction of the sample ionized per pulse can be increased for adsorbable elements such as krypton and xenon by collecting the sample on a cold spot near the ionizing region and liberating these atoms with a laser heating pulse timed with reference to the ionizing pulse. There is a tradeoff between this means of favourably reducing the effective volume of the spectrometer and unfavourably reducing the pulse rate as necessary to provide time for "repumping" of the sample atoms to the cold spot. This modified technique is especially powerful (Chen *et al* 1984; Lehmann *et al* 1985) where the objective is to count a small number of atoms of an isotope such as ^{81}Kr which does not occur in the atmosphere so that memory accumulation in a prolonged run is not a problem. It is proposed (Hurst 1986) to use this technique in an underground solar neutrino observatory similar to that operated by R Davis Jr (Davis *et al* 1968; Bahcall and Davis 1982) except that a bromine-rich liquid would be substituted for Davis's carbon tetrachloride and the RIMS counting of ^{81}Kr ($T_{1/2} = 2 \times 10^5$ years) would replace the radioactive counting of ^{37}Ar. Investigation of RIMS for conventional noble gas mass spectrometry is under investigation at several laboratories (Schneider 1986; Turner 1986). Properly implemented, the laser-induced resonance ionization can be highly element-specific so that the ion pulse is free from background. Disadvantages are that a different, complicated laser scheme is required for each element and these schemes become progressively more difficult as the ionization energy of the element increases with reduced atomic number.

The FTMS technique is an outgrowth and improvement upon ion cyclotron resonance (ICR) mass spectroscopy. The newer technique has become practicable with the development of improved electronic methods for digitizing signals and rapidly obtaining their frequency components by Fourier transformation. The signal is generated in a FTMS cell configured as a cubic box with six electrically-insulated face plates (see figure 1) and placed in a strong DC magnetic field. Small DC potentials

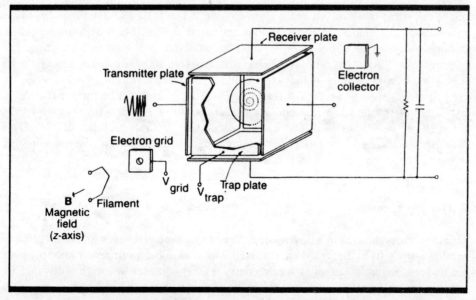

Figure 1. Typical trapped ion cell for Fourier transform mass spectroscopy. Reproduced with permission from Johnston (1987).

are applied to the plates in such a way as to trap ions produced by an electron beam traversing the cell, in a direction parallel to the magnetic field, from a filament just outside one of the plates. If we can refer to the pair of plates which accelerates and catches the electron beam as the longitudinal plates of the cell, one of the transverse pairs of plates is used pulsewise to apply AC power at frequencies which match the cyclotron frequencies of the ions of interest, thereby exciting them into cyclotron orbits of small enough radius so that they remain trapped in the cell after the AC power has been applied. The remaining pair of transverse plates is used, after the excitation pulse has been applied, to measure the voltage induced by the ions oscillating, at their cyclotron frequencies and in phase, along the coordinate perpendicular to these plates. This voltage, which decays with time, is the signal which is digitized and analysed for its Fourier components. A complete run or scan takes only a second or so, so that even improving the digitized signal by accumulating data from multiple sweeps, which is how weak peaks are detected, does not lead to excessively long running times. The technique is capable of both high sensitivity, with as few as 100 ions detectable in typical apparatus, and high mass resolution, indeed enough to differentiate dissimilar chemical species at the same mass number (Gross and Rempel 1984; Comisarow 1985; Marshall 1985). For these reasons FTMS has been a powerful technique in chemical studies of, for example, ion-molecular reaction rates. At the time we undertook this research the method had not, to our knowledge, been applied to the measurement of isotopic abundances.

If FTMS as described above can be successfully adapted to static analysis of noble gases its advantages will be the high sensitivity and high resolution we have quoted plus applicability to all the noble gases with the same equipment. Since the ions are never accelerated to energies high enough for implantation, the method should be quite free from memory effects. An important disadvantage of the method is its limited dynamic range. Space charge effects limit the capacity for ions to about 10^6 in number (McIver et al 1981; Francl et al 1983). Although overabundant ions can be swept out of the cell while the ionizing beam is on in order to increase the dynamic range (see more about this below), in conventional operation the ratio of maximum signal to minimum detectable signal is $\sim 10^4$ and the available precision in measuring, for example, the $^{40}Ar/^{36}Ar$ ratio would be unacceptable in many cases, especially in highly radiogenic samples. (The corresponding ratio of maximum signal to noise in conventional magnetic sector instruments is $\sim 3 \times 10^5$ and can be further increased by orders of magnitude with separate collection of the two ion beams.) Probably FTMS will be impractical for measurements in natural samples of $^3He/^4He$. Nevertheless there are a number of problems, especially in xenology, where the limited dynamic range would not be a serious disadvantage.

3. Our approach

Our laboratory, housed in a physics department and doing its best work when physics graduate students at Berkeley are involved, has always been interested in improving techniques for noble gas mass spectrometry. For this reason we were motivated to explore one or the other of the two methods discussed above. A bit of history accounts for our deciding to try FTMS. A University of California colleague, Prof Robert McIver of the Irvine campus, is one of the world's leading practitioners of FTMS; his laboratory in the Chemistry Department at Irvine contains a number of FTMS set-ups, the expensive components being large ~ 1 tesla magnets and the associated digital electronics, including special hardware for obtaining rapid Fourier transforms (see Sherman et al 1985 for description). After several telephone conversations on the subject he invited one of us (JHR) to visit Irvine and to observe a dynamic run of an atmospheric xenon spectrum. The story has its amusing aspects. The first dynamic xenon run, and as far as we know the first serious attempt to measure isotope abundances by FTMS, was carried out in December 1985 while we were in flight from Oakland to Irvine so that we were presented with a xenon spectrum (figure 2) upon arrival. The experiment had been a simple exercise for the McIver laboratory. All Dr W Bowers, who carried it out, had to do was to attach a xenon flask to the system and crack a valve slightly so as to leak xenon into a continually pumped FTMS apparatus. The leak rate was adjusted so that the number of trapped xenon ions was $\sim 50,000$. The "chirp" or succession of excitation frequencies for the cyclotron resonance was stepped through those frequencies which correspond to the mass range 80 to 200. Twenty scans were combined to form the 9-bit digitized signal which was Fourier-analysed. The results to us seemed, and still do, highly impressive. Our immediate decision, with promise of cooperation from Dr McIver, was to build a static FTMS cell at Berkeley which could be tested in one of the Irvine set-ups. The procedure decided upon was strictly cut and try. If dynamic xenon analysis was so easy, maybe static analysis would also be easy. (It is not.)

While one of us (PM) oversaw the construction of a FTMS cell, based upon Irvine designs, and the associated plumbing, several additional experiments were carried out at Irvine by Dr Bowers using a newly available 12-bit analog to digital converter (ADC). Figure 3 shows the ^{124}XE and ^{126}Xe peaks obtained when the Xe ion population was 500,000. These peaks correspond to approximately 500 ions. They were not seen in figure 2 because the 9-bit ADC had insufficient dynamic range. Figure 3 is valuable for estimating the noise that can be anticipated in an FTMS device. As far as we know the signals in figure 3, aside from the two xenon peaks, can

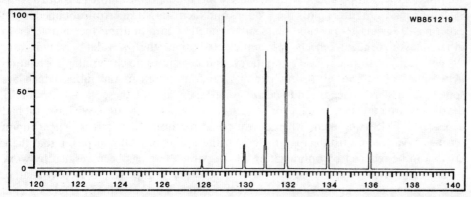

Figure 2. First measurement of Xe isotopic abundances as run dynamically by W Bowers in Prof R T McIver's laboratory at the University of California, Irvine in December 1985. About 50,000 ions were stored in the trap; xenon partial pressure, $P_{Xe} = 4 \times 10^{-10}$ torr; background pressure $P_{BG} \approx 3 \times 10^{-11}$ torr; magnetic field 1·16 tesla; electron emission 1·8 μA for 100 ms; 70 V electrons; mass range excited was 80 to 200; the detect chirp was delayed 100 ms after emission pulse; transient was digitized for 33 ms, 16 k points, 9-bit ADC.

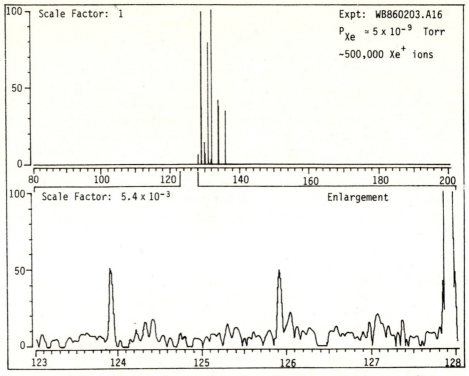

Figure 3. Increased dynamic range with a 12-bit ADC allows detection of $^{124}Xe^+$ and $^{126}Xe^+$ peaks without ejection of the more abundant isotopes. The trapped ion population was 500,000; Xe partial pressure $P_{Xe} = 4.5 \times 10^{-9}$ torr; other parameters similar to those given in caption to figure 2, but the signal was averaged over 1000 scans.

all be attributed to Johnson noise in the preamplifier. In a brief investigation of the accuracy of the isotope abundance measurements for xenon, 15 more runs of 20 scans each were made. Figure 4 plots the δ values and their statistical errors obtained in the experiment. Owing to possible mass discrimination in admitting the sample or to other mass-dependent effects we anticipated that the δ values would vary systematically with mass. To the contrary, figure 4 shows no discernible correlation between δ values and mass, but exhibits random errors in the ratios which are sometimes 3%. No background spectra were taken so that these values are totally devoid of corrections for baseline or for defects in the RF excitation. We have not had access to FTMS equipment whereby we could see how these errors depend upon settings of various adjustable parameters for the equipment. Our suspicion that these errors can be reduced by optimization of the parameters has been verified by work published in two recent abstracts from a research group at the Department of Geological Sciences at the State University of New York at Albany (DeLong *et al* 1988; Spell *et al* 1988). That group, which has been funded so as to provide it with a commercial FTMS system for experimentation with earth science applications, has identified sources of fractionation in dynamic FTMS of krypton and has identified conditions for improved accuracy, achieving agreement with ~1% with accepted values of the isotopic abundances.

BOOK REVIEWS

From Mantle to Meteorites edited by K. Gopalan et al. Indian Academy of Sciences, Bangalore, 1990, 330pp. US $19.00 (ISBN 0-19-81-1; distributed by Oxford University Press).

TWENTY-ONE PAPERS assembled in this volume are as diverse in their subject matters as the interests of the venerable scientist to whom they are dedicated, Devendra Lal of the Tata Institute of Fundamental Research in Bombay. Prof. Lal began his career as a physicist, but was lured early into the problems of detection of cosmic-ray produced isotopes in the natural environment. With this research reference he became, in the words of James Arnold (UC/San Diego), "for a physicist, the best chemist there is." Over the years he and his group at the Tata Institute, as well as with colleagues in the United States, have sought out an impressive list of such isotope tracers: ^{7}Be, ^{28}Mg, ^{38}Cl, ^{38}S, ^{32}Si. Each one of these sequesters itself into different environmental and geological niches, and each such niche has led Prof. Lal into areas of earth and planetary sciences he would never have predicted would involve him when he started his career: oceanography, hydrology, atmospheric science, environmental science, lunar surface studies, meteoritics. His broad interests and engaging personality have created close collegial relationships with many researchers in other parts of the world, especially at the Scripps Institution of Oceanography where he holds a joint appointment. At the Tata Institute he broadened the scope of their research mission instituting programs in plasma physics, climatology, and infrared astronomy, in addition to their original programs in nuclear and atmospheric physics. His successes in these diverse fields of endeavor led to this book, a festschrift commemorating his arrival into the sixth decade of life.

The papers are uniformly well prepared. They range in size from a few as five pages to almost fifty pages. They cover an impressive range of subject matter—theoretical physics, mantle geophysics and geochemistry, spectrometry, experimental techniques, micrometeorites, planetary accretion. It is not possible to offer details about all the twenty-one papers, so a few are singled out here as examples of the great diversity represented in this volume.

Ramanath Cowsik (Tata Institute) leads off with a fascinating history of the search for forces in nature which are weaker than the gravitational force (which itself is a factor of only 10^{-33} the electrostatic force between electron and proton). The mathematics will appear here and there, unfamiliar to many with geochemical backgrounds, but they are used in a descriptive manner to illustrate largely verbal arguments.

D. A. Mendis (Scripps Institution) and Gustav Arrhenius (Scripps Institution) offer a short review of the process of planetary accretion from ionized matter in a plasma, an idea proposed by Alfvén 20–25 years ago, but still not given much consideration in current theoretical models, which start from a neutral gas.

Peter Wyllie (Caltech) offers a neat primer on the several mantle convection models currently competing for favor in the deep earth geophysical community and the respective programs proposed to test them.

Manfred Schidlowski (MPI/Mainz) reviews current thinking and experimentation on a subject your reviewer thought was long ago dead—*panspermia*, the idea that life started on the Earth by "inoculation" from interstellar space. A consequence of this, of course, is that life would exist on any congenial planet like the Earth and that it is likely the process still goes on. Interstellar "things" might be dropping in every day!

G. Cini Castagnoli (Ist. Cosmo. CNR, Torino, Italy) describes the use of thermoluminescence to determine past solar variations.

Michel Maurette (CSNSM, Orsey, France), Claus Hammer (U. Copenhagen), and Michel Pourchet (Domaine Univ., St-Martin-d'Here, France) offer a long and thoroughly absorbing paper on the reasoning that led to the collection of micrometeorites from Greenland and Antarctica. These objects are clearly samples from the interplanetary medium of material not represented in the meteorite collections of the world.

Karl Turekian (Yale) is known for his succinct, tightly reasoned arguments for gleaning the most geological information from the most economical input of geochemical data. He offers here a neat argument concerning the control of degassing of our planet (actually could be applied to any inner planet) using ^{40}Ar systematics. By this he finds there is little or no recycling of crustal carbon; carbon emitted at ocean ridges is largely primordial.

Bruce Bolt (UC/Berkeley) has written an elegant and cautionary essay, "The Turbulence of Geophysical Time," which philosophically treats the problem of long geological time spans and long-term geophysical predictions. Systems are linear and straightforward if phenomena are time-wise delimited, but become nonlinear and chaotic if not delimited.

T. Narasimhan (Lawrence Berkeley Laboratory) contributed a very nice paper on new trends in hydrogeology and its emerging relationship to broad areas of earth sciences.

The diversity of papers is obviously large. Although many of them deal with quite special subjects, they are written for readers who are not themselves specialists in the specific fields. Most readers with a good background in earth sciences will find them quite accessible.

The book is a fitting tribute to a man who has made contributions to many areas of geochemistry. Who, however, is the audience for this book? I would recommend it as a source of additional reading material for early graduate level courses in the earth sciences. Individual papers in it are excellent introductions to a variety of subjects and techniques. Some of the papers would augment material presented in class. It can also be recommended to any earth scientist who finds himself or herself trapped into a narrow speciality and would like to sample ideas and methods in other areas of endeavor. Considering how narrow are many individuals in the earth sciences today, this book fills a not unworthy niche.

Department of the Geophysical Sciences   Edward J. Olsen
University of Chicago
Chicago, IL 60637, USA

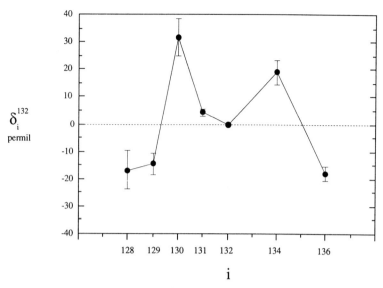

Figure 4. Isotope abundance data for xenon. Results are expressed in permil deviation, δ_i^{132}, of the measured ratio from the accepted atmospheric ratio: i.e.

$$[(i_{Xe}/^{132}Xe)_{meas.}/(i_{Xe}/^{132}Xe)_{atmos.} - 1] \times 1000 \text{ permil}.$$

Measured abundances in 15 runs of 20 scans were averaged. Statistical errors for the means are ~ 5 permil on the average but the errors in the ratios are sometimes ~ 30 permil. The xenon pressure was similar to that in figure 2.

4. Static FTMS

The question to be answered in a test of static FTMS is whether or not the method, which is impressive for 50,000 ions, can be made to work for a sample of about 50,000 *atoms*. The obvious test procedure is to introduce a xenon sample of that size into an FTMS cell and, using a larger electron current than is conventional, explore how fully the sample can be converted into trapped ions. The mean-life of the atoms before ionization will be $eV/(iL\sigma)$ where V is the volume of the cell and the other symbols are as previously defined. Our cell (see figure 5) is a cube 1·25 inches on edge, which determines V and L. Taking 5 mA as a practical upper limit to i, based upon experience with conventional electron impact ion sources, and using a value for σ of 4 square Å (Mathur and Badrinathan 1986), the minimum mean-life is found to be 0·8 seconds, which appears prima facie not to be a prohibitively long pumping time for the trap. The proposed electron current and electron beam time are, however, both very much greater than conventional in FTMS so the question immediately arises as to whether or not trapping and RF excitation will still be efficient under these conditions. We have not yet made adequate theoretical study of this question but the following simple considerations suggest that the trapping will be unimpaired and in fact facilitated:

A 5 mA current of 70 volt electrons puts $2 \cdot 0 \times 10^8$ electrons in our cell while the current flows. This is a much greater charge than the limiting 10^6 positive ions we

have mentioned above, but calculation shows that the maximum electric force on these electrons due to space charge is only 1% of the magnetic force experienced in the 1 tesla field, suggesting that the electron beam should be well confined to the same volume as when the electron current is small. In the foregoing calculation the electron beam was assumed to be confined to a diameter equal to the 0·030" width of the filament ribbon. If the electron beam is dispersed over a greater volume, the 1% figure will be correspondingly reduced. The large negative space charge at the electron beam will be conducive to trapping positive ions within it. Another effect of the volume distribution of negative charge will be to induce a positive charge on the surface of the longitudinal (as defined above) plates so that trapping should also be maintained at the extremities of the electron sheath.

If there are not disruptive transient effects on the trapped positive ions when the electron beam is finally extinguished (a question we have not investigated), the positive ions should be ready for analysis in the same manner as when much smaller electron beams of shorter duration are used in conventional FTMS.

We can expect a relatively higher concentration of multiply charged ions as the result of the increased trapping times, because the singly charged ions will long be confined to a volume where further electron impacts will occur. Trapping occurs in the ordinary Nier source and has been used (Plumlee 1957; Baker and Hasted 1966) to estimate ratios of the cross-sections for progressive ionization to the higher charge states. From the relative strengths of the multi-charged ion currents for xenon in the Nier source, they appear not to drain the singly-charged state excessively, although the trapping times in the Nier source are shorter than we contemplate. In conventional FTMS analysis (W Bowers, private communication) the Xe^{++} signal detected is only 5 to 8% of the Xe^+ signal. Incidentally, one of us in describing years ago (Reynolds 1956) the first static mass spectrometer for noble gases, called attention to highly charged (up to 6 +) tungsten and mercury ions in the background spectrum, without understanding what produced them. Now we can see that trapping of mercury and tungsten ions in the electron beam is responsible for these high charge states.

5. Tests with the Berkeley static cell

The transportable FTMS cell was fitted with a vacuum system and with two noble gas pipettes as shown in figure 6. The reservoir with air xenon was to provide small samples for tune-up and general testing. The ^{128}Xe reservoir was to provide *very* small calibrated samples for tests of the sensitivity which would not be invalidated by xenon background. The system was equipped with two small getters and an ion gauge for pressure measurements.

The schedules at Irvine and the ongoing research on other topics at Berkeley have restricted our tests of the system at Irvine to a single 4-week period in November,

---

Figure 5. The Berkeley FT-cell is a cube 1·25" on edge constructed of 304 non-magnetic stainless steel, MACOR machinable ceramic, OFHC copper wire, and 0·030" × 0·001" rhenium filament. The trimmed mounting flange shown failed to make an adequate seal and was discarded in favour of a custom-made untrimmed flange which fit between the poles of the Irvine magnet.

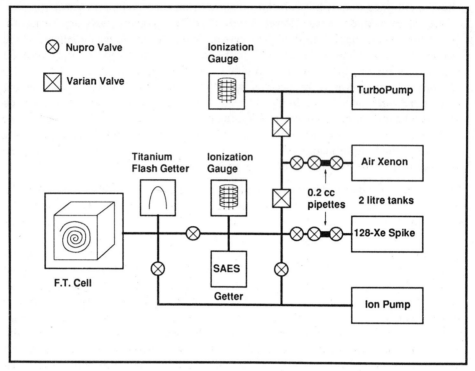

Figure 6. Schematic diagram of the UHV plumbing for the static test FT-cell. See text. The volume of cell, getters, and ion gauge totalled less than 400 cm^3.

1987. Murphy's Law, which provides that anything that can go wrong will, was strongly in force upon that occasion. To begin with, there had been a misunderstanding between Irvine and Berkeley about how much electron current was available from the Irvine electronics so that all experimentation had to be done at far smaller electron currents than have been discussed above: the maximum electron current was 60 μamps. We had absolutely no hope, then, of seeing the extremely small xenon samples provided by the ^{128}Xe pipette. The air xenon pipette, also being of limited usefulness in the situation, was refilled at the Irvine location with flask xenon. Unfortunately our pipettes, even before being refilled, were contaminated with chemically-active gases such as H_2O, CO and CO_2 to an extent that taxed the limited gettering we had provided and that led to trouble. In tests of the vacuum system at Berkeley, without introducing samples from the pipettes, we had been able to achieve vacuum in the 10^{-10} range, despite the low pumping speed built into the small-volume system by the use of small diameter tubing, but after pipette samples were introduced at Irvine, the background pressure degraded to pressures in the 10^{-9} torr range. (The spectra in figures 2 and 3 were obtained with a dynamic background pressure in the 10^{-11} range.) We were immediately reminded of how important it is to have a clean xenon sample in FTMS: other gases which are no problem in a conventional magnetic sector instrument can quickly saturate the allowable ion population of $\sim 10^6$ positive ions. Such ions must be removed by pre-excitation while the ionizing electron current is flowing.

At the end of the Irvine tests, most of which were devoted to diagnostic experiments which were chosen to explore and to understand better the various problems we were

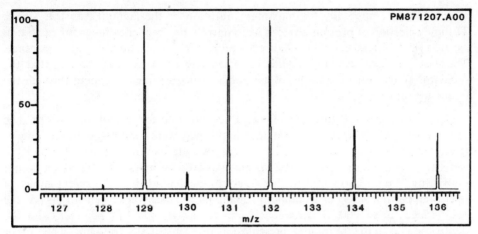

Figure 7. Xenon mass spectrum obtained in a static (i.e. unpumped) system. See text.

encountering, a static xenon run was successfully made, but at a xenon pressure necessarily many orders of magnitude greater than what we originally intended. By increasing the ratio of xenon partial pressure to the background pressure we were, of course, overpowering the contamination from light ions and simulating a cleaner sample. The spectrum from the static runs is shown in figure 7. The conditions and the procedure for obtaining the spectrum, which had been roughly optimized, were as follows:

Conditions: Xenon pressure 2.7×10^{-8} torr, corresponding to 3×10^{10} Xe atoms in the cell. Electron voltage: 70.

Sequence: (i) Electron (ionizing pulse: $4.0\,\mu$A for 10 ms duration).
(ii) "Chirps" for pre-excitation of light ions: *First*: During last half of electron pulse. Mass range was set for 15 to 50 atomic mass units. This prolonged excitation accelerates ions in that mass range, at resonance, into large cyclotron orbits which intersect the lateral plates of the cell and lead to neutralization. *Second*: For 5 additional ms starting 5 ms after the end of the electron pulse. The purpose of this step was to remove any light ions formed *after* the sweep-out of their mass.
(iii) "Detect delay": A dormant period of 130 ms. By lengthening this delay to 6 seconds (the maximum delay provided for in the Irvine electronics) it was established in other runs that xenon ions were trapped for at least six seconds.
(iv) *Excitation pulse*: Mass range 75 to 240, corresponding to a frequency range 2·25 MHz to 0·70 MHz; 1550 steps; 100 Hz frequency change per step; time per step 1 μs; total length of excitation pulse 1·55 ms.
(v) Transient digitized (32,000 points, 12 bits) for 66 ms.
Note: Another technique for removing unwanted ions is to superpose an alternating voltage on the small DC potentials applied to the longitudinal plates as part of the trapping scheme. If the frequency of this signal resonates with the trapping frequency (the frequency with which the ions of a given mass undergo cyclic motion parallel to the electron beam, owing to the approximately parabolic potential produced by the DC trapping scheme) for a given mass, it can eject ions of or near that mass. We employed this "trapping ejection" during the entire electron pulse at a frequency chosen to eject H^+ ions (see below). The frequency and amplitude of the AC signal

were adjusted empirically for maximum duration of the (xenon) transient signal. Trapping ejection of protons enabled us to narrow the frequency range for excitation ejection (15 to 50 mass units), making for more effective ejection in that mass range. This last statement can be understood when one takes note of the fact that the oscillator system for excitation pulses is constrained to linear stepping through the frequency range selected.

Data for masses 128 through 136 were accumulated for 200 of these scans. The xenon isotope abundance ratios obtained in this way were comparable in accuracy to the dynamic runs discussed above except that the rarer isotopes at mass 128 and 130 were strongly discriminated against (as much as 50% for mass 128). We as yet do not know why. We do know that the accuracy can be degraded by a different choice of operating parameters. For example by changing the electron current to $2.9\,\mu A$, the mass range to 80–200, the frequency step to 200 Hz for $1\,\mu s$ per step, and the digitization to a 16,000 point representation of 33 ms of the transient, while other parameters were unchanged, we introduced additional severe discrimination at masses 134 and 136. Even so, those at Irvine who use the FTMS technique daily attributed our poor accuracy basically to a combination of space charge and/or non-uniform excitation across the mass range.

Since our Irvine visit, our hosts there have experimented with using larger values of electron emission in a dynamic cell. Their results are somewhat disquieting: after an initial increase of signal with increased emission, the signal began to fall off with further increased emission and could be made to disappear within the $60\,\mu A$ limits of their present system. The effect is not totally due to disruption of the trapping by a transient occurring when the emission is interrupted, because it was seen in runs when the emission was left on throughout the scan. One must therefore attribute an effect either (i) to loss of trapping, or (ii) interference by the negative space charge with excitation, or (iii) extremely rapid dephasing of the cyclotron orbits by the space charge. In view of the arguments given above for *enhanced* trapping, we favour explanations (ii) or (iii). Effect (ii), but not (iii), would interfere with the sweeping out of impurity ions by pre-excitation during the ionization pulse.

6. Where do we go from here?

Many questions about the utility of static FTMS for the noble gases remain and future research has to be directed at answering them. We discuss them in the order of least progress made until now:

(i) *Re: trapping effectiveness for very small samples*: In the present work, where large ionizing currents were unavailable to us, we have only identified this problem and given very crude arguments for possible success. The next step is either to examine the question with computer simulations (or better yet adequate analytical theory) or by experimentation with a clean system. We say "clean system" because a "dirty system" necessarily involves the additional complexity of pre-excitation removal of background ions. As stated above, it may be possible to trap small samples with large ionization currents but quite impossible to apply pre-excitation removal in the presence of the large negative space charges involved. If so, a definitive test of the method requires clean conditions. In part this question can be tackled by dynamic analysis, i.e. exploring further what happens when the cell is clean (easy for the dynamic system) and the ionizing current is large. Final evaluation should be with a static system.

(ii) *Re: successful ion ejection by preexcitation during ionization with large electron currents*: See above. Here again the approach can be either theoretical or experimental.

(iii) *Re: obtaining adequate cleanliness in the cell*: We characterize this problem as difficult to solve but we see no fundamental barrier. With proper choice of materials, adequate polishing, sufficiently good pumps connected to the cell with high conductance tubing, and high temperature baking, it should be possible to achieve pressures in the 10^{-11} range under static conditions. The present cell includes fairly large areas of a machinable insulator with the trade name Macor. Whether this material can meet the static vacuum requirements is something we are not presently certain about. But if Macor is too gassy, it should be possible to fabricate a cell with some other insulating scheme.

A NASA working group (Planetary Materials and Geochemistry Working Group 1988) has reported on possible advanced analytical facilities relevant to research in that field and the report includes a section on advanced mass spectrometry. The group, in its brief discussion of FTMS, has correctly identified the need for a clean sample and proposes that multiple RIMS pulses be the source of the noble gas ions, because that ionization mechanism is element-specific. We agree that such a composite machine would be a very powerful tool for ultrasensitive noble gas mass spectrometry. We see no fundamental reason why such a spectrometer would not work, but certain disadvantages are obvious: (i) different laser schemes are needed for argon, krypton, and xenon; (ii) it is usually said that RIMS for helium and neon is impossible; (iii) accuracy problems, already present in FTMS, are bound to be aggravated in a machine that requires prolonged ion storage from repeated RIMS pulses; and (iv) the cost and complexity of the instrument would seem to require its development and operation at a national centre rather than in the laboratory of a single investigator.

A FTMS machine of the type we have (barely) tried, if it can be made to work, should not be more expensive or complex than the magnetic instruments currently being acquired by individual investigators. For that reason we deem it sensible to continue exploration of the idea in a modest way and along the lines we have suggested above.

Acknowledgements

We thank our colleagues at Irvine, especially Prof Robert McIver, for sharing their facilities and knowledge and thereby enabling this preliminary work to be carried out on a low budget. Their laboratory is extremely active in applying FTMS to chemical studies. Even the brief testing of our apparatus was a significant intrusion into their work and is much appreciated. This work was supported by NASA under Grant No. NAG 9-34 and bears code # 152.

References

Bahcall J N and Davis R Jr 1982 An account of the development of the solar neutrino problem; in *Essays in nuclear astrophysics* (ed.) C A Barnes, D D Clayton and D N Schram (Cambridge: University Press) pp. 243–285

Baker F A and Halstead J B 1966 Electron collision studies with trapped positive ions; *Philos. Trans. R. Soc.* **A261** 33–65

Chen C H, Kramer S D, Allman S L and Hurst G S 1984 Selective counting of krypton atoms using resonance ionization spectroscopy; *Appl. Phys. Lett.* **44** 640–642

Comisarow M B 1985 Fundamental aspects and applications of fourier-transform ion-cyclotron resonance spectrometry; *Anal. Chim. Acta* **178** 1–15

Davis R Jr, Harmer D S and Hoffman K C 1968 Search for neutrinos from the Sun; *Phys. Rev. Lett.* **20** 1205–1209

DeLong S E, Mitchell D, Spell T L and Harrison T M 1988 Isotope ratio measurements by fourier transform mass spectrometry. I. General fractionation effects; *Eos* **69** 1501–1502 (Abstr.)

Drake R E, Deino A L, Curtis G H, Swisher C S, Turrin B, McCrory M and Becker T 1988 Applications of $^{40}Ar/^{39}Ar$ dating of single-crystals by laser-fusion; *Eos* **69** 1502 (Abstr.)

Francl T J, Sherman M G, Hunter R L, Locke M J, Bowers W D and McIver R T Jr 1983 Experimental determination of the effects of space charge on ion cyclotron resonance frequencies; *Int. J. Mass Spectrom. Ion Processes* **54** 189–199

Gross M L and Rempel D L 1984 Fourier transform mass spectrometry; *Science* **226** 261–268

Hurst G S 1986 Feasibility of a $^{81}Br(v,e^-)$ ^{81}Kr solar neutrino experiment; in *Resonance ionization spectroscopy* (eds) G S Hurst and C Grey Morgan (Bristol: Institute of Physics) **71** 283–288

Hurst G S and Payne M G 1988 *Principles and applications of resonance ionization spectroscopy* (Bristol: Adam Hilger) p. 417

Johnston M 1987 Fourier transform mass spectrometry; in *Spectroscopy* (Oregon: Spring Shield) **2** 14–20

Kelley S, Turner G, Butterfield A W and Shepherd T J 1986 The source and significance of argon isotopes in fluid inclusions from areas of mineralization; *Earth Planet. Sci. Lett.* **79** 303–318

Kennedy B M, Lynch M A, Reynolds J H and Smith S P 1985 Intensive sampling of noble gases in fluids at Yellowstone: I. Early overview of the data; regional patterns; *Geochim. Cosmochim. Acta* **49** 1251–1262

Kirschbaum K 1988 Carrier phases for iodine in the Allende meteorite and their associated $^{129}Xe_r/^{127}I$ ratios: A laser microprobe study; *Geochim. Cosmochim. Acta* **52** 679–699

Kirschbaum C, Irwin J J, Böhlke J K and Glassely W E 1987 Simultaneous analysis of halogens and noble gases in neutron-irradiated quartz veins: a laser microprobe study; *Eos* **68** 1514 (Abstr.)

Lehmann B E, Oeschger H, Loosli H H, Hurst G S, Allman S L, Chen C H, Kramer S D, Payne M G, Phillips R C, Willis R D and Thonnard N 1985 Counting ^{81}Kr atoms for analysis of groundwater; *J. Geophys. Res.* **90(B13)** 11547–11551

McIver R T Jr 1971 A trapped ion analyzer cell for ion cyclotron resonance spectroscopy; *Rev. Sci. Instrum.* **41** 555–558

McIver R T Jr., Hunter R L, Ledford E B Jr., Locke M J and Francl T J 1981 A capacitance bridge circuit for broadband detection of ion cyclotron resonance signals; *Int. J. Mass Spectrom. Ion Phys.* **39** 65–84

Mamyrin B A and Tolstikhin L N 1984 *Helium isotopes in nature*, Developments in geochemistry (Amsterdam: Elsevier) **3** 273

Marshall A G 1985 Fourier transform ion cyclotron resonance mass spectrometry; *Acc. Chem. Res.* **18** 316–322

Mathur D and Badrinathan C 1986 On the ionization of xenon by electrons; *Int. J. Mass Spectrom. Ion Processes* **68** 9–14

Ott U, Begemann F, Yang J and Epstein S 1988 S-process krypton of variable isotopic composition in the Murchison meteorite; *Nature (London)* **332** 700–702

Ozima M, Zashu S, Mattey D P and Pillinger C T 1985 Helium, argon, and carbon isotopic compositions in diamonds and their implications in mantle evolution; *Geochem. J.* **19** 127–134

Planetary Materials and Geochemistry Working Group 1988 *Advanced Analytical Facilities Report* LPI Tech. Rpt. 88-11 (Houston: Lunar Planetary Institute) p. 38

Plumlee R H 1957 Space charge neutralization in the ionizing beam of a mass spectrometer; *Rev. Sci. Instrum.* **28** 830–832

Reynolds J H 1956 High sensitivity mass spectrometer for noble gas analysis; *Rev. Sci. Instrum.* **27** 928–934

Rison W and Craig H 1983 Helium isotopes and mantle volatiles in Loihi Seamount and Hawaiian Island basalts and xenoliths; *Earth Planet. Sci. Lett.* **66** 407–426

Schneider K 1986 Resonance ionization spectroscopy with xenon; in *Resonance ionization spectroscopy* (eds) G S Hurst and C Grey Morgan (Bristol: Institute of Physics) **84** 67–73

Sherman M G, Kingsley J R, Hemminger J C and McIver R T Jr 1985 Surface analysis by laser desorption of neutral molecules with detection by Fourier Transform Mass Spectrometry; *Anal. Chim. Acta* **178** 79–89

Spell T L, DeLong S E, Mitchell D and Harrison T M 1988 Isotope ratio measurements by Fourier Transform Mass Spectrometry. II. Optimization of experiment parameters; *Eos* **69** 1502 (Abstr.)

Staudacher Th and Allegrè C J 1982 Terrestrial xenology; *Earth Planet. Sci. Lett.* **60** 389–406

Tang M, Lewis R S, Anders E, Grady M, Wright I P and Pillinger C T 1988a Isotopic anomalies of Ne, Xe and C in meteorites. I. Separation of carriers by density and chemical resistance; *Geochim. Cosmochim. Acta* **52** 1221–1234

Tang M, Lewis R S and Anders E 1988b Isotopic anomalies of Ne, Xe and C in meteorites. II. Interstellar diamond and SiC: carriers of exotic noble gases; *Geochim. Cosmochim. Acta* **52** 1235–1244

Tang M, Lewis R S and Anders E 1988c Isotopic anomalies of Ne, Xe and C in meteorites. III. Local and exotic noble gas components and their interrelations: *Geochim. Cosmochim. Acta* **52** 1245–1254

Thonnard N, Payne M G, Wright M C and Schmitt H W 1984 Noble gas atom counting using RIS and TOF mass spectrometry; in *Resonance Ionization Spectroscopy* (eds) G S Hurst and M G Payne (Bristol: Institute of Physics) **71** 227–234

Turner G 1988 Application of RIMS to the study of noble gases in meteorites; in *Resonance Ionization Spectroscopy* 1986 (eds) G S Hurst and C Grey Morgan, *Inst. Phys. Conf. Ser.* (Bristol: Institute of Physics) **84** 51–58

Turner G 1988 Hydrothermal fluids and argon isotopes in quartz veins and cherts; *Geochim. Cosmochim. Acta* **52** 1443–1448

Wilbrans J R, York D and Schliestedt M 1987 Laser probe $^{40}Ar/^{39}Ar$ age spectra of single crystals of phengite from the Cycladic blue schist belt, Greece; *Eos* **69** 518 (Abstr.)

Multidisciplinary investigations of new collections of Greenland and Antarctica micrometeorites

MICHEL MAURETTE, CLAUS HAMMER* and MICHEL POURCHET**

Centre de Spectrométrie de Masse et de Spectrométrie Nucléaire, 91406-Orsay, France
* Institut of Geophysics, University of Copenhagen, DK-2200 Copenhagen, Denmark
** Laboratoire de Glaciologie, Domaine Universitaire, 38402-Saint-Martin-d'Here, France

Abstract. There is no collection of extraterrestrial material on the Earth or in the interplanetary medium, which is free from artifacts and biases. Studies of cosmic dust grains recently collected on the Greenland and Antarctica ice sheets, allow a better assessment of such limitations. Grains with a "chondritic" composition are overabundant in these two new collections. Isotopic measurements (^{10}Be, ^{26}Al, Neon), as well as grain size distribution, indicate that they are genuine micrometeorites. The unexpected high ratio of unmelted to melted chondritic grains runs contrarily to classical predictions of frictional heating in the atmosphere. A wide scatter in their contents of major, minor, and trace elements suggests that they are different from the major classes of meteorites. Potential terrestrial applications of micrometeorite studies range from the storage of nuclear wastes to prebiotic synthesis on the early Earth.

Keywords. Greenland micrometeorites; Antarctica micrometeorites; mining of cosmic dust; Blue Ice-I expedition; micrometeorites; cosmic dust; chondritic composition; Blue-Ice sediments; terrestrial weathering; terrestrial contamination; unmelted micrometeorites; carbonaceous meteorites; micrometeorite flux; composite-pyrolyzable micrometeorite; frictional heating; thermal shielding; prebiotic evolution; ablation constraints; COKE computation; biogenic microlithography; etch canals; nuclear wastes.

1. Introduction

In the 1970's Brownlee and collaborators (Brownlee 1985) initiated very successful operations to collect cosmic dust grains in the stratosphere and in deep sea sediments, with regard to previous attempts yielding unrealistic high values for their accretion rate by the Earth (see Hodges 1981 for a summary of earlier works). Major breakthroughs resulted from a better understanding of contamination problems, the application of new criteria of extraterrestrial origin, and the development of ingenious schemes for collecting large quantities of grains on a semi-industrial basis. As of now about 1000 tiny cosmic dust grains with an average size of $\sim 10\,\mu m$, labelled as *interplanetary dust particles* (*IDPs*), have been collected in the stratosphere, and several 100,000 cosmic grains with much larger sizes (up to a few 1 mm), have been dragged with a huge magnetic rake (~ 300 kg) from deep sea sediments at the centre of the Pacific (Clanton *et al* 1982; Brownlee 1985).

In the stratosphere about 80% of the IDPs are "unmelted" grains with a "chondritic" composition (see §3.1), in which the proportion of Fe/Ni grains is low,

---
* To whom correspondence should be addressed

and $\sim 20\%$ of the grains appear as "melted" spherules. In contrast, in previous studies of deep sea collections, at least 99% of the cosmic dust grains were described as "stony" and "iron" melted spherules of a roughly similar abundance ($\sim 50\%$), and the abundance of extraterrestrial particles quoted as "largely unmelted" was smaller than 1%.

This drastic loss of large unmelted grains could be understood in terms of the classical "ablation" model (Whipple 1950), describing the deceleration of hypervelocity particles in the atmosphere, assuming that they are micrometeoroids made of "compact-inert" materials, such as chunks of meteorites. These models predict the existence of a critical size of $\sim 100\,\mu m$, beyond which most micrometeoroids are melted (and thus spherulized) during this impact. Then mass loss (ablation) occurs through aerodynamical shearing.

These models suggested the occurrence of a "screening" effect of the Earth's atmosphere, that would preferentially transmit micrometeoroids with low impact velocity, representing debris of asteroids or short-period comets. On the other hand, the most interesting micrometeoroids originating from long-period comets or from the interstellar medium, should have highly eccentric orbit, and consequently much higher impact velocity. They should be preferentially destroyed upon impact with the atmosphere, and consequently strongly depleted in terrestrial collections.

In this paper we shall discuss mostly our recent studies that deal with new collections of cosmic dust grains on the Greenland and Antarctica ice sheets, and which have not been considered in previous "cosmic dust" reviews (Brownlee 1985; MacKinnon and Reitmeijer 1987; Bradley 1987). There is no collection of cosmic dust (including ours), either in *space* or on the Earth, that is free from artefacts and biases. Our work was initiated by some limitations of previous collections of stratospheric and deep sea cosmic dust grains, that we had hoped to successfully tackle. In particular we believed that the comparison of very different collections of micrometeorites*, covering a wide range of grain sizes should end up identifying some of their "invariant" characteristics, that are independent of collection procedures, terrestrial weathering and heat metamorphism during atmospheric entry. Such characteristics should reveal new filiations between a variety of solar system objects, and yield new clues about the early history of the solar system.

We present below the new schemes that we started to apply in 1984, for using the ultra-clean ice caps of Greenland and Antarctica as giant collectors of micrometeorites, and explain why the Greenland and Antarctic collections are remarkably complementary. We next discuss several discoveries made during the investigation of these new collections of micrometeorites over the last 3 years, that either reveal new features of the cosmic dust complex, or deal with unexpected applications of these micrometeorite studies to other fields of science and technology. In this discussion we shall rely extensively on results obtained by our French colleagues, E Robin, Ph Bonny, G Raisbeck and F Yiou, and G Callot, M Christophe-Levy, C Jéhanno, C Jouret and P Siry.

---

* As we show in §§3.2 and 6 most cosmic dust grains collected on the Earth in the size range $>100\,\mu m$ originate from small-sized micrometeoroids; we shall frequently quote them as micrometeorites without distinguishing any further between melted and unmelted particles.

2. The mining of cosmic dust on the margin zones of the Greenland and Antarctica ice caps

2.1 Limitations of previous cosmic dust collections

One of the great advantages of the stratospheric collection of IDPs is its high abundance ($\sim 80\%$) of unmelted cosmic dust grains, that should reflect the small size of their parent bodies (about 10–20 μm), preventing their strong heating during atmospheric entry (see §7). Another marked advantage of IDPs is their short residence time in the stratosphere, that should minimize to a considerable extent their terrestrial weathering, although some problem of silicone oil contamination might complicate important issues such as the determination of the "cosmochemical" Mg/Si ratio.

Conversely these small masses present several drawbacks. First they are smaller than the limit of detection of powerful analytical techniques of isotope analysis. The IDPs are also too small for our own studies, including the characterization of the micropore structure, the dispersion of elemental abundances in the grains *at different scales of magnification of the analysed volume*, and the search for tiny micrometeorite impact craters on their surface. Moreover, they cannot generate measurable trails of ionized gases in the atmosphere, that would help in ascertaining their origin. Such trails delineate the trajectory of much larger micrometeoroids, yielding *radar* and *visual* meteors, that correspond to particles with masses in excess of $\sim 10^{-6}$ g and a few mg, respectively. Statistics of the best determined trajectories (Millman 1972), indicate that the vast majority of micrometeoroids ($> 90\%$) have cometary orbits, quite distinct from the asteroidal orbits of the parent bodies of meteorites. When a big cosmic dust grain (i.e. masses 10^{-6} g), *is first identified in the laboratory* as the residue of a micrometeoroid (size < 1 cm) then it can be statistically affiliated with cometary debris. This important "statistical cometary connection" is not available for the much smaller IDPs.

In contrast deep sea collections contain already a large number ($> 100,000$) of big micrometeorites with size up to a few mm. However magnetic extraction combined with a strong weathering of the chondritic grains in their relatively old ($\sim 100,000$ y) host sediments, act as to considerably enrich deep sea collections in strongly heated up cosmic dust grains loaded with magnetite crystals ("stony" and "iron" spherules), to the detriment of the weakly magnetic unmelted grains, that are quite rare in this collection (Brownlee 1985).

Over the last few years several attempts have been made to find cosmic dust grains in Antarctica ice (see references in Koeberl *et al* 1988). However they only yielded a small number of cosmic spherules, or a few unmelted grains that did not show up a chondritic composition. Moreover the interesting collection of spherules of Koeberl *et al* might be strongly weathered, like Antarctica meteorites (Gooding 1986).

Our first "Blue Lake I" expedition was an optimistic attempt to collect an ideal type of terrestrial sediment, very enriched in cosmic dust, from which we could extract by simple mechanical means (thus excluding magnetic extraction) a large number of unmelted micrometeorites with masses in excess of 10^{-6} g (i.e. sizes $> 100 \mu$m). We further required these grains to be in a much better state of preservation than those extracted from deep sea sediments. This rationale looks logical now, but in reality, this polar hunt for micrometeorites was plagued with "calculated risks" (i.e. bold assumptions), that fortunately ended up as a success.

2.2 Speculations about cosmic dust placers on the Greenland ice cap

Polar ices might be the most favourable sediments for cosmic dust collection. In fact most of the host sediment (i.e. the ice) is readily eliminated upon melting, without requiring a preliminary disaggregation procedure that might further destroy the most friable grains. Moreover the grains should be kept most of the time in deep freeze conditions, and this should markedly minimize terrestrial weathering.

The major problem with the ices is their very low "maximum" concentration of cosmic dust grains, that can be computed assuming a 100% transmission efficiency of the micrometeorite flux of Grün et al (1985) in the atmosphere, and evaluating the ice accumulation rate, V(cm/y). At the site of our expeditions on the Greenland and Antarctica ice caps $V \sim 50$ and ~ 20 cm/y, respectively. The concentrations of $> 100\,\mu$m-size cosmic dust grains would thus reach low values of ~ 3 and ~ 10 grains per ton of ice, respectively.

Our concept of "giant" cosmic dust collector, intended to get a large number of micrometeorites, requires the melting of a huge amount of ice. It originated from a set of aerial stereoview photographs of the ablation zone of the west Greenland ice cap (latitude of Sondrestromfjord) taken by Bauer in 1958, at an elevation of 5000 m. Small seasonal "blue lakes" are formed on the melt zone during the short Arctic summer. Each year these shallow lakes are formed at about the same locations, because they somewhat delineate the much steeper and invariant "hill/valley" structure of the basement rocks on which the ice flows.

These stereoviews clearly showed irregular patches of dark sediments ("cryoconite") on the bottom of the lakes. They also indicated that the shallow collection basin associated with each lake has a diameter of a few km. We made the following bold assumption: Each year, during the short arctic summer (July–August), a ~ 1 m-thick superficial layer of ice is melted over the surface of the collection basins. The running waters are captured in the lake, and the micrometeorites initially deposited in the ice much further inland, finally settle down at favourable "retentive" sites on the lake bottom.

The amount of melt ice water thus cycled yearly through a lake reaches an enormous value, of about 10^8 tons per year! Moreover at the site of our first expedition ("Blue Lake I"), at about 20 km from the margin of the Sondrestromfjord ice field, the ice flow model developed by Neels Reeh indicates that the ice surface is about 2000 year-old. This yearly process of melting was in fact repeated there over a time scale of ~ 2000 y, providing a "gigantic" amount of ultra-pure melt ice water in excess of 10^{10} tons. Although major losses of grains and/or water did occur (for example when a big crevasse intersects the lake), it might be that cryoconite deposits accumulated through this "placer" type mechanism are the richest mine of cosmic dust grains on Earth.

2.3 Two "Blue Lake" expeditions in Greenland

In July 1984 a Danish-French expedition with five members* was organized to check the validity of this speculation. We settled down at a base camp at ~ 20 km from the margin of the Sondrestromfjord ice field. Our major task was to collect ~ 20 kg of

---

*C Hammer, M Maurette, K Ramussen, N Reeh and H H Thomsen

cryoconite on the bottom of a blue lake with a length of about 600 m. The subsequent analyses of this sample revealed that these deposits are indeed the richest and best preserved mines of cosmic dust grains known on the Earth yet (Maurette et al 1986b). In §6·3 we estimate that ~ 1 kg of cryoconite can be assimilated to a cosmic dust collector of ~ 500 m^2, exposed for one year in space (the largest collectors of micrometeorites placed on board of satellites have an effective area of ~ 1 m^2).

To fully exploit this amazing "space" equivalence we launched in July–August 1987, a new "Blue Lake II" expedition, to disaggregate right in the field ~ 1 ton of cryoconite, that might end up being equivalent to a giant space collector of 500,000 m^2 exposed for one year in space at 1 a.u. Our objective was to extract only the biggest micrometeorites (size $> 500 \mu$m) that are extremely rare in space (about one per 250 m^2 per year) and that have never been investigated before.

This search for large unmelted micrometeorite was again a calculated risk. First the classical model of micrometeorite ablation (Whipple 1950), predicts that big micrometeorites should be preferentially destroyed in the atmosphere. But they might also be preferentially lost during transport to the placer type deposits. Cryoconite is made of cocoons of filamentary siderobacteria (Callot et al 1987), with average sizes of $\sim 0.5 - 1$ mm. The vast majority of grains with sizes up to a few 100μm are tightly encapsulated in the cocoons, that easily float in running waters, carrying with an astonishingly high efficiency cosmic dust grains to their final deposits. But mm-size grains might not be encapsulated in the cocoons. Consequently their individual transport "trajectory" to the deposits might be quite different from that of the cocoons, and major grain losses might occur.

We lauched the "Blue Lake II" expedition with 13 members*, establishing two base camps at ~ 10 km and ~ 25 km from the ice margin. With a view to minimizing grain losses we only conducted a direct extraction of micrometeorites with sizes $> 500 \mu$m. In this procedure, illustrated in figure 1B, ~ 10 kg sample of cryoconite are placed on a large stainless steel sieve with an opening of 500μm, which is half-immersed in running water. A pressurized jet of water is then fired on this sample, that get sieved and disaggregated in about 10 minutes, yielding a few tens of mg of mineral grains.

A small helicopter was also used during 3 full days for sampling ~ 15 kg samples of cryoconite at 10 distinct sites between ~ 1 km and ~ 50 km from the ice margin. This peculiar sampling was made to investigate the past activity of the ancient micrometeorite flux over a time scale of about 10^4 y. Another important operation, still to be fully exploited, was to try to detect the acid fall-out of major historical volcanic eruptions, for estimating the age of the ice surface of atleast one site. This determination would severely constrain the ice flow model of Reeh, used for dating purposes.

In §3 we shall see that most of our objectives were again fulfilled. Why then launch an additional expedition to Antarctica, to collect small micrometeorites that are very abundant in cryoconite?

2.4 The Blue Ice I expedition in Antarctica: a search for "invariant" characteristics of micrometeorites

The preliminary analyses of the Blue Lake I samples suggested the occurrence of

---

*C Hammer, M Maurette, M Pourchet, N Reeh, H H Thomsen and others

Figure 1. Deposits of cryoconite on the melt zone of the west Greenland ice cap.

biases in our collections of small micrometeorites, that have been briefly outlined elsewhere (de Angelis *et al* 1988). We currently disaggregate cryoconite samples held in a stainless steel sieve with a hard nylon brush, recuperating about 100 cosmic dust grains per $\sim 100\,g$ of cryoconite. In this procedure the most friable grains might be preferentially destroyed. A gentle melting of the ice, followed by the filtering of melt ice water on a sieve on which the grains can be directly handpicked under a binocular microscope, would minimize the destruction of the most friable grains.

Moreover Callot *et al* (1987) showed that biogenic corrosion is effective in cryoconite, that might preferentially *alter* the most interesting porous cosmic dust grains, probably loaded with good nutrients for sustaining biogenic activity (i.e. sulphur and C-rich material). It is likely that micrometeorites just released during the melting of Antarctica blue ice should be much less corroded than Greenland micrometeorites, hosted for long periods in cryoconite.

A comparison between Antarctica and Greenland micrometeorites might thus reveal such biases of the Greenland collections. The "Blue Ice I" expedition with 3 members*, that was proposed in 1985, was finally accepted in 1987. In December 1987 we thus settled down a camp of "water-pocketing" *to melt in the field ~ 100 tons of blue ice.* For this operation we used ~ 4000 litres of fuel, two generators of pressured steam (delivering jets of hot waters), two generators of electricity, and a variety of pumps.

We made pockets of melt ice water right under the ice surface (figure 2), directing a

Figure 2. Water pocketing at $\sim 1\,km$ from the margin of the Antarctica ice cap, near "Cap Prudhomme" at about $6\,km$ from the French station of Dumont d'Urville. A participant penetrates into an empty $\sim 2\,m^3$ pocket, to assess its geometrical shape.

---

*M de Angelis, M Maurette and M Pourchet

jet of hot water into a small drill core with an initial diameter of ~ 20 cm. Each pocket of water slowly expended in lateral extension, until it reached a volume of a few m^3, in about 5 hours of operation. The average depth of the top surface of these pockets, could be adjusted between a few tens of cm and ~ 5 m. This well-controlled water pocketing under the ice surface shielded efficiently melt ice water from contaminations originating from our activities, such as fuel burning. It was a key factor in our successful operation.

On each day of good weather about 2 to 4 individual pockets were obtained, corresponding to a daily volume of water of about 5 to 10 tons. This water, as well as the sediments deposited on the bottom of the host holes, were vacuum-cleaned with the pumps, and collected on a stack of 3 stainless steel sieves with openings of 50, 100 and 400 μm, respectively. The total volume of water was measured, directing the filtered water into a plastic container of 1 m^3.

With a field microscope we counted the total number of well-defined* cosmic spherules in each sieve fraction. We thus estimated their concentration in the ices, as well as some major characteristics of their size distribution, that was very helpful. This daily control of the outcrop of cosmic spheres allowed us to rate them from "good" to "bad" (in terms of terrestrial contamination), but also to discover a peculiar dust accumulation process effective in the top ~ 1.5 m-thick layer of the ice field. In this enriched layer, found at depths ranging from ~ 0.2 up to 1.5 m, the concentrations of spherules with a size > 50 μm (about 100 per ton of ice), increase up to a factor of ~ 20 with regard to the corresponding background level measured at a depth of ~ 5 m. This process allowed us to find the richest and best preserved collection of unmelted micrometeorites (about 5000 grains) and cosmic spherules (~ 5000 sperules) found in Antarctica, yet.

3. Individual or statistical identification of micrometeorites

3.1 *Validation of chondritic compositions*

A high concentration of Ni in an "iron-rich" grain is considered as a good criteria of extraterrestrial origin. When they are analysed with an energy-dispersive X-ray spectrometer (EDX) the primitive fine-grained carbonaceous meteorites with a *chondritic composition* show also high concentration of iron and nickel, as well as high magnesium and sulphur contents, and measurable concentrations of Ca and Cr (see the 3 first EDX spectra from the top in figure 3). The finding of spherules or "unmelted" polycrystalline fragments presenting spectra somewhat related to this composition is highly suggestive of an extraterrestrial origin.

However in the Greenland and Antarctica collections, there is a wide variation in element concentrations around this composition (see the 5 other EDX spectra in figure 3). A top priority was to establish definitively the validity of this broad class of composition as a criteria of "extraterrestriality", in particular for grains that contain undetectable amount of Ni in the EDX spectra, such as the glassy spheres and some of the most interesting unmelted grains.

---

*See the "chondritic-barred" spheres and the "glassy-chondritic spheres described in §4.3 and illustrated in figures 4A and 4C, respectively.

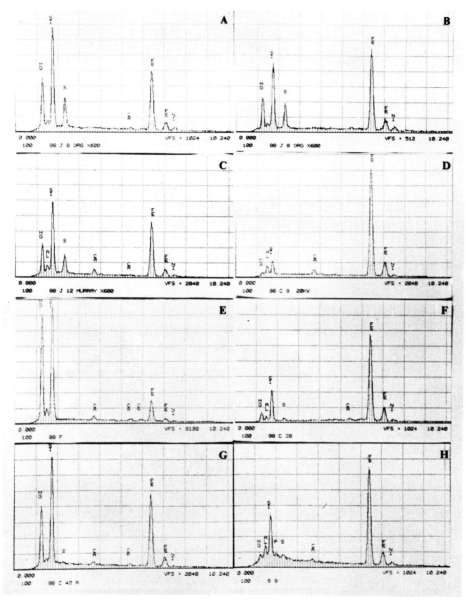

Figure 3. Semiquantitative broad beam analysis of major element composition with an energy-dispersive X-ray spectrometer. All analyses were performed on internal fracture surfaces, at the same scale of the analysed volume. The largest variations observed for a set of ~ 30 analyses of the Orgueil meteorite have been reported in A and B. Orgueil and Murray (C) show high concentrations of S. This element is quite depleted in the set of analyses corresponding to porous unmelted micrometeorites (D to F). The specific choice of spectra in D to H illustrates a drastic variation in major element abundances, which is not observed in carbonaceous chondrites (Greenland micrometeorites in E, H; Antarctica micrometeorites in D, F, G). Courtesy of P Siry.

During the course of this work two independent methods were available at CSNSM to identify the extraterrestrial origin of *individual* chondritic grains, based on either a search for solar flare tracks in unmelted grains of any size, or measurements of ^{10}Be and ^{26}Al cosmogenic nuclides in both spherules and unmelted grains with sizes $> 400\,\mu$m. Moreover the powerful method of instrumental neutron activation analyses (INAA) was available at CEFR (Bonté *et al* 1987a), to search for high iridium concentrations in individual grains with sizes $> 100\,\mu$m.

Our search for latent tracks in about 10 sulphur-rich grains with a high voltage electron microscope (HVEM) was unsuccessful yet, but this does not preclude an extraterrestrial origin of the grains*. In 1983 Raisbeck *et al* did use the powerful technique of accelerator mass spectrometry to measure the concentrations of ^{10}Be and ^{26}Al in about 10 *individual* deep sea spheres with a minimum size of $\sim 400\,\mu$m. At this time only the two varieties of spherules that are most resistant to terrestrial weathering in deep sea sediments were analysed, namely the chondritic-barred spherules (figure 4A) and spheres made of oxidized alloys of Fe/Ni (figure 4D). All the spheres showed high values of the concentrations of the two nuclides, yielding a strict *individual* criteria of extraterrestrial origin.

Next this method was applied to a selection of 7 big Greenland spherules (extracted from ~ 2 kg of cryoconite), including new varieties of spheres, such as glassy spheres (figure 4C), and a "largely-unmelted" grain containing large relics of mineral grains. All grains showed high concentrations of cosmogenic nuclides (Raisbeck *et al* 1986). Recently these measurements were extended to a set of 24 unmelted chondritic

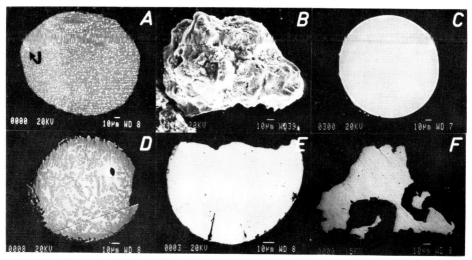

Figure 4. Major families of Greenland micrometeorites: *A*, chondritic barred spherule; *B*, unmelted chondritic fragment; *C*, homogeneous glassy spherule; *D*, spherule made of an oxidized alloy of Fe/Ni; *E*, unoxidized Fe/Ni spherule; *F*, unmelted-unoxidized Fe/Ni alloy. Courtesy of "Nature".

---

*The same method was applied in 1970 to discover solar flare tracks in micron size lunar dust grains (Borg *et al* 1970). In 1977 we searched unsuccessfully for tracks in a typical IDPs, lent to us by D É Brownlee (Bibring *et al* 1978). These tracks were recently discovered by Bradley *et al* (1984) in similar IDPs. This unreliability in fossil track detection results probably from a process of track "maturation" that would stabilize tracks against an ill-understood process of electron beam "ionization annealing" in the microscope. Tracks in lunar dust grains have rightly matured in the lunar regolith, whereas tracks in both IDPs and PUMs would rather behave like fresh tracks, fading away quickly during TEM observations.

fragments from Greenland, extracted from the 100–200 μm size fraction, analysed at once as a single grain with a mass of ∼ 130 μg. This "grain" also showed high concentrations of ^{26}Al and ^{10}Be (Raisbeck and Yiou 1987).

The validation of the chondritic composition was realized for the first time by Robin (1988) for a much larger variety of grains (about 130 spherules and 60 unmelted fragments), relying on iridium measurements in *individual* grains, preselected from both optical microscope observations* and EDX analyses. With the only exception of a small proportion of "homogeneous" glassy spherules (figure 4C), all grains, including unmelted chondritic fragments with no detectable Ni in their EDX spectra, did show the high iridium concentration expected for extraterrestrial grains with chondritic and/or Fe/Ni alloy compositions.

3.2 *Individual or statistical identification of micrometeorites*

Beside the validation of the chondritic and Fe/Ni compositions high ^{10}Be/^{26}Al ratios (> 5) gives a strict criteria to identify a genuine micrometeorite, originating from a parent micrometeoroid with a size smaller than about 1 cm. This method has some threshold of sensitivity for the analysis of individual grains, that corresponds to both a minimum mass of ∼ 100 μg (size > 400 μm) and a minimum residence time of ∼ 50,000 y in the interplanetary medium.

For a set of ∼ 20 smaller grains from the 100–200 μm size fraction all analysed at once as a single particle, this individual criterion switches to a "statistical" criterion. This set might still contain some terrestrial grains and/or debris of much larger meteorites and/or micrometeorites with undetectable exposure ages. Moreover while estimating a very important upper limit on the size of the parent micrometeoroids of the grains, the method cannot fix the exact preatmospheric size of micrometeorites, or their transmission efficiency in the atmosphere.

In §6 we describe a measurement of the mass distribution of the Greenland cosmic dust grains, that is amazingly similar (in shape) to that of the micrometeorite flux in the interplanetary medium at 1 a.u., as compiled by Grün *et al* (1985). This measurement first *statistically* extends the conclusion of Raisbeck and Yiou (1987) down to the very small sizes of the IDPs (∼ 10 μm). Moreover it demonstrates for the first time the unexpectedly high survival rate of micrometeoroids in the atmosphere, that constitutes a very strong constraint about ablation and fragmentation mechanisms (see §7).

4. New micrometeorites and their host Blue ice sediments

4.1 *High concentrations of micrometeorites in blue ice sediments*

In wet cryoconite samples extracted at the site of the "Blue Lake I" expedition (at about 20 km from the ice margin), the concentration of cosmic spherules with size

---

*The grains were first directly hand-picked under a binocular from various size fractions of the "Blue Ice" sediments, relying on simple characteristics such as being "dark and colourless", weakly magnetic, friable, and showing slow sedimentation rate when released in water. In Antarctica the sediments were directly obtained by filtering melt ice water on a stack of stainless steel sieves. In Greenland a mechanical disaggregation of cryoconite, conveniently performed on the sieves with a hard nylon brush, was required to extract these grains.

> 100 μm, per kg of wet cryoconite (about 700 spheres) is already much higher than that found in ~ 100,000 y old deep sea sediments extracted at the centre of the Pacific (about 1 to 10 per kg of clays). But after a simple mechanical disaggregation of cryoconite a minute mineral fraction, representing about 100 ppm of the initial mass of cryoconite is collected on the 100 μm sieve. In this fraction the proportion (number) of cosmic spherules can reach an amazingly high value of ~ 10% in the two richest cryoconite samples collected near Jakobson. Samples collected at the Sondrestromfjord sites are much more heavily contaminated with quartz grains. But these grains constitute a white "substrate" on which the dark-colourless cosmic dust grains can be easily spotted and handpicked.

At a given site on the ice cap the concentration of cosmic spherules with sizes < 300 μm is rather independent (within a factor of 2) on either the disaggregation procedure or the type of deposits. These deposits can be observed as mm-thick layers in the countless cryoconite holes that constitute most of the ice surface (figure 1, top), as ~ 5 cm-thick dried deposits right on the top surface of the ice field, as extensive ~ 1 cm-thick placers found under running water, along the banks of small torrents (figure 1, bottom), or on the bottom of the lakes.

In sharp contrast the concentration of big spherules (sizes > 500 μm) is now dependent on both parameters. At the site of Camp II (~ 25 km from the ice margin), for our best disaggregation procedure (see §2·3), this concentration varies from ~ 1 per 10 kg of cryoconite for the worst "washed out" deposits under running waters, up to ~ 1 per kg for the richest ~ 2 cm-thick deposits found on the bottom of the most "mature" cryoconite holes, reaching sizes of few meters.

In Antarctica we extracted a few grams of sediments from 100 tons of blue ice. This small concentration, amounting to a few tens of ppb, illustrates the ultra-clean characteristic of the Antarctica ice collector. In the richest size fraction (50–100 μm) of our best daily collection the proportion of cosmic dust grains reaches an amazingly high value of ~ 30%.

4.2 *Terrestrial contamination and weathering trends:*
An unexpected complementarity in Greenland and Antarctica

In Greenland terrestrial contamination is typical of wind-borne dust, as the 50–100 μm size fraction contains about 200 times more terrestrial grains than the > 400 μm fraction. In the "purest" > 400 μm size fraction about 20% of the grains are extraterrestrial. But it becomes almost impossible to handpick unmelted micrometeorites in the most contaminated 50–100 μm size fraction. The exact opposite trend is observed in Antarctica blue ice sediments, where terrestrial contamination now results from a turbulent upward motion of moraine debris (still poorly understood), that originate from a parent bed lying ~ 200 m below the ice surface. Now the 50–100 μm size fraction is the purest fraction, and most of the mass of terrestrial debris is found in the > 400 μm fraction, in which it will be very difficult to handpick rare unmelted micrometeorites.

With regard to their "richest" size fractions the two collections are thus remarkably complementary. The Antarctica collection should allow the first meaningful comparison of the 50–100 μm size micrometeorites with stratospheric IDPs, for searching for a possible filiation between the two collections. On the other hand Greenland cryoconite is the only exploitable deposits of big unmelted micrometeorites, that are extremely rare in space.

Another complementarity of the two collections, is their very distinct weathering patterns (see §8), that should be most useful to define "invariant" characteristics of micrometeorites not seriously affected by this process. Weathering in Greenland results from biogenic etching. In Antarctica, the grains exhibit a very different weathering pattern due to "ordinary" chemical corrosion of natural waters. Another consequence of these distinct weathering patterns is the specific contamination of porous unmelted fragments in trace elements (Robin 1988) observed in Greenland but not in Antarctica.

A top superficial $< 1 \mu$m-thick layer is frequently observed on the unmelted micrometeorites, that has an overall composition in major elements slightly distinct from that of "fresh" internal surfaces obtained by fracturing. In Antarctica this layer shows some enrichment in both Al and Fe with regard to that observed in similar grains extracted from Greenland cryoconite. Such "skins" might result either from weathering or ablation heat metamorphism, or a combination of both. But there is a peculiar skin on a specific family of unmelted fragments, that look like puffing up, and which is observed in both Greenland and Antarctica (see the two top micrographs in figure 5). We thus believe that it rather represents an ultrathin skin of ablation, very different from the much thicker fusion crust of meteorites, that contains new information about the ablation process. This skin is probably slightly altered during terrestrial weathering, while keeping its overall aspect.

4.3 *The new families of micrometeorites, and the unexpected high abundance of unmelted chondritic fragments*

In the 6 SEM micrographs (A to F) reported in figure 4 we illustrate the major types of cosmic dust grains with sizes $> 100 \mu$m found in a typical sample of cryoconite collected at ~ 20 km from the margin of the Sondrestromfjord ice field. With regard to grains previously reported in the deep sea collections the following major differences were observed: the dominant stony grains ($\sim 50\%$) are again barred-chondritic spheres (A); there is an unexpectedly high proportion of unmelted chondritic micrometeorites (B) of about 25%, that was discovered by E Robin, and which is much larger than that previously quoted ($< 1\%$) for "largely-unmelted" grains in deep sea collections; the proportion ($\sim 2\%$) of grains made of an oxidized alloy of Fe/Ni (D) is about 25 times smaller than in deep sea sediments.

New families of grains were discovered. First, besides their high abundance, there are two broad varieties of unmelted chondritic micrometeorites, namely: the crystalline unmelted micrometeorites (CUMs) with large single crystals and very little residual porosity and/or matrix material; the highly porous unmelted micrometeorites (PUMs), made of friable aggregates of very tiny grains, that cover a great diversity of particles. We also discovered new glassy chondritic spherules (C), and Robin found spherules of unoxidized Fe/Ni alloys (E), as well as very rare *unmelted* grains of unoxidized Fe/Ni alloys (F), that were never reported before. In the next sections we shall focus on the two broad types of unmelted chondritic micrometeorites.

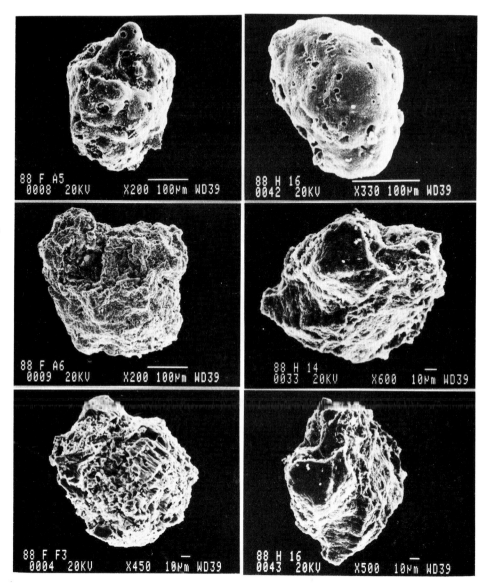

Figure 5. Variety of unmelted micrometeorites. The most external < 1 μm thick layer of the two grains in the top micrographs start to be melted, yielding a typical "puffing up" habit. Most of the micropores in the bulk volume of these specific grains show jagged contours yet, indicating that they have not reached melting temperature inside. Greenland and Antarctica micrometeorites are on the left and right hand sides respectively. The grains in the middle and bottom micrographs refer to porous and crystalline micrometeorites, respectively.

5. Search for invariant characteristics of unmelted micrometeorites

The EDX and INAA analyses of Robin (1988)* have been very useful to identify elemental abundances that have not been affected by terrestrial weathering in Greenland cryoconite. The list of "unweathered" elements includes most of the major elements (Mg, Si, Fe, Ca), that have already been measured on a much finer scale of the analysed volume ($< 1\,\mu^3$) in IDPs and meteorites (see Bradley 1987), or tiny dust grains from comet Halley (Langevin et al 1987). The concentration of important minor and trace elements (Ir, Sc, Cr, Ni, etc.) is also an "invariant" characteristic of the grains, that was measured at a much larger scale ($\sim 10^7\,\mu^3$) of the analysed volume.

5.1 *Crystalline unmelted micrometeorites from Greenland: About 50% of them have no clear analogs in the meteorite collections*

The mineralogy of 14 CUMs with size of $\sim 200\,\mu m$ from the Greenland collection, already analysed for their trace element contents (Robin 1988), has recently been investigated by Christophe-Levy and Bourot-Denise (1988). This work was extended to 15 CUMs from the 100–200 μm size fraction of the Antarctica collection, that have not been analysed yet for their trace element contents. Their major objective was to relate the CUMs to major types of chondritic meteorites, by means of the classical methods of mineralogy and petrography. Besides the absence of a terrestrial contamination in trace elements, the advantage of these crystalline grains is that they contain large crystals of olivine and pyroxene, that can be meaningfully analysed for their major and minor element content with an electron microprobe.

In figure 6 the compositions of these minerals have been plotted on a binary diagram, altogether with the shaded area that delineate the corresponding trends observed for chondritic meteorites. About 12 grains might be related to ordinary chondrites. In this first subset one grain can be strictly related to the H6 chondrites and another grain to the L chondrites; 2 grains are clearly more oxidized and 2 grains more reduced than in ordinary chondrites, and; one grain shows a disequilibrated assemblage of crystals. But all the other grains of this preliminary selection (i.e. 43% of the total) contain Fe-poor silicates, *and this proportion is very different from the smaller value* ($\sim 1\%$) observed in the groups of iron poor meteorites (enstatite chondrites and/or achondrites). Finally two of the grains of this second subset cannot be related to known classes of meteorites.

Even the crystalline chondritic micrometeorites, that look like a collection of chunks of well-crystallized meteorites and/or chondrules, seem to contain new materials with no clear analogs in meteorite collections. Their trace elements contents should soon yield important clues about their filiation with meteorites.

---

*The detailed analyses of ~ 35 major, minor and trace elements in about 200 *individual* Greenland micrometeorites with sizes $\sim 200\,\mu m$ (~ 130 spheres and ~ 60 unmelted fragments) have been reported in the Ph.D thesis of Robin (1988). This impressive set of analyses is available upon request at the Centre des Faibles Radioactivités du CNRS, Avenue de la Terrace, 92120 Gif-sur-Yvette, France.

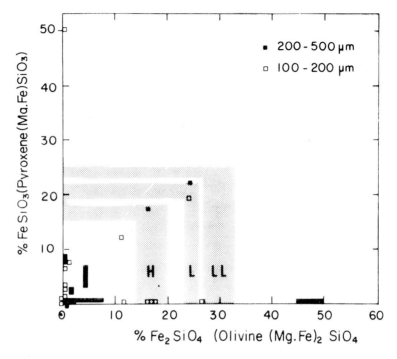

Figure 6. Binary plot of the mineralogical composition of 14 crystalline unmelted micrometeorites (Courtesy M Christophe-Levy).

5.2 *Porous unmelted fragments: they are not simple chunks of primitive carbonaceous meteorites*

A preliminary comparison, based on Si/Mg/Fe ternary diagrams obtained from broad beam EDX analyses (magnification $\sim 1000\,\mu^3$), already suggested that the evolutionary "track" of the PUMs on such diagrams is rather similar to that of primitive carbonaceous meteorites, with the exception of a much broader dispersion of abundances (Bonté et al 1987a). This trend was quite distinct from those of the few IDPs analysed at that time by others, at a much smaller scale of analysed volume ($\sim 1\,\mu^3$). This apparent similarity with the primitive meteorites is no longer observed, considering other characteristics.

First the fine-grained PUMs microstructure is quite different. SEM micrographs of internal fracture surface of chunks of Orgueil (top micrograph on the left, figure 7), and PUMs from the Greenland and Antarctica collections (figure 7) illustrate the striking porosity of the PUMs, which is not observed in Orgueil, Murray and Mokoia. In about 80% of the PUMs, all micropores have very irregular "jagged" contours. In the remaining fraction the PUMs show a typical "puffing up" habit (top micrographs in figure 5), and the jagged pores now coexist with tiny vesicles with smoother habits (last micrograph in figure 7), indicative of a partial melting of the least refractory components. Both types of PUMs are very distinct from a family of "largely unmelted" chondritic micrometeorites, that contain much larger vesicles and large "relic" crystals, that we shall not consider in this paper. This porosity is also consistent with the low density (1·5–2 g/cm^3) of a few large PUMs,

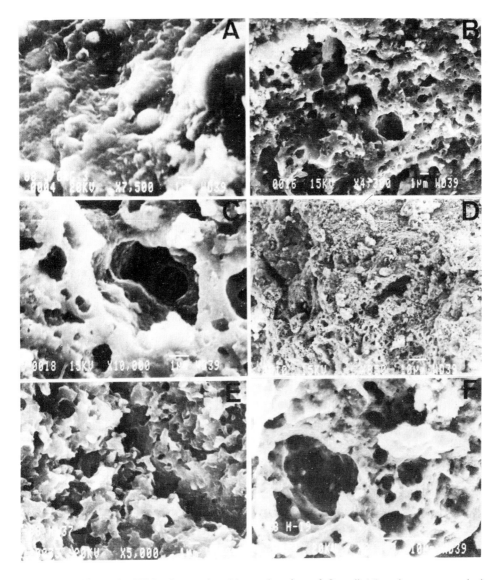

Figure 7. SEM micrographs of internal surface of Orgueil (A) and porous unmelted micrometeorites, including two 1 mm-size big micrometeorites (B and C) from the Blue Lake II expedition, as well as ~ 200 μm-size grains from Greenland (B, C, D) and Antarctica (E, F). One of the grains (D) was loaded with a sticky component that gets spread on the surface during microfracturing with a sharp surgical blade. G corresponds to the most heated-up grains of this series of unmelted micrometeorites, showing puffing-up habits.

inferred from direct weight measurements by Ph. Sarda. We have checked that carbonaceous meteorites do not develop such highly porous structures upon either pyrolysis up to ~ 1000°C, or accelerated leaching tests in hot water and steam.

About major, minor and trace elements, one of the most striking differences with the corresponding data measured for meteorites is the much broader dispersion of elemental abundances, that frequently prevents the strict definition of *average* values.

For example the frequency distributions of Fe/(Fe + Mg) and Ir/Si ratios observed for 31 PUMs show a very asymmetric shape, that peaks a value slightly higher than the chondritic average. Nevertheless the arithmetic mean for both ratios just corresponds to the chondritic values!. (On the same plots crystalline fragments show very different distributions, displaced to much smaller values). In the next section we shall further discuss these dispersions in the abundances of Ir, Ni, Fe and Si, for both crystalline and porous unmelted chondritic micrometeorites, and a large variety of chondritic spherules.

5.3 *Scatter diagrams and their power transformations*

Binary and ternary plots of elemental abundances have been used to classify and correlate diverse solar system objects, including various components of meteorites, stratospheric IDPs, and submicron size grains from comet Halley, that impacted the ionization mass spectrometers on board the Giotto and Vega spacecrafts. In these previous attempts ratios of elemental abundances were generally used to plot these scatter diagrams. The most extreme example of scatter was observed for Halley dust, analysed at a very small scale of the analysed volume ($\sim 10^{-4} \mu^3$), while ultramicrotome sections of C1 carbonaceous chondrites yielded the least scattered plots on the same scale.

We also observed a large scatter of elemental abundances for the new Greenland and Antarctica micrometeorites, analysed at a much larger scale of analysed volume ($\sim 10^7 \mu^3$). We argue elsewhere (Maurette and Passoja 1988) that these scatters and

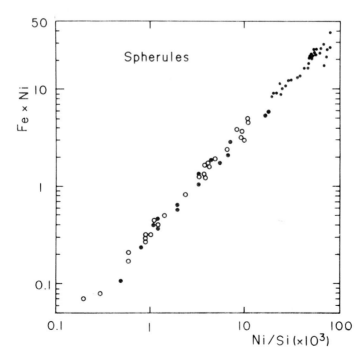

Figure 8. Variations of the Fe × Ni product with Ni/Si ratios for 74 chondritic spherules (log scale).

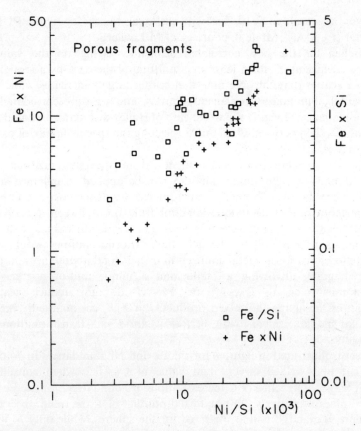

Figure 9. Variations of the Fe × Ni product with Ni/Si ratios for 31 porous unmelted micrometeorites (log scale).

their evolution in diverse solar system objects might be related to the earlier "fabric" of the constituent material of micrometeorites, that might have involved some "fractal" aggregation process. Indeed, in material sciences, aggregation processes frequently lead to the formation of fractal clusters, which are self-similar over a large size range. Grain aggregation processes in the early solar system might have also followed fractal scaling. This might appear as a peculiar scatter of elemental abundances, that would not simply follow the rules of ordinary statistical fluctuations (R Rammal, personal communication). These information might have been partially altered during the subsequent history of the grains, that possibly included their "weathering" in damp asteroidal regolith (McSween 1979), and/or their more recent turbulent reprocessing by fast jets of water and volatile species on the dark surface of cometary nuclei.

We analysed data obtained at different scales of analysed volume for 74 chondritic spheres with sizes $\geqslant 100\,\mu m$ (INAA analyses at a scale of $\sim 10^7\,\mu^3$; courtesy of E Robin); 30 anhydrous and 30 hydrated IDPs (broad beam EDX analysis at a scale of $\sim 10^2\,\mu^3$; courtesy of D Brownlee and L Schram); 30 ultramicrotomed sections of each one of the following samples: one hydrated and one anhydrous IDPs, and chunks of the Orgueil and Murchison micrometeorites (point count analysis at a scale of $\sim 10^{-3}\,\mu^3$; courtesy of J Bradley and D Brownlee); 32 tiny grains of comet Halley that were analysed at a scale of $\sim 10^{-4}\,\mu^3$ with the highest accuracy by the PUMA mass spectrometer (PUMA 1, Mode 0; courtesy of M Lawler).

We searched for the best correlated elements, relying on the value of the correlations coefficient, C, for a least square fitting of the data by a straight line on binary y vs x scatter diagram. We note that scatter largely increases while switching from elemental abundances to elemental ratios, and we only considered plots of elemental abundances. Even in the worst case of Halley dust data, we could find clear correlations ($C \geqslant 0.9$) between O, Si and S selecting the specific family of oxygen-rich grains (O/Si $\geqslant 1$).

When the best correlations have been identified statistical methods such as factory analysis and stepwise determinant analysis can be applied to enhance similarities and/or differences between families of micrometeorites at the cost of computer time (see an application of these techniques for chondrules, in McSween *et al* 1983). Power transformations ($X \to X^n$ have also been used in statistics to reduce scatter in data analysis) (Hoaglin *et al* 1983). We felt these transformations would be more appropriate to search for a fractal connection in elemental abundances, in particular in scatter diagrams involving a major and a minor and/or a trace element. Transformations such as Y vs X^n or Y^n vs X do not reduce scatter. But transformations involving the mixed product, $P = Y \cdot X$, can markedly increase the value of C for specific elements, with the P vs X^2 and P vs X transformations yielding the same values of C.

This effect is illustrated in figure 8 for the Fe and Ni abundances in 74 chondritic spheres, that first yielded a very bad value of $C(\sim 0.5)$ when considering the Fe/Si vs Ni/Si plot. The value of C increases up to 0.77 and 0.992 for the Fe vs Ni and Fe·Ni vs Ni plot, respectively. On the mixed product plot, the trend observed for 31 PUMs (figure 9) exactly fits that observed for the spheres, while only $\sim 50\%$ of the CUMs roughly follow this trend. We noted a similar improvement on the scatter diagram of the Ni and Ir concentration measured for 31 PUMs, with the value of C increasing from 0.3 (Ni/Si vs Ir/Si) up to 0.992 for the mixed product.

We further enhanced the smoothing effects of power transformation considering the numerical integral, $F(X) = \int Y dX$, computed from the Y vs X scatter diagram. The resulting curves obtained for the spheres, PUMs and CUMs from Greenland are now quite separated. However upon a simple vertical translation the PUMs curve exactly fits the relationship noted for the spheres, including all residual fluctuations. A similar "translational" fit cannot be obtained for the CUMs.

We just started to play the risky game of power transformations, and we still do not fully understand their properties. A much less dramatic improvement of C is noted for combinations of major elements, and the improvement is quite spectacular when one elemental abundance widely fluctuates.

Nevertheless odd dispersions in element abundances can now be defined from two numbers, the slopes of the log P plots, and the corresponding values of C. We have still to relate these characteristics to the geometry of grain aggregation, and check whether they depend on the micropore structure. This "linearization" of elemental abundances might allow more meaningful comparisons between various families of micrometeorites.

The 74 spheres considered in figure 8 include very different particles, such as barred chondritic spheres, transparent homogeneous glasses and "largely unmelted" highly vesicular objects with relic grains. The spheres also greatly differ with regard to

Figure 10. Typical "mature" cryoconite hole used as micrometeorite detector.

their concentrations of Ir (10 ppb up to 1500 ppb) and Ni (10,200 ppm to 0·02 ppm). They constitute a very homogeneous family of solar system objects, that should derive from a common reservoir of material. The astonishingly good match observed for the

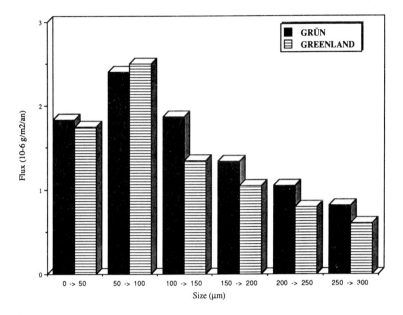

Figure 11. Comparison of the differential mass distributions of Greenland cosmic dust grains and micrometeorites in the interplanetary medium at 1 a.u.

plots of the spheres and the PUMs strongly suggests that the parent material of the spheres are the porous fragments.

One of the most amazing feature is that a strong metamorphic event, such as ablation melting, does not obliterate the chemical signature of the parent material, as inprinted in log P. Preliminary results indicate that the corresponding trend observed for P(Fe, Ni) in primitive carbonaceous meteorites would be quite distinct, showing in particular a smaller slope. This again supports the view that PUMs are not simple chunks of primitive meteorites, unless some "aqueous" weathering in a damp regolith (McSween 1979; Bunch and Chang 1980), did remobilize very efficiently the element distributions. An alternative explanation would be that a high proportion of heavy elements in micrometeorites are initially trapped in some volatile phases, that get preferentially lost during ablation*. It might be interesting to conduct some pyrolysis experiments and/or accelerated leaching tests in hot waters with chunks of either carbonaceous chondrites or the largest PUMs available (size ~ 1 mm). The initial slope of log P(Ni, Fe) might vary during such treatments.

This fit between the plots of spheres and PUMs has other implications. It constitutes a new constraint on the nature of ablation process, and indicates that terrestrial weathering does not markedly affect the Fe, Ni, Ir and Si concentrations in the PUMs. Indeed, as many of the spherules reported in figure 10 (i.e. the barred chondritic spheres) are hermetically sealed against weathering, the plots observed for the PUMs and the spheres should not match.

---

* A possible constraint about the composition of this mass loss is the analysis of "metallic" layers in the upper atmosphere by laser sounding. These layers show an overabundance of alkali metals and iron with regard to calcium. If this enrichment is associated with a volatile component of micrometeorites the P(Fe, Ca) products for the spheres and the PUMs should scale somewhat differentially on a log scale.

5.4 Unexpected exposure ages of micrometeorites in the interplanetary medium

Another striking difference between micrometeorites and meteorites was discovered by Raisbeck *et al* (1983): the ^{10}Be exposure age of both spherules and unmelted grains (about $5 \cdot 10^5$ y) corresponds to their residence time in the interplanetary medium. These high values have been very recently reconfirmed from ^{21}Ne measurements in *individual* PUMs with sizes $\sim 200\,\mu$m, by C Olinger and C Hohenberg. They are much larger than previous estimates of the lifetimes ($\sim 10^4$ y) of $100\,\mu$m-size micrometeoroids in conventional orbits around the sun in the zodiacal cloud*, that assume destruction rate of about $\sim 1\,\text{cm}/10^5$ y.

If micrometeorites follow conventional low eccentricity orbits in the zodiacal cloud, the processes of grain destruction should have been much smaller over the last 500,000 y than today, and the present-day interplanetary medium is much dustier than in the past. An alternative explanation is that solar flare activity was much higher in the distant past (Raisbeck, personal communication), producing over a short exposure of 10^4 y the required amount of ^{10}Be. But micrometeoroids might follow more eccentric orbits, corresponding to long-period comets of the Oort clouds, thus acquiring long exposure ages. Along such orbits their collision and sputtering lifetimes should be much larger, and this might explain our unsuccessful search yet for micrometeorite impact craters on the surface of unmelted micrometeorites. The high ^{10}Be exposure age of the grains would require a substantial flux of *low energy* galactic cosmic rays outside the heliopause.

These long exposure ages, that are already very different from the corresponding distributions reported for stony meteorites, indicate that the new micrometeorites can hardly be related to major asteroidal events generating meteorites.

6. Micrometeorite flux on the ice caps

6.1 Mature cryoconite holes, and their saturated layer of sediments

Our new method to evaluate the micrometeorite flux (Maurette *et al* 1987; Robin 1988; Bonny *et al* 1989) starts with the measurement of the size distribution of about 2000 individual chondritic spherules (both glassy and barred) with sizes $> 50\,\mu$m, that can be easily transformed into a mass distribution. Such spheres represent about 70% of the cosmic dust grains deposited in the ice. Corrections have to be applied to take into consideration a major contribution of grains smaller than $50\,\mu$m, as well as the relatively minor contribution of the other types of larger grains (mostly unmelted grains).

Then we rely on field measurements, yielding the maximum amount of cryoconite deposited per m^2 (~ 2 kg) on the bottom of the largest and *most mature* cryoconite holes (figure 10). These specific holes did survive for the longest period during their chaotic flow right on the top surface of the ice field. During this motion, that lasted for a duration, ΔT, they captured cosmic dust grains from two sources: the direct infall of the micrometeorite flux over the "last" ΔT time window, during which the holes have moved on the surface of the ice cap, from their birth place (near the onset of the melt

---

* For example with a flux of $\sim 10\,\mu$m-size impacting micrometeorites of about $1/\text{m}^2/\text{day}$, a $200\,\mu$m-size grain in space would suffer a direct destructive hit every 10^4 y

zone) to the present collection site and the melting of the successive annual ice layers formed much further inland (over a longer period $\sim 2 \times \Delta T$), which fed the holes with cosmic dust grains initially trapped in these layers*.

This model, including the maximum value of $2 \times \Delta T$ corresponding to the age of the ice surface (~ 2000 y), was inferred from the ice flow model of Reeh (1988). This is a strict upper limit as the holes have probably a shorter lifetime against destructive events such as flooding, opening of crevasses and so forth (see §6·3). We thus deduced that 2 kg of cryoconite picked up in the most mature hole found at 20 km from the ice margin are equivalent to a detector of $1\,m^2$ exposed during ~ 3000 y in the micrometeorite flux transmitted by the Earth atmosphere. With this equivalence we directly deduced the flux of Greenland cosmic dust grains from their mass distribution.

6.2 *Similarities between differential mass distributions of Greenland cosmic dust and micrometeoroids in the interplanetary medium*

In figure 11 we give as a shaded histogram the differential mass distribution of the Greenland cosmic dust grains, reported earlier (Maurette *et al* 1987). Its shape is independent on the exact value of the residence time of cryoconote holes on the ice surface, ΔT, used in our modelling exercises, that can only produce a *vertical shift*. It is very different from distributions expected from volcanic dust, wind-borne contamination, and regolith reprocessing by repeated impact cratering.

In fact it fits the distribution directly computed for the micrometeorite flux in the interplanetary medium at 1 a.u. (Grün *et al* 1985), before its transmission through the atmosphere. This distribution, delineated by the full line histogram in figure 12, was recently improved by Bonny *et al* (1989)**, with regard to our first estimate (see figure 2 in Maurette *et al* 1987). This fit is observed for the peak position as also for the slopes of the distribution. In contrast the "absolute" value of the total mass of micrometeorite accreted each year by the Earth (about 5000 tons/y in the size range 50–300 μm), would be about 4 times smaller than the estimate deduced from Grün *et al* (1985).

A closer look reveals slight differences in shapes. There is a misfit in the coincidence of the right wings, that get amazingly restored by a small shift ($\sim 50\,\mu$m) of the Greenland distribution to the left. A more accurate definition of the size increment of $\sim 50\,\mu$m used to plot the size distribution is probably responsible for this effect, unless the ablation process is highly specific, removing a constant $\sim 50\,\mu$m-thick layer of the grains, independently on their initial size. There is a more pronounced misfit on the left wing, corresponding to an important deficit of small micrometeorites. This deficit might correspond either to a slight change of the exponent of the mass distribution for sizes $< 60\,\mu$m, or more likely to a preferential loss of weakly magnetic

---

*This factor 2 represents the slower flow rate of the deep buried annual layers (that intersected the collection site at shallow angle during their upward motion), with regard to the faster flow of the ice field surface, carrying the cryoconite holes.

**The new distribution takes into consideration a two-fold increase of the micrometeorite flux incident on the Earth's atmosphere, and that represents the gravitational focusing effect of the Earth. Next a third degree polynomial fit to the spectrum of Grün was used, to replace the sharp kink in the micrometeorite spectrum, generally postulated in all previous studies around $\sim 10^{-7}$ g, by a smooth transition.

micrometeorites during the magnetic extraction. The position of the peak is rather sensitive to the approximation of a smooth transition around 10^{-7} g, and the sharp kink postulated in all previous work tends to shift the peak to a smaller size.

This comparison allows several inferences. We generalize on a *statistical* basis, to a smaller mass range where most of the micrometeorites are found, the earlier conclusion of Raisbeck and Yiou (1987), namely: the parent bodies of the vast majority of cosmic dust grains found in terrestrial sediments are indeed genuine micrometeorites. This conclusion would be valid in particular for grains in the size range $< 50 \mu m$, corresponding to stratospheric collections. Furthermore we infer for the first time that the ablation (i.e. mass loss) of micrometeoroids in the atmosphere is indeed very weak, even when the residues are melted spherules. Indeed in the case of strong ablation a clear shift in the peak of the distribution and/or some discontinuity around a size of about 100 μm should be observed. This condition of "weak" ablation also agrees with the high relative abundance of unmelted micrometeorites, observed up to ~ 1 mm. Both observations represent new constraints on ablation models, that decided Bonny et al (1987; 1988) to propose a new modelling of this process (see §7). A major problem is to understand the lack of any major fragmentation of micrometeoroids in the atmosphere that can be deduced from the same argument, and that runs somewhat contrary to indirect evidence inferred from studies of radio and visible meteors.

For the Antarctica ice cap only a rough value of the cosmic dust flux can be given. Only the concentration (about 5 grains with size $> 50 \mu m$ per ton) and mass distribution of the "background" cosmic dust grains, that we found well outside the enriched superficial layer of the ice field (i.e. at depths of ~ 5 m), can be used for a flux estimate. If we take a reasonable value of 20 cm/y for the sedimentation rate of the host blue ice of the grains, then the Antarctica flux agrees within a factor of 3 with the micrometeorite flux.

6.3 *Effective collection efficiency of the Greenland ice cap for micrometeorites*

In the 50–300 μm size fraction, we found that the total mass influx of the Earth (about 5000 tons/y) is about 4 times smaller than the value deduced by multiplying the flux of Grün et al in space (10,000 tons/y) by the factor ($\times 2$) of gravitational focusing of the Earth. During a preliminary disaggregation of ~ 40 kg of cryoconite in the field, intended to define the "best" procedure, we have recently extracted from the $\gtrsim 500 \mu m$ size fraction 6 chondritic spherules and our first 2 "big" PUMs (no CUMs have been found yet in this large size fraction). The total mass of these big grains (about 5 mg) then yields a flux of cosmic dust which is about 15 times smaller than the predictions inferred from Grün et al (1985).

The factor of overall collection efficiency of the 50–300 μm micrometeorites ($\sim 25\%$) might represent a loss of grains during their transmission through the atmosphere and subsequent collection/disaggregation. But we believe that it rather reflects the finite lifetime of the "mature" cryoconite holes, that is smaller than the value, $\Delta T \sim 1000$ y, used in this preliminary evaluation. This lifetime should be dependent on such destructive events as flooding, opening of crevasses and so forth. The value of ΔT should be about 4 times shorter than the maximum value of ~ 1000 y considered yet in our modelling exercises, to restore a 100% collection efficiency. As both the shape of our micrometeorite mass spectrum and the high proportion of unmelted micrometeorites strongly suggest a very good transmission of micrometeorites through the atmosphere, we rather favour this interpretation of $\Delta T \sim 250$ y, that we should soon check from our field studies.

With $\Delta T \sim 250$ y, we deduce that 1 kg of cryoconite used for collecting grains with sizes $< 300\,\mu m$, is equivalent to a 100% efficient space collector of $\sim 500\,m^2$ exposed for one year in space. The space equivalence of one ton of cryoconite used only for collecting the big micrometeorites (size $> 500\,\mu m$) would then rate in the same unit of "100% efficiency", as a giant collector of $100,000\,m^2/y$, when we consider a further decrease in the overall collection efficiency noted in our work (a factor ~ 3).

About 2 months of effort in the field (by 3 participants) was required to melt ~ 100 tons of blue ice, yielding ~ 5000 cosmic dust grains with a size $> 50\,\mu m$. For comparison the same amount of grains is contained in ~ 1 kg of Greenland cryoconite, that can be collected in a few minutes on the Greenland ice cap, and disaggregated within a few hours. This shows that the melt zone of the Greenland ice cap is yet a unique collector to hunt for big micrometeorites that are extremely rare in space. But collection sites in which these micrometeorites are uncontaminated by REE have still to be identified (see §8·5).

7. Improved thermal protection of a composite-pyrolyzable micrometeorite, giant thermal flux, and prebiotic evolution

7.1 New ablation constraints and the inadequacy of classical ablation models

Bonny et al (1988, 1989) did observe a high ratio, $U/M \sim 30\%$, of unmelted fragments to melted spherules, which looks rather constant from $\sim 100\,\mu m$ up to ~ 1 mm. They stated that this result cannot be explained by classical ablation models, that only consider micrometeorites made of "compact-inert" materials, such as chunks of meteorites.

The proportion of micrometeorites that are not heated up beyond melting temperature, can be computed with a model developed by Fraundorf (1980) for the specific case of IDPs up to sizes of $25\,\mu m$. This model, which directly originates from the classical theory of Whipple (1950), also integrates the distributions of speed and entry angle of micrometeorites. Robin (1988) argued that this proportion is equivalent to the ratio, U/M, of unmelted fragments to melted spherules measured in the new collection of micrometeorites.

We did believe at first (Bonny et al 1987) that U/M represents a strong constraint on the ablation mechanism, when considering melting temperature. Indeed starting with a reasonable value of $T_f \sim 1600°K$, and values of density ($\sim 2·5\,g·cm^{-3}$) and average speed ($V \sim 20\,km/s$) of micrometeorites given by Grün et al (1985) U/M sharply drops from 10% down to 1% and 0·1% for sizes of $100\,\mu m$, $300\,\mu m$ and 1 mm, respectively. Thus very shallow entry angles are required for the survival of $\sim$ mm-size micrometeorites as unmelted fragments. This corresponds to an overall transmission efficiency throughout the atmosphere ($\sim 0·1\%$), that is much smaller than that ($\sim 25\%$) measured in Greenland. Classical modelling of the ablation process was considered as inadequate.

But predictions of U/M ratios are uncertain. On the one hand the values of U/M are very difficult to measure over a large size range. This results from the friability of both glassy spheres and unmelted micrometeorites, inducing their preferential destruction upon mechanical disaggregation*. Moreover in this modelling U/M scales as T_f^8, V^{-6} and ρ^2. Consequently minor changes in the values of these parameters trigger drastic changes in the predictions. We selected favourable values of these parameters, to check whether they might fit our data. Taking the most recent speed distribution of micrometeorites ($V \sim 14$ km/s), that was corrected for various effects of the Earth (Southworth and Sekanina 1973), a low density of $1\,\text{g}\cdot\text{cm}^{-3}$, and $T_f \sim 1600°$K, we restore a reasonable fit with our data up to sizes of $\sim 300\,\mu$m, as U/M scales as 69%, 10% and 1% for sizes of 100 μm, 300 μm and 1 mm, respectively.

However these classical predictions become worst when T_f is replaced by a more realistic critical temperature of "degradation" of micrometeorites, T_c. During in-situ pyrolysis of micron size PUMs in the hot stage of the HVEM (heating rates of $\sim 10°$C/min) a high proportion of the PUMs start to develop rounded habits at $T_c \sim 900$–$1000°$C. We believe that melting temperatures used in classical ablation models should be approximated by a lower value, corresponding to the development of a clammy state of the grains, which is a few hundreds °C lower than T_f. Moreover our recent observation of high concentrations of pyrolyzable carbon-rich material in the PUMs by high voltage electron spectroscopy (Bonny et al 1989) indicates that the temperature of the PUMs has not reached the stage of graphitization ($\sim 900°$C). A lower limit of the temperature of micrometeorites might be soon deduced from their strong depletion in sulphur. All data should help in bracketing the critical temperature of "degradation" of micrometeorites at values, $T_c \sim 1000°$C. With this value, classical models can hardly account for the very good transmission of micrometeorites, as the predicted variations of U/M for the same set of favourable parameters scale as 10%, 1% and 0.1% for sizes of 100 μm, 300 μm and 1 mm, respectively. This implies the occurrence of efficient cooling mechanisms, that have to be clearly identified.

The detection of carbon in PUMs (Bonny et al 1989), as well as previous studies of IDPs, suggest that a large proportion of micrometeorites have a porous aggregate structure of silicates, loaded with some volatile C-rich material. Bonny et al (1987) noted that this material is somewhat similar to the "composite-pyrolyzable" materials used for the thermal shielding of reentry vehicles in the atmosphere. They did apply a complex computer program ("COKE")**, to check whether a synthetic "*composite-pyrolyzable*" micrometeorite made from this material (silica fibers bounded with 30% of phenolic resin), would be much better thermally shielded during impact than "*compact-inert*" micrometeorites (i.e. composed of a chunk of meteorite), only considered in the classical computations.

---

* We first disaggregated at once 140 g of cryoconite held on a 100 μm sieve, in about 30 minutes of operation with a hard nylon brush. Next we successively disaggregated several ~ 10 g samples of cryoconite, relying on a more gentle procedure, where a jet of water is fired on cryoconite held under running water in the sieve. We found that both the value of U/M ratio and the proportion of glassy spherules increased up to a factor ~ 2 during the gentle procedure.

** This code was developed by Darmon and Balageas (1986) at the Office National de Recherches Aéronautiques (ONERA), to model the ablation of much larger man-made thermal shieldings, used on rockets entering the Earth's atmosphere at speeds < 5 km/s, that are smaller than the average speed of meteorites (~ 20 km/s).

7.2 The COKE computations and giant heat flux

The COKE programme accounts for new complex cooling mechanisms, that are all activated by the pyrolysis of the volatile component upon heating, such as: endothermal chemical reactions; percolation of pyrolysis gases in the porous structure, that requires a *connected* pore structure; formation of a protective gas "gasket" around the grains. An improved version of the code should soon assess the additional cooling effects of melting and volatilization.

These very lengthy computations have only been performed yet on two test particles, with sizes of 60 and 500 μm. In figure 13 we just report for the 60 μm particle, the variation of the residual proportion of the pyrolyzable component as a function of time. For the synthetic micrometeorite an important proportion of this component is still preserved at the end of the heat pulse. This thermal protection does improve with micrometeorite size up to at least the maximum value of 500 μm considered by Bonny.

The most efficient cooling process might be the "blocking" effect of the gas gasket formed by the pyrolyzed gases, that somewhat deflects the destructive incoming supersonic air flow. Unfortunately this very complex process which involves the theory of turbulent "chaotic" flow, has been poorly modelled yet. Bonny is attempting to improve our understanding of this key process.

This new modelling of the ablation of tiny micrometeorites suffering enormous heat flux in the 100 MW range, will probably find applications in the development of thermal shieldings of new generation of rockets. Such rockets might be much faster than seen today, suffering "giant" thermal fluxes, in the 100 MW range, that cannot be simulated in the laboratory.

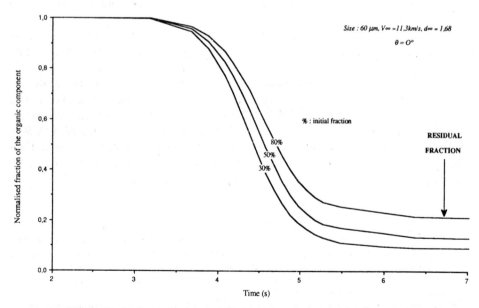

Figure 12. Residual concentration of pyrolizable organic component in a synthetic micrometeorite, during a pulse heating evolved during hypervelocity impact at 20 km/s with the upper atmosphere (normal incidence). The ordinate gives the "normalized" proportion of the pyrolizable component that has not been transformed yet into "graphite".

7.3 Composite-pyrolyzable micrometeorites and prebiotic evolution

The primitive Earth was bombarded by the accretionary tail, composed of micrometeorites, meteorites and planetesimals. The earlier flux of micrometeorites was certainly much higher than today's (possibly by a factor $> 10^4$). The results outlined in the previous section strongly suggest that the composite-pyrolyzable micrometeorites are preferentially transmitted in the atmosphere, thus keeping an important residual concentration of the initial organic component. The homogeneous and intense rainfall of such micrometeorites could have reached very localized and isolated spots on the Earth, such as hot ponds of highly mineralized waters, hot pools of dense oil spread on early seas, etc.

A previous scenario of exobiology proposed by Californian biologists postulates that extraterrestrial organic matter was raining from the sky. This material ended up being adsorbed on the high specific area of terrestrial clays, and the "catalytic" properties of the clays helped in the making of complex prebiotic molecules. The porous unmelted micrometeorites with their high porosity and fine-grained structure are already "cosmic" clays, containing their own load of organic molecules. As soon as they quit the very hostile interplanetary medium to fall in highly favourable terrestrial environments they might start behaving as individual "minicentres" of prebiotic synthesis.

This scenario is presently investigated in collaboration with A Brack (Centre de Biophysique Moléculaire du CNRS, Orléans). The first experiment is to check whether the PUMs have any measurable catalytic activity for the synthesis of simple proteins from mixtures of amino-acids. It would be interesting to investigate whether iron hydroxides and/or silica gels loaded with trace elements, might synergetically enhance prebiotic synthesis.

8. Terrestrial weathering of micrometeorites and related applications

Gooding and collaborators (1986) have reported on quantitative weathering studies of Antarctica meteorites collected in a very dry and cold environment. The first attempt to discuss the weathering of much smaller micrometeorites was presented by Callot *et al* (1987), who showed that the most abundant "barred" chondritic spherules behave like reliable "weathering microprobe". This opened up a new field in weathering studies described below, with applications ranging from biogenic "microlithography" to the long-term storage of nuclear wastes.

8.1 *From etch canals in chondritic barred spherules to biogenic microlithography*

The most abundant cosmic dust grains are the "barred" chondritic spheres, made of parallel bars of olivine and interstitial glass (figure 4A). The glass bars, loaded with bright magnetite crystals, are much thinner than the slightly brighter olivine bars. In deep sea sediments glass is etched out at a rate much faster than olivine. This delineates a typical pattern of etch canals, that are remarkably free of reaction products, and that frequently extend up to the centre of large grains, at depths $> 100 \mu m$.

We have also observed a pattern of etch canals in the Greenland spheres, that is also remarkably free of reaction products, but that extends to much shallower depths ($< 10\,\mu$m). This weak weathering is very different from that observed in deep sea spheres, because olivine is now etched out at a rate much faster than glass (see figure 3 in Callot *et al* 1987). For grains extracted from Antarctica ices the etch pattern reverses to the trend observed for deep sea spheres (figure 13). However the etch canals penetrate only to very shallow depths of a few μm, demonstrating that weathering in Antarctica ices is even weaker than in Greenland cryoconite.

Callot *et al* (1987) conducted a series of experiments to both identify the nature of cryoconite, and understand its peculiar weathering effects. First they found that cryoconite is made of cocoons of filamentary *sidero*bacteria, in which the grains are tightly encapsulated. Slabs of glasses and olivine were microfractured and exposed to the chemical action of various types of waters, including sea water and a 10^{-3} N solution of citric acid, that simulates the strongest acid excretions from living microorganisms. One series of slabs was exposed to microcolonies of *sidero*fungi* (*Penicillium nostatum*). These fungi, known to be surviving in very harsh and cold environments, did reproduce the distinct weathering trend observed in the cocoons of *sidero*bacteria, where olivine is etched out at a rate much faster than glass.

Very recently J M Robbez Masson confirmed this biogenic etching trend for another variety of fungi, *Aspergillus amstellodami*. In figure 14 we report the striking biogenic corrosion prints of their filamentary structure, as observed with a Nomarsky phase contrast microscope (figures 14A and B) and a SEM (figure 14C). These biogenic "microlithographies" yield new information. There is a central groove where biogenic etching acts as a smooth chemical polishing of the surface. Then the next adjacent shallower groove corresponds to a rough chemical etching of the surface, that enhances the contrast of polishing scratches. Finally the most external etched rims appear as a pattern of adjacent darker circular area. These micrographs decorate the various etching processes effective along a *single living* micro-organism, that might be related to specific properties of micro-organisms, minerals, soils and climates.

8.2 *Terrestrial contamination of Greenland micrometeorites in trace elements and the problem of iron hydroxides*

Another development in biogenic corrosion should follow the preliminary INAA analyses of Robin (1988), pertaining to the severe contamination in trace elements of the Greenland PUMs. This contamination is no longer observed for the same family of grains extracted from a minute fraction of sediments found in ~ 100 kg of Antarctica ice core, or for the Greenland crystalline unmelted fragments.

Jéhanno and Rocchia (locally quoted by Robin 1988) postulate that the carrier of the trace element contamination is iron hydroxide generated by biogenic activity, that end up filling up partially the pore structure of the PUMs. It is important to check the validity of this conclusion, because the PUMs might be hopefully washed out of their iron hydroxides, using recipes already developed by soil scientists to remove such hydroxides from clay minerals, that are also characterized by high specific area.

---

* Such micro-organisms synthetize in their cell membranes very efficient "siderophores" (ferric-ion-specific ligands), that have already been documented by cell physiologists in interpreting the fast transport of iron in cells. This process might be responsible for the biogenic corrosion trends noted in this section.

New collections of micrometeorites 117

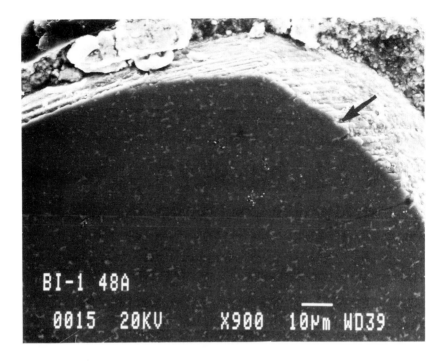

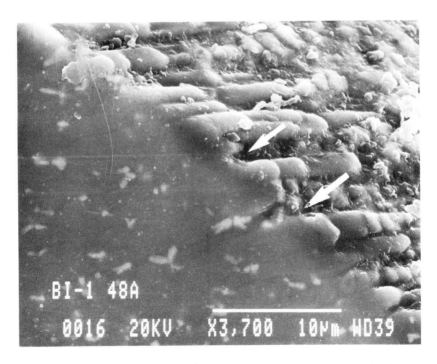

Figure 13. SEM micrographs of an Antarctica barred chondritic sphere showing the intersection of internal and external surfaces. Interstitial glass (white arrows) has been weakly etched out up to depths of a few μm at the most.

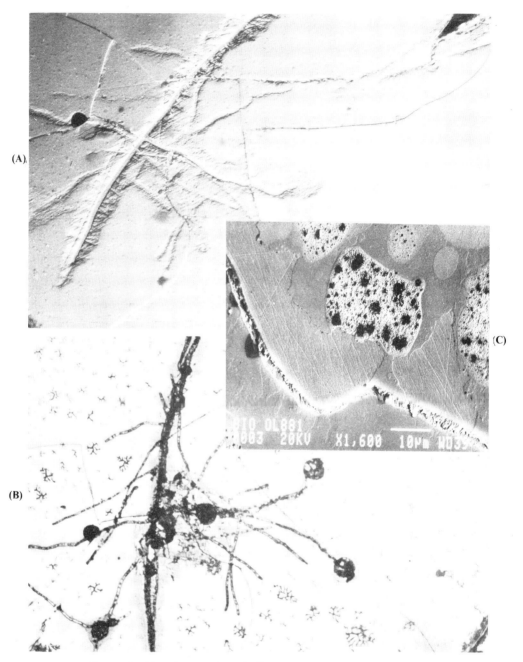

Figure 14. Biogenic corrosion of olivine by *Aspergillus amstellodami*. **A**. Shows the preferential etch pattern observed after cleaning olivine in warm alcohol, and **B**. The original surface with filamentary fungi. The SEM micrograph reported in **C** complete the Nomarsky phase contrast observations reported in **A**.

We have some doubts about the nature of this host phase. Iron hydroxides generally adsorb very selectively elements such as Sc, with an ionic radius similar to that of iron. This is not compatible with the pattern of the trace elements noted by

Robin either in the PUMs, or in grains of iron hydroxides extracted from cryoconite. Indeed the REEs show no marked fractionation with regard to terrestrial crust abundances, and Sc is one of the least anomalous element as it falls right on the Orgueil line, like Mg and Si. Moreover we argued in §6·4 that the original iron concentration of the PUMs has not been markedly modified upon weathering.

This contamination, that was investigated yet in only one sample of cryoconite (JAK3), might not be of general occurrence on the Greenland ice cap. This sample, that was not collected by us, was stored at room temperature for about 3 months before being lent to us. Biogenic activity was certainly accelerated with regard to our own samples kept in deep freeze conditions most of the time. Another alternative carrier phase might be related to biogenic excretions of Al-rich silica gels (that are not as highly selective for the adsorption of elements than Fe-hydroxides), that compose the Si-rich shells of the siderobacteria. It might be very difficult to remove selectively this host phase without altering the grain composition.

It is amazing that rare gases such as ^{21}Ne can still be measured in the tiny micrometeorites of Greenland, while they have been widely redistributed in Antarctica meteorites. This confirms that micrometeorites of the new collections have been much better shielded against terrestrial weathering than Antarctica meteorites, exposed for long durations in a very dry and cold environment.

It is a great challenge for soil scientists and microbiologists to understand how biogenic activity can preferentially extract trace elements from minerals and soils, and redistribute them in various types of colloidal excretions, that might be easily transported, contributing to the formation of mineral ore deposits.

8.3 *Long term disposal of high level nuclear wastes*

Applications of the present studies to the storage of nuclear wastes are straightforward. Such wastes are encapsulated in specific matrices, such as the French glass R7T7 and the Australian ceramic "SYNROC". These radwaste matrices, together with their appropriate canisters and technological barriers, will be disposed in deep-seated geological repositories. Their durability has been scaled from thousands of accelerated "leaching" tests in boiling waters with no biogenic activity. These tests show that amorphous silicates (i.e. glass) are corroded at much faster rate than crystalline silicates and ceramics.

The simple observation of the exact opposite trend in both cosmic dust weathering and the associated simulation experiments shows that these earlier scales of durability have to be reassessed if living micro-organisms can grow in radwaste geological repositories. In addition the trace element studies of Robin *et al* (1988) suggest that any biogenic activity in these repositories might redistribute actinides of the U and Th families in colloidal forms of iron hydroxides and/or Al-rich silica, that might be transported to the biosphere.

9. Suggestions about future collections of micrometeorites on the Antarctica and Greenland ice caps

In Antarctica the $< 50\,\mu$m residue *is not to be thrown away* on the wrong premise that it should be heavily contaminated with wind-borne dust. If our stack of sieves had included an additional sieve with an opening of $\sim 20\,\mu m$, we should have easily collected a huge number of IDPs! In Antarctica water pocketing can only be applied in fields of consolidated blue ice, that only outcrop up to a few km from the ice margin. Moreover, areas where ice is *under compression* have to be carefully selected because most of the blue ice field is frequently loaded with barely visible cracks, in which melt ice water gets quickly lost, preventing the completion of the water pockets. Further inland the ice field is made of permeable firn up to depths of $\sim 100\,m$, and the pocketing operations become very difficult (but not impossible) to achieve.

Favourable cryoconite samples, that contain a minimum amount of *wind-borne* terrestrial grains (and consequently the highest concentrations of cosmic dust), can be found at favourable locations in Greenland. In fact the two richest samples of cryoconite were collected on two ice fields with margin right on the sea shore, exposed to much cleaner winds blowing from the sea*. As the margin of the Sondrestromfjord ice field is at $\sim 30\,km$ from the sea shore, the finest mineral fraction is at least 100 times more polluted than in the Jakobson sample.

It might be possible to avoid problems related to both biogenic weathering and the friability of the porous unmelted aggregates, conducting collections at higher latitudes on the melt zone of the Greenland ice cap. At these latitude ice fields can be found, that look much cleaner and freer of cryoconite. One might direcly collect $\gtrsim 100\,\mu m$ micrometeorites, *not encapsulated in the cocoons of siderobacteria*, just collecting right on a clift of the ice margin the water of a small torrent (see the dark arrow in figure 16), that would flow through stainless steel sieves. The direct hand-picking of large grains not encapsulated in the cocoons of siderobacteria on the sieves, should drastically minimize the two major problems quoted above. We argue that the low density PUMs might be even preferentially enriched in this future collection. With an expected debit of $> 5\,l$ per second per stack of sieves, and 10 stacks of sieves, about $\sim 100,000$ tons of melt ice water might be cycled over a one-month interval, yielding a quantity of grains similar to that trapped in $\sim 500\,kg$ of cryoconite.

In future collections it is important to preserve any sample of "Blue Ice" sediments. in a freezer, as soon as they have been collected. Indeed Greenland cryoconite at room temperature is extremely active, quickly yielding red-coloured interstitial waters. Moreover we have observed that Antarctica PUMs can be very quickly populated by micro-organisms, when they are left unprotected on a laboratory bench. This is bad for the characterization of "primitive" organic compounds in the grains!.

10. Summary

These preliminary investigations were aimed at searching for a new variety of cosmic dust grains, establishing their origin, and assessing their extent of terrestrial weathering and/or heat metamorphism during atmospheric entry. Most of the work has still to be done for relating the most interesting *unmelted* Greenland and Antarctica micrometeorites to meteorites and stratospheric IDPs, and to identify

---

* The JAK-3 sample was collected in 1984 by K Eichelmeyer and W Harrison, at 35 km from the margin of the Jakobson ice field about 500 km to the North of Sondrestromfjord. The "Blue Lake III" sample ($\sim 6\,kg$ of cryoconite), was collected for us in July 1988 by a team of 12-year-old scouts lead by a guide A Drouin, at 9 km from the margin of another ice field, at about 100 km to the north of Jakobson.

New collections of micrometeorites 121

Figure 15. For caption, see p. 126

invariant characteristics of these grains, that are independent of terrestrial weathering and ablation heat metamorphism.

The melt zone of the Greenland ice cap is the best "giant" ice cap collector of *big* micrometeorites (sizes $> 500\,\mu$m), where one ton of cryoconite can be assimilated to a 100% efficient space collector with a "minimum" geometrical exposure factor of $\sim 100{,}000\,\text{m}^2/\text{y}$. On the other hand the water-pocketing procedure used in Antarctica, that requires about 50 litres of fuel to collect ~ 100 micrometeorites with sizes $> 50\,\mu$m, constitutes our "best" method to gently collect ~ 2000 *small* unmelted micrometeorites from about 100 tons of blue ice. These micrometeorites are essentially unweathered and/or undestroyed by the collection procedure, and the size range of the collection can be extended from $< 300\,\mu$m *down to the very small size of stratospheric IDPs.*

The most interesting porous unmelted micrometeorites, probably loaded with C-rich material, correspond to a new class of low albedo fluffy material, that seem to be of a general occurrence in cometary environment (Cruishank 1987; Dollfus 1988) and in the zodiacal cloud (Levasseur 1988). These very fine-grained micrometeorites are unfortunately the most difficult to analyse. Fortunately, methods of analytical microscopy and ion microprobe analysis already developed for the study of much smaller stratospheric IDPs, are available for their "high spatial resolution" studies. New "fractal" methods of looking at the dispersion of abundances over a large scale of the analysed volume should be developed. Such invariant measurements might yield new information about both the early history of the grains, and their filiation with other solar system objects.

In Greenland the high specific area of the PUMs seems to be responsible for their contamination in trace elements. Fortunately this contamination, that is not observed in Antarctica, does not noticeably affect important elements such as Mg, Si, Fe, Ni and Ir. It is also likely that some sampling sites in Greenland are uncontaminated. If the carrier phases of this contamination are iron hydroxides, it might be selectively washed out from the grains. Even if the grains have been weathered to some extent, they still behave as closed system for rare gases, that have been drastically remobilized in Antarctica meteorites.

The preliminary results obtained yet suggest that most of the unmelted chondritic grains, with the exception of a small proportion of the CUMs, are different from meteorites, including carbonaceous chondrites. A quantitative comparison of these new micrometeorites with the tiny stratospheric IDPs has still to be done at a much smaller scale of magnification of the analysed volume ($\sim 10^{-4}\,\mu^3$). Indeed both types of materials might derive from different parent bodies. First, a kink at $\sim 10^{-7}\,\text{g}$ (corresponding to sizes of $\sim 60\,\mu$m) is observed in the micrometeorite flux at 1 a.u. Next, different dynamical processes are required to transfer large and small micrometeorites to earth-crossing orbits. Finally, survival in the atmosphere should be highly dependent on porosity, existence of a pyrolyzable component in the grains and so on.

We have deliberately presented some speculations that run contrary to previous thoughts. In particular, the terrestrial atmosphere might rather well transmit fast micrometeoroids on highly eccentric orbits, originating from long period comets or the interstellar medium, and arriving at near normal incidence. Such grains would suffer a short pulse heating at high temperature, that might be less destructive than a longer heating at moderate temperatures, associated with slow micrometeoroids,

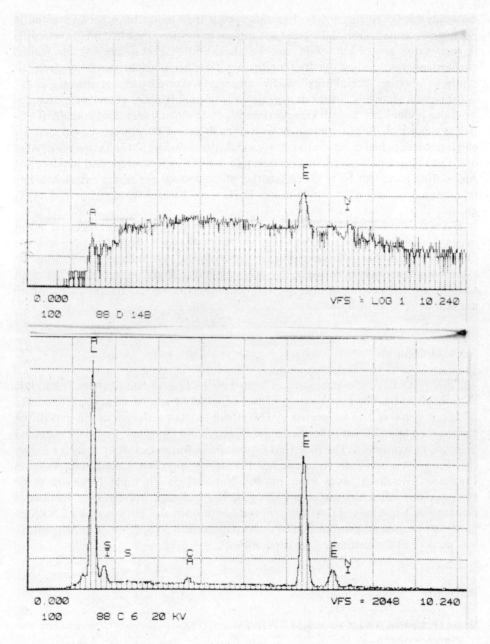

Figure 16. EDX spectra of odd grains. The top grain, that contains a height concentration of organic material, shows a clear association of Fe and Ni.

originating from short period comets or the asteroidal belt. Measurements of the distribution of ^{10}Be and ^{21}Ne exposure ages of micrometeorites, might further constrain the nature of their orbits.

The "reactive" constituent material of the PUMs might improve their thermal shielding during atmospheric entry. They should thus be preferentially enriched in the cosmic dust complex found on the Earth with regard to chunks of "compact-inert" materials similar to meteorites. For this reason they might have been involved in prebiotic evolution. One of the great advantages of the new unmelted micrometeorites is their large size. Each grain can be shared into several chunks, for fruitful consortium-type studies, or for conducting simulation experiments in prebiotic synthesis, ablation "metallurgy", and weathering in damp extraterrestrial regoliths.

It might be that new classes of *non-chondritic* extraterrestrial grains were excluded from our preliminary chondritic selection. We have already identified some Al-Ti-Fe rich grains which do not contain Ir, as well as dark and porous aggregates that are either much enriched in sulphur and organic matter, or which show a clear association of Fe and Ni in an Al- and C-rich matrix (see the EDX spectra reported in figure 16). An exciting topic will be to check whether such unusual grains are extraterrestrial, relying on measurements of ^{21}Ne and D/H ratios by the Washington University group.

The meaningful analyses of the new collections of micrometeorites reported in this paper, leading to both the definition of their *invariant characteristics*, and the understanding of their full implications in terms of solar system history, will require an extensive collaboration between the fields of mineralogy, astrophysics, materials sciences, the engineering of instrumental developments, and.... fractal chaos, which has still to be vigorously initiated.

Acknowledgements

The first "Blue Lake I" expedition was funded by the Institut National des Sciences de l'Univers (INSU). The participation of two of us (M M and M P) to the "Blue Lake II" expedition was also supported by INSU, but most of the coast of this expedition was granted by the Council for Scientific Research of Danemark, and the Greenland Scientific Organization. The Blue Ice I expedition in Antarctica was jointly funded by the Terres Australes et Antarctiques Françaises and the Expeditions Polaires Françaises. Research funds from Institut National de Physique Nucléaire et de Physique des Particules are acknowledged. The technical assistance of Institut of Geophysics, University of Copenhagen, and Laboratoire de Glaciologie du CNRS (in particular the helpful assistance of M de Angelis in the field operations in Antarctica), was decisive in the success of the expeditions.

References

Bibring J P, Brownlee D A and Maurette M 1978 HVEM and HVES observations of U-2 stratospheric dust grains; *Lunar Planet. Sci.* **15** 439

Bonny Ph, Balageas D, Devezeaux D and Maurette M 1987 Atmospheric entry of micrometeorites containing organic materials; *Lunar Planet. Sci.* **18** 118

Bonny Ph, Balageas D and Maurette M 1988 Entrée atmosphérique de micrometeorites poreuses chargées en matière organique, *Rapport ONERA T.P.* 1988–110

Borg J, Dran J C, Durrieu L, Jouret C and Maurette M 1970 High voltage electron microscope observations of fossil nuclear particle tracks in extraterrestrial matter; *Earth Planet. Sci. Lett.* **8** 379

Bonté Ph, Jéhanno C, Maurette M and Robin E 1987a A high abundance and great diversity of unmelted cosmic dust grains on the West Greenland ice cap; *Lunar Planet. Sci.* **18** 105

Bonté Ph, Jéhanno C, Maurette M and Brownlee D E 1987b Platinum metals and microstructure in magnetic deep sea cosmic spherules; *J. Geophys. Res.* **92** E64'

Bradley J P 1988 Analysis of chondritic interplanetary dust thin sections; *Geochim. Cosmochim. Acta* **52** 889

Bradley J P, Brownlee D E and Fraundorf P 1984 Discovery of nuclear tracks in interplanetary dust; *Science* **226** 1432

Brownlee D E 1985 Cosmic dust: collection and research; *Annu. Rev. Earth Planet. Sci.* **13** 147

Brownlee D E, Wheelock M M, Temple S, Bradley J and Kissel J 1987 A quantitative comparison of comet Halley and carbonaceous chondrites as the submicron level; *Lunar Planet. Sci.* **18** 133

Bunch T E and Chang S 1980 Carbonaceous chondrite-2. Carbonaceous chondrite phyllosilicates and light element geochemistry as indicators of parent bodies processes and conditions; *Geochim. Cosmochim. Acta* **44** 1543

Callot G, Maurette M, Pottier L and Dubois A 1987 Biogenic etching of amorphous and crystalline silicates; *Nature (London)* **328** 147

Christophe Michel-Levy M and Bourot-Denise M 1988 Comparaison mineralogique de grains de poussière cosmique collectés au Groénland et en Antarctiques, Compte Rendus des 'Journées de Planétologie 1988', Observatoire de Besancon

Clanton U S, Nace G A, Gabel E M, Warren J L and Dardano C B 1982 Possible comet samples: the NASA cosmic dust program; *LPSC* **13** 109

Cruikshank D P 1987 Dark matter in the solar system; *Adv. Space Res.* **7** 109

Darmon G and Balageas D 1986 *Rev. Gen. Therm. Fr.* **192** 197

De Angelis M, Maurette M and Pourchet M 1988 Assessment of biases in collections of Greenland micrometeorites; *LPSC* **19** 255

Dollfus A 1988 Ejection des poussières par les comètes: charactéristiques par polarimétrie, Comptes Rendus des 'Journées de Planétologie 1988', Observatoire de Besancon

Fraundorf P 1980 The distribution of temperature maximum for micrometeorites decelerated in the Earth's atmosphere without melting; *Geophys. Res. Lett.* 10 765

Gooding J L 1986 Clay-mineraloid weathering in Antarctic meteorites; *Geochim. Cosmochim. Acta* **50** 2215

Grün E, Zook H A, Fechtig H and Glenn R H 1985 Collisional balance of the meteoritic complex; *Icarus* **62** 244

Hoaglin D C, Mosteller F and Tukey J W 1983 Understanding robust and exploratory data analysis, (John Wiley)

Hodge P 1981 Interplanetary dust, (Gordon and Breach)

Koeberl C, Hagen E and Faure G 1988 Chemical composition and morphology of meteorite ablation spherules in neogene till in the dominion range, transantarctic mountains; *Lunar Planet. Sci.* 18 625

Langevin Y, Kissel J, Bertaux J L and Chassefière E 1987 First statistical analysis of 5000 mass spectra of cometary grains obtained by PUMA 1 (Vega-1) and PIA (Giotto) impact ionization mass spectrometers in the compressed modes; *Astron. Astrophys.* **187** 761

Levasseur-Regourd A C and Dumont R 1988 Après IRAS: propriétés locales de la poussière interplanétaire, Comptes Rendus des 'Journées de Planétologie 1988', Besancon

Mackinnon I D R and Rietmeijer F J M 1987 Mineralogy of chondritic interplanetary dust particles; *Rev. Geophys.* **25** 1527

Maurette M, Passoja D E, Maas D and Laurence M 1986a An improved fractal characterization of projectile impact microcraters. Preliminary implications to micrometeorites; *LPSC* **17** 1002

Maurette M, Hammer C, Brownlee D E, Reeh R and Thomsen H H 1986b Placers of cosmic dust in the blue ice lakes of Greenland; *Science* **233** 869

Maurette M, Jéhanno C, Robin E and Hammer C 1987 Characteristics and mass distribution of extraterrestrial dust from the Greenland ice cap; *Nature (London)* **328** 699

Maurette M and Passoja D E 1988 Scatter diagrams of elemental abundances in micrometeorites: a non statistical interpretation; *Lunar Planet. Sci.* 19 642

McSween H Y 1979 Are carbonaceous chondrites primitive or processed? *Geochim. Cosmochim. Acta* **43** 1761

McSween H Y, Fronabarger A K and Driese S G 1983 in *Chondrules and their origins* (ed.) A King (Lunar and Planetary Institute) p. 201

Millman P M 1972 Cometary meteoroids, in *Nobel symposium No. 21: From plasma to planet* (ed.) A Elvius (New York: John Wiley) p. 157

Raisbeck G M, Yiou F, Klein J, Middleton R and Yamakoshi Y 1983 ^{26}Al and ^{10}Be in deep sea stony spherules; evidence for small parent bodies; *Lunar Planet. Sci.* **14** 622

Raisbeck G, Yiou F, Bourles D and Maurette M 1986 ^{10}Be and ^{26}Al in Greenland cosmic spherules; evidence for irradiation in space as small objects and a probable cometary origin; *Meteoritics* **21** 487

Raisbeck G and Yiou F 1987 ^{10}Be and ^{26}Al in micrometeorites from Greenland ice; *Meteoritics* **22** 485
Reeh N 1988 (in press)
Robin E 1988 *Des poussières cosmiques dans les cryoconites du Groénland: Nature origine et applications*, Thèse de Doctorat de Sciences, Orsay
Schramm L S, Brownlee D E and Wheelock M M 1989 Major element composition of stratospheric micrometeorites; *Meteoritics* **24** 99
Southworth R B and Sekanina Z 1973 Physical and dynamical studies of meteors; NASA CR 2316
Whipple F 1950 The theory of micrometeorites, Part 1: In an isothermal atmosphere; *Proc. Nat. Acad. Sci. USA* **36** 687

Figure 15. Clean and accessible ice margin at a latitude of 80°N, on the East Greenland ice cap. Cryoconite deposits are not formed at these high latitudes, where the arctic summer is quite short (Courtesy of J F Loubières).

Cyclic behaviour of the solar-terrestrial system

G CINI CASTAGNOLI

Istituto di Fisica Generale dell'Università, Istituto di Cosmogeofisica del CNR Torino, 10133 Corso Fiume 4, Italia

Abstract. The thermoluminescence (TL) profile in a recent sedimentary core of the Ionian Sea reveals two evident periodicities on the decennial time scale at 12·06 yr and 10·80 years and two long periodicities at 137 and 59 years. The frequency difference between these couple of lines is the same. Two principal carrier waves are therefore envisaged with periods respectively 11·4 and 82·5 years, both with an amplitude modulation with a period of 207 years. The similarity of the carrier periods to the Schwabe and to the Gleissberg solar cycles respectively favours the point of view that the TL profile is essentially modulated by the solar output. Moreover the amplitude modulation is shown to have the same period of the principal wave present in the atmospheric decadal record of $\Delta^{14}C$ in tree-rings in the last 4·500 years.

For all these considerations it is conceivable to propose the measurement of the TL profile in uniform sediments as a new physical method that can be used in order to advance in the quantitative understanding of the geophysical cycles, driving the solar-terrestrial system.

Keywords. Solar-terrestrial relationships; solar cycles; thermoluminescence in sea core; cyclic behaviour; Ionian Sea.

1. Introduction

The search of the solar imprint in different time series of the data available concerning the physical processes occurring in the solar-terrestrial system has the aim of obtaining information from the past which may be useful for understanding the present and possibly for predicting the future.

Unbroken series of data through a succession of equal intervals of time have been the object of many investigations by a great number of scientists in the field of natural sciences. Information reservoirs like trees, sediments, polar ice, stalagtites from underground caves, corals and nodules are the precious and unique monitors of the past-time terrestrial environment. They preserve in strict chronological sequence the records of chemical and isotopic composition, climate variations and other information relevant to studies of cyclic phenomena induced by the solar-terrestrial relationships. Moreover, exceptional climatic conditions and interesting astronomical phenomena have been historically recorded in written documents; for example, aurorae, sunspots and eclipses are described in ways that may provide very useful proxy data for the solar activity and the solar corona output.

In order to understand and, therefore, to forecast within, so to say, 50 years the Sun-Earth forcing effects, time series of at least a hundred times longer are necessary, and this is not easy to be achieved. Trends, cycles, periodicities and recurrences were thoroughly searched in different records. Douglass's discussion of the climatic dilemma "of cycles having a very confusing tendency to stop" in the tree-rings widths is one of the classic examples of the results obtained. The signals are often buried in the noise and *ad hoc* techniques of numerical analysis have to be investigated in order to separate signal from noise.

By a suitable numerical analysis of the time series it is possible to infer in some cases the solar-terrestrial forcing effect. As a classic example we can refer to the discovery of the 27-day recurrence of the geomagnetic effects related to the "M regions" of the Sun made by Chapman and Bartels, using the harmonic analysis of the "international magnetic character, figure C data" plotted not in the usual way, but in a sequence of vectors called "summation dial". Since then, in recent years progress has been achieved in the field of solar-terrestrial physics, at least for what concerns the direct observation of solar periodicities. Measurements of the solar-activity indices, of the solar constant, of the emission at different wavelengths, of the interplanetary magnetic field and plasma performed on board of spacecrafts have shown that the solar properties are not as constant as had been thought before. However the short length of time over which these records are taken still does not allow a full understanding of the complex physical processes visible on the Sun and of how their variations influence the terrestrial environment. Therefore from the point of view of the determination of long periodicities and recurrences to be found imprinted in the terrestrial records the most useful solar observation is still the yearly sunspot number (R_z): its best-known feature is the 11-year cycle which was discovered by Schwabe in 1843. This pattern of variability should be accurately memorized in the terrestrial phenomena controlled by the solar activity. Long running terrestrial series showing this prominent cyclicity may provide valuable information on other periodicities relevant to human life time.

In this paper I will present the results obtained by measuring the thermoluminescence (TL) profile as a function of depth in a recent sea sedimentary core, spanning the last two millennia (Cini Castagnoli *et al* 1988a, b).

2. TL profile of recent sea sediments

This new physical method has been devised because the natural TL of polyminerals forming the sediment is sensitive to light and ionizing agents and a natural equilibrium dictated by the solar output level was thought to be responsible for the TL level of the great number of grains statistically collected in each sample. This investigation has been undertaken in order to identify in sediments global recurrences and periodicities of solar origin, which were detected in terrestrial archives by other methods. For instance the 11-yr solar cycle was found in the concentration of ^{10}Be in Milcent ice core, between 1180 and 1800 A.D. (Beer *et al* 1985).

Before sedimentation the crystalline mineral grains which form the uniform core are exposed to ionizing radiation and to direct sunlight. Free electrons and holes are produced at an equilibrium value, which depends on the year of deposition. When these crystals are heated in the laboratory, these trapped charges are released and are able to recombine generating the TL signal by a radiative process. The TL level accumulated in the crystals before their deposition into the sediment is augmented by the TL acquired *in situ* (after deposition) from the local radioactivity concentration in the core. This second effect is superposed on the first one and it is an increasing function of time. In very recent sediments the predepositional TL may be predominant over the TL acquired after sedimentation and it is considered mainly responsible for the signal which is investigated here.

The GT14 core was drilled by means of a gravity corer in the Ionian Sea (lat. 39°45'55" N, long. 17°53'302 E) at a water depth of 166 m on the Italian continental shelf. The core is a fossiliferous calcareous mud with terrigenous clasts. No sign of sedimentary stratification is visible (even under the microscope) through the whole core. The only marker of stratification surfaces is given by preferred alignment of mica flakes and/or by the long axes of bioclastics. This core is thus a very monotonous sediment whose texture and microfossil associations are uniform.

The sedimentation rate has been determined by a radiochemical method based on the measurement of the ^{210}Pb activity "in excess" (with respect to the *in situ* decay of ^{226}Ra) as a function of depth (Krishnaswamy *et al* 1971). A sedimentation rate of $S = (6.3 \pm 0.3) 10^{-2}$ cm y^{-1} has been determined. The least-square fit of the ^{210}Pb excess activity with respect to an exponential decay function has a correlation coefficient $r = 0.98$, indicating that the sedimentation rate is significantly constant. A peak of the ^{137}Cs activity, due to testing of nuclear weapons, has been found in the first centimetre of the core indicating that the top layers of the core were not lost in the coring operations and that bioturbation is negligible.

The constant sedimentation rate of the core gives the possibility of transforming the depth profile into a time profile. TL measurements were obtained on sequential sediment layers of width 2·5 mm and a first estimate of the equivalent time interval was $\Delta t = 4$ y. To better determine the value of Δt we observed that a large amount of clinopyroxenes was found in the layer 174. The clinopyroxenes form needle-like crystals, with rhythmic zoning similar to that found in the grandmass pyroxenes of volcanic rocks. The association of this layer with the Ischia volcanic event in 1301 A.D. allows one to determine the time interval with an error of approximately 1%, giving $\Delta t = (3.87 \pm 0.04)$ y.

To measure the TL core profile, one gram of material was taken from each layer.

The samples were prepared in red light treating the wet material by successive dispersion, washing in NaOH, water and acetone, using the centrifuge after each step to preserve the original composition of the polyminerals. After drying in oven at 40°C overnight, the sediment powder was gently sieved and the fraction $> 44 \mu$m was divided in samples of 15 mg for the TL measurement. Glow curve measurements were then obtained by using a TL analyser described by Miono and Ohta (1979). The TL signals were recorded in a neutral atmosphere of ultrapure nitrogen (flow 5 l/min) at a heating rate of 5°C/s and read at 340°C, where the glow curves are flat (see figure 4 of Cini Castagnoli and Bonino 1988). A mean TL value was obtained over 4 or 5 samples per layer in order to construct the TL profile. By this procedure 467 points were obtained for the whole length of the core, for a total of more than 2000 glow curves. The TL signal for each sample is given with a relative error smaller than 5%. The experimental TL data have an average intensity of 58 arbitrary units (a.u.) and a standard deviation of 6 a.u.

In figure 1 the time series of 427 points (with zero mean and standard deviation 5·0 a.u.) obtained by taking the differences of the raw data from a running average over 41 data points (a running average over roughly 160 y) is shown. The TL profile in figure 1 covers the time interval 247 A.D.–1900 A.D. and shows a non-random behaviour, with equispaced minima. TL fluctuations are seen to range from 1% to

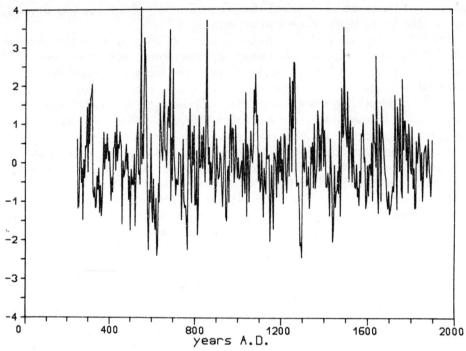

Figure 1. Time series of the TL signals, for 427 layers. The data have been detrended from low periodicities (differences from a running average over 41 data points of the 467 raw data) and are plotted in units of the standard deviation.

10% of the mean value of the raw data. It can be noted, in particular, that the Maunder, the Sporer and the Wolf minima of solar activity are coincident with TL minima. This suggests that the TL series is modulated by the solar activity. However, TL fluctuations are much larger than the variations of the total solar irradiance, suggested to be of the order of 0·4% from 1967 to 1985 as part of a 22 y modulation (Frohlich 1987). Several forms of solar variability, however (γ, X, UV, low-energy solar-wind particles), may be important in determining the solar effects on the Earth environment and consequently on the TL profile. For example, the solar-irradiance fluctuations in the UV range are known to be of the order of 5% on time scales of months and of the order of 10% on the entire solar cycle (Lean 1987), and are thus comparable to the TL fluctuations.

3. TL periodicities: their similarity to those found in the solar activity and in terrestrial archives

In order to detect the periodicities of the TL series the power spectrum of the data was computed through a standard discrete Fourier transform routine from the detrended 427-point time series and it is shown in figure 2. Two strong peaks corresponding to periods of 137 and 59 y, emerge in the low-frequency range. Other two peaks are evident in the high-frequency range with periods of 12·06 and 10·80 y years.

Smaller peaks can also be observed at 43·5 and 22 y periods. In the same figure 2, a second low level curve is also reported. This spectrum has been computed from a

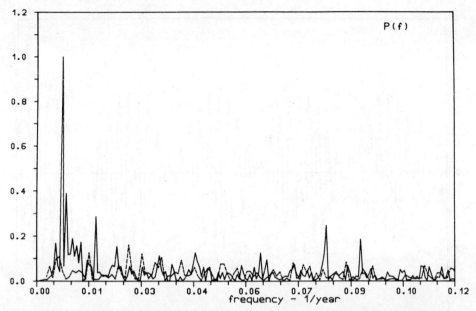

Figure 2. Power spectrum of the 427 data points of the TL series shown in figure 1.

"bleached" TL profile. One sample of each layer of the core was exposed to remove bleachable TL a standard sunlamp for 420 minutes before measuring the glow curve. The time series thus obtained has an average value of 36 TL a.u. and a standard deviation of 7 a.u. It does not show any evident periodicity in the spectrum, neither on decennial nor on secular time scales. The bleached TL profile is important because it provides a background noise level for the original TL spectra. In this respect we note that the secular and the decennial peaks of 137, 59, 12·06 and 10·80 years are all significant at 95% and all clearly emerge from the background noise represented by the bleached spectrum.

The existence of the high-frequency peaks, corresponding to typical frequencies of the solar cycles, suggests a connection between the TL signal and the solar activity. The carrier period of the beat of these two high-frequency waves is, in fact, 11·4 y, a period which appears both in the sunspot record as the average value between 1824 A.D. and 1903 A.D. (Attolini et al 1985) and in the ^{10}Be cosmogenic isotope record in the Milcent ice core during the interval 1180 A.D.–1500 A.D. (Attolini et al 1988). The 10·8 y period is, on the other hand, evident in the same ice core from 1500 A.D. to 1800 A.D. (Cini Castagnoli et al 1984). Likewise interesting are the two low-frequency peaks, suggesting that the TL signal is connected with long-term fluctuations in the Sun-Earth system. The 137 y period is, in fact, quite similar to the period of 133 y recently found from historical aurorae records (Attolini et al 1988).

By the method of superposition of epochs (summation of subseries with length T) the amplitude and phases of the waves which are the best fit to the folded data were determined, for the four principal periods T indicated by the power spectrum. The sum of the fundamental waves takes care of at least 50% of the standard deviation of the entire time series.

In figure 3 the modulated train obtained by the beat of the two high-frequency waves with amplitudes and phases determined by the experimental procedure

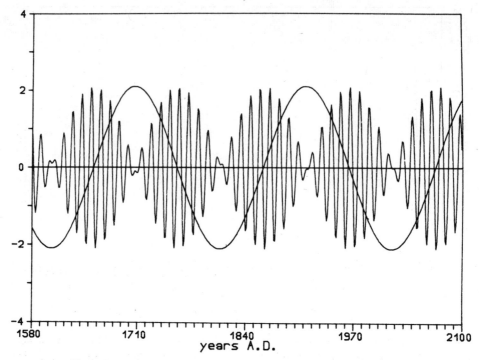

Figure 3. Modulated wave-train generated by the beat of the 12·06 yr and the 10·80 yr components. The carrier wave has a period of 11·4 yr and the secular modulation has a period of 206·7 yr. Superposed is the sinusoid of period 206 yr resulting from the $\Delta^{14}C$ series of figure 5, obtained as discussed in the text. Notice that maximum and minimum values of the ^{14}C fluctuations appear at secular minima of Sunspots.

mentioned above is shown. The 11·4 y wave is modulated by a wave of period approximately of 207 y, the distance between the amplitude modulation minima thus being approximately 103·5 years. In figure 4 the squared amplitude modulation obtained from the beat of the 12·06 and 10·80 waves is shown with the time series of the annual sunspot number R_z as given by Waldemeier (1961) and by Solar Data, World data centre A, Boulder. It is an interesting result that this modulation, deduced from the TL data, apparently describes the general shape of the secular variation of the sunspot number with the right period and phase. We note in particular that the minima after the 5th and the 14th small sunspot cycles, recorded in 1810 A.D. and 1913 A.D., coincide with two modulation minima. These results thus suggest that the high-frequency TL waves may have a solar origin. The beat of two waves of 12·06 and 10·80 years may be responsible for at least one aspect of the sunspot secular variations. Of course at the moment no answer can be given to the question whether two distinct decennial cycles are really active in the Sun or if there is only one decennial cycle which is amplitude-modulated by a secular cycle.

Another interesting result can be deduced from the power spectrum of figure 2. The distance between the two low-frequency peaks is identical to the distance between the two high-frequency peaks. The distance is $\Delta = (v_1 - v_2)/2 = 0·00483$ corresponding to $T = 207$ yr, while the mean frequency $\Delta = (v_1 + v_2)/2 = 0·012$ corresponds to $T = 82·5$ yr.

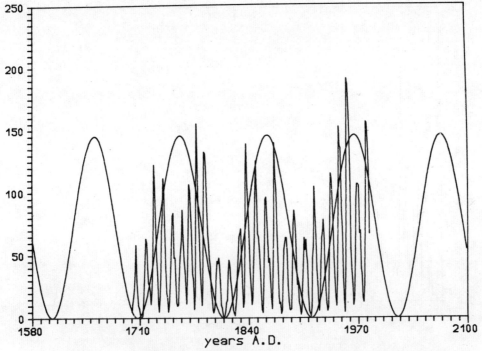

Figure 4. Time series of the yearly sunspot number in the last three centuries. Superposed to this, the squared amplitude modulation of the wavetrain reported in figure 3 is shown.

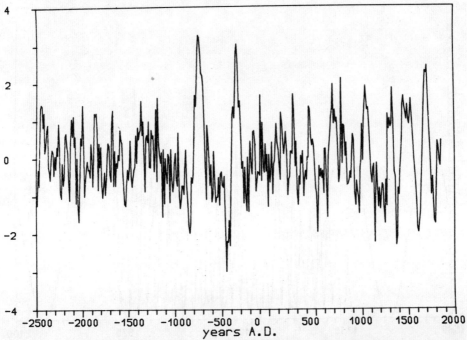

Figure 5. A 4·400 yr record of decadal atmospheric (tree-ring) $\Delta^{14}C$ after removal of the long term trend as approximated by a spline similar to a 400 yr moving average after Stuiver and Braziunas (1988).

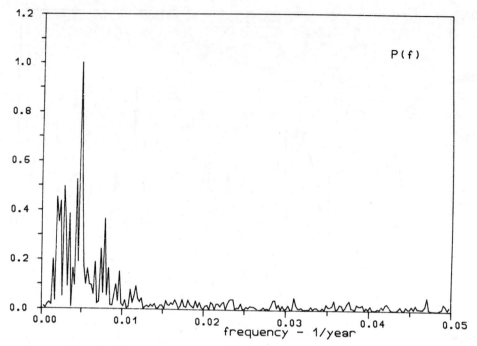

Figure 6. Power spectrum of the $\Delta^{14}C$ series shown in figure 5.

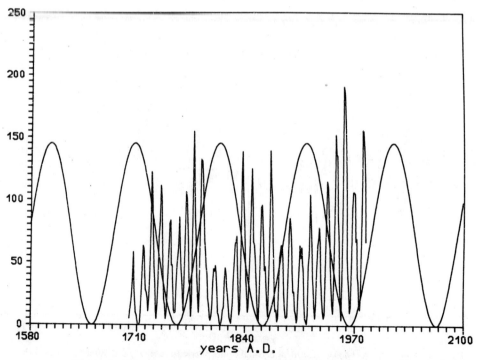

Figure 7. Time series of the yearly sunspot number and the squared $\Delta^{14}C$ sinusoid of figure 3: notice that this sinusoid is opposite to the TL sinusoid shown in figure 4.

It is interesting to note that the TL-modulating wave of period $T = 207$ yr is also the principal wave that we can find in the 4500 yr decadal $\Delta^{14}C$ time series as measured in tree rings (Stuiver and Becker 1986). The series, shown in figure 5, is given by Stuiver and Braziunas (1988) in their figure 1, the long-term trend being removed from the decadal record by fitting a cubic spline to the data.

The power spectrum of the time series of figure 5 is shown in figure 6. The dominant period is the 207-yr peak. By the same method of the superposition of epochs used for the TL waves, the amplitude and phase of the $\Delta^{14}C$ wave have been determined. The $\Delta^{14}C$ wave is shown in figure 3 superimposed on the TL high-frequency wave train. Maximum values of positive and negative fluctuations in $\Delta^{14}C$ are coincident with the TL nodes. The square of the $\Delta^{14}C$ wave is shown in figure 7 together with sunspot series. A comparison between figures 4 and 7 clearly shows that the behaviour of TL and $\Delta^{14}C$ are opposite.

4. Conclusion

From the previous analysis we may conclude that, whatever the controlling agents of the TL profile may be, definite periodicities have been obtained in a recent carbonate sea sedimentary core.

Two high-frequency peaks at 12·06 and 10·80 years and two low-frequency peaks at 137 and 59 years are the main features of the TL power density spectrum.

The splitting $\Delta = v_1 - v_2$ is the same in the two cases.

The amplitude modulation of two principal carrier waves is therefore envisaged.

The periods of the carrier waves are respectively 11·4 and 82·5 years corresponding to the Schwabe and the Gleissberg solar cycles.

The period of the amplitude modulation wave is 207 years which is found to be the main periodicity of the high precision (2 per mil) atmospheric $\Delta^{14}C$ decadal record in the last 4500 years.

The similarities between the periodicities found in the TL profile and the solar activity cycles favour the point of view that the TL profile itself is essentially modulated by the solar activity. It is difficult to envisage other sources of periodical variations in the samples preparation or in TL acquired *in situ* from local radioactivity. Changes in the predepositional transport of the sediment material due to climatic effects cannot be excluded.

In conclusion a new physical method, the measurement of the TL profile in uniform sediments, is proposed in order to advance the quantitative understanding of the major geophysical cycles by which the complicated terrestrial machine is driven, which could thus be useful in predicting future changes on the time scales of the human life.

Acknowledgements

The author acknowledges the most valuable contribution of Dr G Bonino to this research. She is grateful to Dr A Provenzale for the analysis of the data and to Prof C Castagnoli for helpful encouragement and discussion. Technical assistance during the present experiment was provided by Mr P Cerale and Mr A Romero.

References

Attolini M R, Galli M and Cini Castagnoli G 1985 On the R_z sunspot relative number variations; *Solar Phys.* **95** 391–395

Attolini M R, Cecchini S, Cini Castagnoli G, Galli M and Nanni T 1988 On the existence of the 11-year cycle in solar activity before the Mauander Minimum; *J. Geophys. Res.* **93** 12729–12734

Attolini M R, Galli M and Nanni T 1988 *Proc. NATO Adv. Rec Workshop, Durham* (eds) R Stephenson, R A Wolfendale and J A Eddy (Dordrecht: D. Reidel) pp. 49–68

Beer J, Oeschger H, Finkel R C, Cini Castagnoli G, Bonino G, Attolini M R and Galli M 1985 Accelerator measurements of ^{10}Be: the 11-year solar cycle from 1190–1800 AD; *Nucl. Instrum. Methods* **B10/11** 415–418

Cini Castagnoli G and Bonino G 1988 *Thermoluminescence in sediments.* Course XLV of the International School of Physics Enrico Fermi on; *The solar terrestrial relationships and the earth environment in the last millenia* (ed.) G Cini Castagnoli (Amsterdam: North Holland) p. 323–328

Cini Castagnoli G, Bonino G and Provenzale A 1988a On the thermoluminescence profile of an Ionian Sea sediment: evidence of 137, 118, 12·1 and 12·8 y cycles in the last two millenia; *Nuovo Cimento* **C11** 1–12

Cini Castagnoli G, Bonino G and Provenzale A 1988b The TL profile of a recent sea sedimentary core and the solar variability; *Solar Phys.* **117** 187–197

Cini Castagnoli G, Bonino G, Attolini M R, Galli M and Beer J 1984 Solar cycles in the last centuries in ^{10}Be and δ^{18}O in polar ice and in thermoluminescence signals of a sea sediment; *Nuovo Cimento* **C7** 235–244

Frohlich C 1987 Variability of the solar "constant" on time scales of minutes to years; *J. Geophys. Res.* **92** 796–800

Krishnaswami S, Lal D, Martin J M and Meybeck M 1971 Geochronology of lake sediments; *Earth Planet. Sci. Lett.* **11** 407

Lean J 1987 Solar ultraviolet irradiance variations: A review; *J. Geophys. Res.* **92** 839–868

Miono S and Ohta M 1979 *Proc. Sixteenth Int. Cosmic Ray Conf.* (Kyoto: IUPAP) Vol. 2, pp. 263–265

Stuiver M and Becker B 1986 *Radiocarbon* **28** 863–890

Stuiver M and Braziunas F 1988 *Proc. NATO Adv. Rec. Workshop, Durham* (eds) R Stephenson, A Wolfendale and J A Eddy (Dordrecht: D. Reidel) pp. 245–266

Waldmeier M 1961 *The sunspot activities in the years 1616–1960* (Zurich, Switzerland: Schulthess and Co.)

The turbulence of geophysical time

BRUCE A BOLT
Seismographic Station, University of California at Berkeley, Berkeley, California 94720, USA

"The heavens themselves, the planets and this centre
Observe degree, priority and place,
Insisture, course, proportion, season, form: but when the planets
In evil mixture to disorder wander,
What raging of the sea! shaking of earth!
Take but degree away, untune that string,
And, hark, what discord follows!"

Shakespeare, Troilus and Cressida

Abstract. A critique is given of the representation of terrestrial evolution as either a strictly linear or circular process. The danger of the recent emphasis on this dichotomy is to shift attention from the actual complexity of the development of geophysical events. Given that the main task in earth sciences is the establishment of general algorithms for prediction of both past and future planetary constitutions, the descriptive functionals of geophysics and geology for the non-conservative terrestrial system must be considered as multiparameter, stochastic solutions of nonlinear equations. Explanations in terms of reversible time series or of tectonic cycles are, at best, rough simplifications of short-term geological episodes and, at worst, distractions from a fuller understanding of the basic non-stationary mechanisms.

The argument is carried forward in terms of fundamental nonlinear geophysical problems with illustrations such as the geomagnetic dynamo, the tidal equation, and the problem of direct temperature estimation in the Earth. Models involving the characterization of fluid turbulence are considered and the role of regularization of ill-posed problems is discussed. The evolutionary pattern of damped non-stationary mixing of linear, harmonic and random (quasi-chaotic) terms puts severe limitations on geophysical prediction of processes in deep (geological) time but useful predictions of bounds in the short run are a realistic goal.

Keywords. Prediction; geodynamics; planetary evolution; earth structure; terrestrial engine; turbulence; geophysical time; thermodynamic; non-Newtonian world; causality and uniqueness.

1. Introduction

A welcome aspect of the charge to contributors to this volume in honour of President Devendra Lal was that consideration be given to aspects of the time scales of planetary and interplanetary processes and that speculation ("in a restrained way") about the evolution of planetary sciences was in order. In taking up the challenge, this essay discusses, from the viewpoint of an applied mathematician, physical time-dependent processes within the Earth taken as a prototype for the terrestrial planets. My approach is in the spirit of the geophysical work of Sir Harold Jeffreys and Professor K E Bullen, under whose guidance my interest in earth physics was stimulated. In applied mathematics, the observed universe is commonly represented by models defined by differential or integral equations in which time is one of the independent variables. This type of modelling has dominated geophysics since the seminal contributions of Newton, Laplace, Kelvin, Jeffreys, and others. It is expected that methods are available to solve these equations—most often by means of a linearized theory. Yet, as Jeffreys has warned, in geophysics, where the endeavour is to describe the past and predict the future, mathematical methods are no more than a tool and their choice, or even present availability, must not be allowed to dominate the basic physics and chemistry.

The initial task is to clarify the terms of my title. In the first place, my discussion is not about the curvature of time or "time warps" as conceived in general relativity. It is about geological time. The phrase 'turbulence of time' is a shorthand description for the complex temporal variation of *physical events* that constitutes the evolution of the Earth. Time itself is taken to be an infinite ordered sequence of points in a physical continuum which stretches from the origin of the solar system into the indefinite future. Each time point can be associated with physical events or spatio-temporal functions or functionals which may or may not form a reversible causal sequence of repeatable events.

In this connection, my description differs from the contrasting systems, i.e. "Time's arrow" and "Time's cycle", discussed by the geologist Stephen Jay Gould in a widely-read recent book (Gould 1987). Gould compares the interpretation of geological history as a temporal series of unrepeatable events with that of a repetition of fundamental physical states. While Gould recognizes explicitly the oversimplification in these poles of the dichotomy, his emphasis at least gives the impression that it is time itself which has curvature; it may have a fixed directionality or be twisted in repeating cycles. Of course, it is not the time variable that can be expressed in a linear equation or harmonic series but rather, it is the event functionals that can be conceived as having a vector or Fourier form.

One helpful description used by Gould is that of *deep time*. In contrast to *local time*, i.e. the duration of human history (and modern scientific observatories), deep time is the temporal span associated with the age of the Earth, or even the time to the formation of the solar system. While the result is nowadays scientifically uncontested, the arguments that the Earth is more than a hundred million years old sparked a profound controversy between physicists and geologists; the conductive heat transfer equation when integrated through deep time proved incapable, even at the hand of Lord Kelvin, of accommodating the evidence found in the rocks by geologists. (Radioactivity was the then inconceivable missing ingredient in the model.) Nevertheless, with careful selection of the scale of the problem, integration of the geophysical equations of motion over local time is found to be often successful. Our concern in this essay are the speculative limitations of integration into deep time.

The use of the term "turbulence" in the title also needs explanation. My intention is to convey the picture of unstable motion, made up of irregularities of various scales and displaying randomness of speed and orientation. I believe that this phenomenon is fundamental to description of the physical world.

The simplest model of turbulence is based on the elegant experiments of Osborne Reynolds. He passed a thin jet of coloured liquid into the centre of a tube through which the same but uncoloured liquid flowed. He altered the velocity of flow, and, at each speed, measured the resistance of the tube on the liquid within it. At low speeds, the coloured jet of liquid was unbroken and continuous. At a certain critical speed, it became irregular and more unorganized until its continuity was destroyed by eddies. This experiment enabled Reynolds to show that the resistance of a fluid to a moving object passes from being linearly dependent on the velocity during laminar flow to dependence on the squared velocity during turbulent flow.

My selection of this topic was motivated by the continuing puzzles presented by the almost universal representation of the Earth as a mechanical engine. The wide speed and storage capacity of digital computers in recent years has reinforced for many scientists the conception of an algorithm of mechanical creation, coded from Newtonian mechanics and Maxwell's equations. The algorithm is supposed to constitute a computational model of the reversible engine. When started, this computational engine, given the present physical conditions, can be run backwards to the creation much as a videotape can be rewound to produce, frame by frame, the previous state of affairs. The complimentary claim is that if we set $t = 0$ at the present, and adopt the present boundary conditions, the computational engine will run forward to predict the future of the Earth. The implications of this model are so profound that any doubts about it deserve close attention and analysis. For example, recent discussions of this question derive from a deeper awareness of the extent in nature of chaotic motions (e.g. Keilis-Borok 1987), nonlinear behaviour, and non-causality. Rather than a reflection of a "dark side" of the usual scientific optimism, critical rethinking of the design of the evolutionary engine may not only produce a more realistic description but prevent much wasted effort. In what follows, the examples chosen to illustrate the problem serve to focus on the complexities and point to a more realistic model for explanation and prediction (Bolt 1976).

2. The terrestrial clock

In order to explain the depth of geological time, the most skillful geological writers have turned to metaphors. Sir Charles Lyell, for example, stated his comprehension of the newly-emerging concepts of deep time as: "Worlds are seen beyond worlds immeasurably distant from each other and beyond them all innumerable other systems are faintly traced on the confines of the visible universe (Lyell 1830)."

When I was a student, the time used in geophysics was Greenwich Mean Time, fixed at Greenwich in terms of the Earth's rotation. With experience, it becomes clear that specification of observed time is really extremely complicated, involving the many components of the terrestrial wobble (Munk and MacDonald 1960). Nowadays, atomic clocks have been adopted (not without protest) to provide absolute times and are severed from the Earth's rotation. It is then essential to correlate atomic time with Universal Time or time kept by the rotating Earth. The difficulty involved can be seen from one example: the ephemerides of the Sun, moon and planets, based on Newtonian mechanics, show discrepancies with observed positions. Prediction of these discrepancies is close in local time but very uncertain in deep time.

There is no space here to define the various time scales currently adopted to measure the flow of physical processes along the time continuum. One primitive measure of time's passage is worthy of mention, namely, the cycles of ocean and bodily tides. The applied mathematical description of the interactions of gravitational forces between Sun, moon and Earth is among the most attractive and challenging in all of science. Yet there are few simple discussions that are satisfying and very few geologists and geophysicists can claim a deep understanding of the intricacies of the system.

Kant in 1714 was the first to suggest that tidal friction could be a cause of observed discrepancies in the predicted motions of the Sun and moon. Much later, G I Taylor demonstrated the importance of tidal dissipation in shallow seas and Harold Jeffreys linked this effect with changes in the rate of the Earth's rotation, i.e. the length of the day.

Tidal friction and viscous conditions in the Earth's interior give rise to irreversible changes in the directions of the rotations of the bodies involved. As Jeffreys has remarked, straightforward mechanical arguments soon lead to error if the complications due to the irregular forms of the oceans are not included. Although the relevant equations of motion are nonlinear, it can be established with certainty that significant energy dissipation occurs in shallow seas because of the friction of the water flow.

A gripping prediction into deep time related to Earth tides can be found in "The Earth" (1976) where Jeffreys discusses the effect of tidal friction in the future. The bodily tides on the moon are known to affect the Earth-moon system such that each body keeps the same face fixed toward the other with a period of about 47 days. Solar tides lengthen this period of rotation so that the high lunar tides will eventually retard the moon's revolution and cause the moon to approach the Earth. Jeffreys predicts that "This process will continue till the moon is at last dragged down to such a distance that it will be broken up by the action of tides raised in it by the Earth."

While prophecies of this scale send the mind soaring into deep time, the many premises and assumptions underlying the argument indicate caution. Will conditions in the Earth be maintained which allow shallow seas to exist or even grow over geological epochs? How can we be confident of such extrapolations into the future until predictions on the Earth-moon system into the past are verified? How can we avoid the non-uniqueness of the scientific efforts to model the creation of the planetary system? In what follows, it is argued that there are reasons why, in a strict deterministic sense, the past is not the key to future geophysical processes. At least, studies of the length of the day and the role of tidal dissipation establish that a damping term must be included in the description of evolutionary processes. Transfer of external energy by interplanetary torques, perhaps due to cosmic plasma, may reduce but not remove the effect.

The geochronology of the most recent revolutionary geological paradigm also requires a mention. The decisive boost for plate tectonics was the perception of regularity in the geomagnetic reversal patterns on the ocean floor, symmetrically created at the oceanic ridges and eventually evanescent at the deep ocean trenches. In Allan Cox's metaphor (1973): "The Earth's magnetic field acts as a chronometer, accurately [SiC] timing these events. The heart of the timing system is located in Earth's liquid core, where a magnetic field is generated by electrical currents."

Two comments are in order. First, it should be remembered that the mechanism for the reversals themselves is not fully understood at the present time, despite study by outstanding physicists. The hydromagnetic engine involves quite complicated nonlinear processes. Secondly, the future stability of the reversal clock is anything but assured. The dynamo theory for the generation of the Earth's magnetic field is peculiar in that while it purports to be an explanation of one of the major geophysical fields, it has not predicted a major phenomenon which has *subsequently been detected*. Because prediction is the mark of a strong theory, such a situation is somewhat unsettling.

In my own subject, seismology, the terrestrial clock enters in the attempt to predict future earthquakes from past geological evidence. While most of seismology deals with the present and provides, in a quantitative way, sharp spatial boundary conditions for the predicted departures of plate movements, earthquake prediction for reduction of long-term seismic risk must be based on long geologic records. Most critical structures rely on the establishment of two earthquakes: the maximum Holocene earthquake and the maximum Pleistocene earthquake. The first is defined as the maximum earthquake to have occurred in the region in the last ten to eleven thousand years and the second is the maximum earthquake deemed to have occurred in the region in the last 10,000–100,000 years. Establishment of these past earthquakes is a geological exercise involving field structural relationships along faults. The main ingredient is the establishment of the activity of the fault-slip rates and structural offsets. Very recently, striking advances have been made by dating, using radiocarbon methods, the time of liquefaction of sand layers located in stratigraphic sequence in alluvial deposits along active faults. The seminal example is that of Pallett Creek in central California, where inter-occurrence times between large earthquakes have been established at about 130 years for the last one thousand years (Sieh 1984).

3. Nonlinearity

Most solutions of geophysical problems obtain linearity by imposing perturbation techniques or restricting the time interval so that a linear approximation is adequate. A great advantage of a linear system is that a solution can always be constructed by superposition of two other solutions. In stark contrast, in nonlinearity systems, there are no limits to the field of activity, which is unbounded and involves rapid transitions from one type of solution to another. Generally, the problems are extremely intricate and much mathematical effort has been expended on them. The discussion is advanced at this stage if we examine one of the most thoroughly described physical systems related to geophysics, namely, turbulence.

In turbulent flow, there is a spectrum of eddies and motions which convert the kinetic energy of the fluid molecules to heat, but the motion of the molecules cannot be described by the continuum equations. Following Newton, the dissipation is treated usually through the introduction of turbulent viscosity as a defining parameter dependent on the bulk speed.

This spectrum of scales and energies of motion has been the basis of a successful treatment of fluid turbulence at the hand of G K Batchelor (1967) and others as statistical ensembles. Roughly, we can say that in turbulent motion there is no ability to remember the motion that gave birth to the irregularities. Steady turbulence is when the average part of the potential field depends only upon position but not on time. In contrast, when the ensemble average depends on time but not on position, the turbulence is inhomogeneous and averages are taken over spatial coordinates. Turbulence is inherently nonlinear. A deterministic description of such motion starts with the Navier-Stokes equation. This equation governs convection in the Earth's mantle and the dynamics of the fluid core of the Earth where it is fundamentally linked with the generation of the Earth's magnetic field. In any attempt to find the cause of the field, according to P H Roberts (1971), the initial question to be asked is: "Assuming a physically reasonable driving force F, will a flow (of the core fluid) of the right magnitude and character be created? And what will be the size, strength, and nature of the field created, bearing in mind that Lorentz forces will affect the flow?"

This hydromagnetic dynamo problem, in which the fluid displacement and the field are deduced from known F, is nonlinear and difficult. Roberts argues that there is scant hope for a full solution because F is not known *ab initio*. Even so, it is remarkable that ingenious work has led to theorems and partial results which limit the range of solutions, given a plausible state of conditions in the Earth.

Another example of a basic geophysical nonlinear problem is the Laplace tidal equations, which apply to motions of the oceans and the atmosphere (Platzman 1971). Many attempts at partial solutions have been made for geophysical models of interest. The usual simplification is to seek low-pass filtered approximations that limit the span of frequencies involved. The resulting equations are approximations for long-wave equations, where the word "long" refers to both wavelength and period. Nevertheless, efforts to compute the amplitude and phase of ocean tides along the coastlines of the world have shown that the nonlinearities involved make prediction hazardous even at the scale of decades. New approaches may replace Euclidean geometry with a *fractal* basis. Certainly, geophysical growth systems reveal fractal behaviour (Mandelbrot 1982) but its predictive power is still not substantiated.

An alternative but related treatment of nonlinear systems to turbulence is chaos theory. This active branch of physics describes unpredictable behaviour in nonlinear dynamical systems (Hao 1984). In certain experiments, it has been demonstrated how to map nonlinear behaviour so that transitions between quasi-periodic and chaotic motions can be detected.

4. Thermodynamics and temperature estimation

The notion of irreversible processes is the basis of the second law of thermodynamics. The best known process is that of an ideal gas, where after expansion, the gas, while possessing the same amount of energy, loses some of its capacity for doing work. In such irreversible processes there is thermodynamic "degradation," or in other words, the entropy increases. The degree of reversibility in the process is a measure of the increase in entropy of the system. Indeed, it is accepted that all natural processes are irreversible so that quantitatively we can adopt the formulation of Clausius that the entropy of the universe is increasing.

The second law of thermodynamics therefore appears consistent with the idea of the turbulence of geophysical time and helps in defining its scope. The planetary system and its constituent physical properties will never be the same as now, nor have they been quite the same before now. Unfortunately, there are great practical difficulties in using the model of heat engines with reversible and irreversible cycles to determine the detail of the geophysical evolutionary process.

Seismology has provided an apparently robust way of computing the distribution of elastic parameters in the Earth from measurements of travel times of seismic waves between seismic sources and seismographic stations. The tomographic method has been developed extensively since its primitive use in 1906 by R D Oldham to discover the core of the Earth. While the basic problem is nonlinear, the linearized functionals involved usually provide rapid convergence.

Among the assumptions, however, that are involved in this inversion, are some involving temperature. Strictly the elastic parameters derived from the P and S seismic velocities are neither adiabatic nor isothermal and a full thermodynamic treatment requires the absolute temperature to be assigned (see Bullen and Bolt 1985). Some years ago, I derived an index of state θ, which gives a measure of the inhomogeneity in the Earth. The original form was later extended by Bullen to explicitly involve the temperature gradient in the Earth. This work illustrated that it is theoretically possible to infer the temperature distribution in the Earth from the inversion of seismic travel times. We must first, of course, achieve more refinement in the present inversions which essentially ignore the effects of temperature variations. Precise observations of eigenvibration frequencies and travel times must be amassed and then high resolution inversion may produce residuals that can be related to the purely temperature effects inside the Earth.

An alternative formulation is to use a thermoelastic theory for the seismic wave motion, which takes account of these temperature variations explicitly, such as that developed by M A Biot. In the future, inversion to Earth properties using this more general form may be worthwhile. In the meantime, estimates of temperature in the Earth depend on less direct arguments, which assume the mineralogy of the deep interior, and the extrapolation of the results of small-scale experiments in the laboratory to conditions in the bulk volume of the Earth.

5. A non-Newtonian world

Modern geophysics grew out of Newtonian mechanics. Newton himself contributed directly to geophysics through his estimation of the density of the Earth and his treatment of tides. His gravitational theory makes the precise measurement of the universal constant of gravitation G a basic requirement of geophysics, but it is not yet as well determined as other fundamental physical constants.

Recently, a fifth force has been discussed as a near-range adjustment to Newtonian gravitation. So far, however, the experimental results have not been sufficiently error-free to convince many physicists that any discrepancy in measured attraction cannot be fully explained by systematic experimental errors within the expected uncertainties. Although the fifth force is not likely to be important to the prediction of evolution in the short run, it does have significance to the regularity of terrestrial time over geological epochs. Indeed, the fifth force might stem from the effect of the position of all masses of the gravitating planetary system at any given time. As these positions change nonlinear interactions would occur, perhaps involving turbulence at certain scales. The result would be a non-causal geophysical system over deep time.

In 1988 there was a new attempt to measure the constant gravitational with a high precision at the International Bureau of Weights and Measures in Paris. The aim of the experiment is to test whether the gravitational attraction differs between masses of different composition such as carbon, copper and lead. If gravitation depends on the composition of attractive bodies, further chaotic processes arise in the working of the evolutionary engine, including cyclic variations in the radius of each planet.

If the gravitational constant varies or if the fifth force is confirmed, an improved gravitational theory to compute the emphemeris for the solar system is required. Changes in the length of the day might be expected with a consequent hydromagnetic response of the fluid core.

6. Causality and uniqueness

As we have seen in the above discussions, the evolution of physical processes inevitably involves the causal connections between events. Causality, however, remains a central issue of logical investigation and its exact distinguishing features are a matter of debate. Normally, by cause is meant a necessary and sufficient precedent condition. In this sense, inquiry into causes in geophysics is only a special case of the application of scientific inquiry in general, but involving the time factor as an essential variable.

If our causal beliefs are true, we expect to infer that, given a cause, the effect will occur. In this way, we recognize *stages* in the development of processes, i.e. they have a regular succession of properties. It is, however, not logically always the case that the earlier stages *cause* the latter ones because it is possible to have a succession of links in a chain, but the chains are not causally linked. To demonstrate this result, we need only remark that later development is unavoidable. Eruptions of a volcano may be necessary but not sufficient for its extinction. Because, the evidence suggests that causation is partly an external action it is best to discard altogether the assumption of causal chains. It seems sounder, from the basic logical viewpoint, to recognize that in geological evolution there is no unilineal form of development but, rather, interactions between causes and events at each point of time.

Given the above geophysical and logical discussion, what appears to be necessary is to recognize that evolutionary physical processes have unbounded complexity; analysis in terms of a number of simple factors can be wide of the mark. Indeed, ultimately, it may be hopeless to reduce a physical system to a finite set of simple connecting "laws." Many different laws are involved in the same process and processes proceed through time in inter-related ways. For example, it is likely that there are many causes for the acquisition of the present internal structure of the Earth and that the same structure may be reached by different paths with different complexities of inter-related causes.

The conclusion would seem to be that a geophysical property cannot be relied upon to retrace the steps of the process backwards in time. The reason is that all interactions of external causes may not be included in the computations. Similarly, for computational extrapolations into the future, inability to specify all possible interactions at any stage means that, at least when any nonlinear reaction is present, a new crucial branch in the process could occur. Even with the introduction of mathematical theory, to deal with a particular nonlinear problem, the effect of some other nonlinear property can be made more severe by the adopted approximations.

One of the great contributions of the development of inverse theory of the geophysical context in the last decade has been the emphasis on nonuniqueness (Tarantola 1987). Nonlinear inversion, where the data kernels are not linear, means that the solution obtained is dependent on the assumptions about the zero iterate. It is essential in all geophysical problems of this type, such as seismic travel-time and free-oscillation problems, to explore the effect of the variations in the first iteration. An early attempt to do this for models of internal Earth structure was made by Frank Press (1970), who used Monte Carlo methods to determine the range of solutions in the Earth, which satisfied the seismological data. Press used body wave travel-times, surface-wave dispersion data and terrestrial free-oscillation data, in addition to specification of the Earth's mass and moment of inertia. He used a computer to generate and test over 5 million models. The few models that met Press's criteria were rather complex and no simple model emerged. There are weaknesses, however, in a Monte Carlo technique. K E Bullen has argued that the procedure tends to favour over-parametrized models because a simple random walk would be automatically weighted to have a low probability.

A number of approximate methods for solving nonlinear ill-posed problems have been described (e.g. Vogel 1987). By ill-posed is meant that the nonlinear operator does not have a closed range so that the solution is not unique, nor does it depend continuously on the available data (Wahba 1987). The key mathematical strategy at present is first to regularize the problem and then to proceed via a sequence of sub-problems, each of which depends on a particular parameter. It has been found that robust solutions depend upon each sub-problem being well posed, having efficient numerical solutions, and allowing inclusion of *a priori* data about the desirability of the solutions. One of the favoured methods is to use the penalized least-squares, where penalty functions that establish desirability are applied. My own experience with this scheme (Bolt and Uhrhammer 1975) has been in the calculation of a series of Earth models. I found that the algorithm works well in this context but doubts arise concerning the subjectivity of the penalty weights chosen. In any event, the resulting Earth models have not found the favour given to the standardized Earth-model PREM, although the reason for geophysicists' preference has not been made clear (see Bullen and Bolt 1985 for a comparison). In short, the method involves the setting down of a judgemental function.

7. Conclusions

The thrust of the analysis given here is that, in geophysical forecasting through deep time, the model of geophysical turbulence is perhaps close to reality. The simplification of geological evolution as either "Time's arrow" or "Time's cycle" must be replaced by a full appreciation of the complexity of the spatio-temporal processes in geophysics. These processes occur over large ranges of scales in both space and time and involve both internal and external force systems. There is continual input of energy from various sources which produces mechanical responses through feedback loops which often involve nonlinearities. Both locally and globally the evolutive system is not conservative but kinetic energy at every scale is damped by frictional dissipation—ultimately at the molecular level.

Observations of the motions of the atmosphere and the oceans show an enormously broad spectrum of fluid turbulence. The behaviour of the geodynamo is close to this process with motions in the core being likened to weather in magneto-meteorology. The fluid motions are fed across the whole spectrum from the largest to the smallest scale by draining energy by means of a variety of mechanical processes. There are energy cascades occurring from eddies to large turbulent vorticities. At the same time, the large vorticities break down into smaller ones.

The mathematical framework which has been set up to consider the dynamics of these turbulent motions in fluid dynamics may hold, in the long run, the promise of a fruitful approach to solving the broadest class of geophysics problems. Consider the task of inferring the evolutionary history of geological structures in deep time, i.e. the Pre-Cambrian rocks that span 80% of all geological time. Do their physical and chemical properties contain remnant evidence of their past histories? Would such evidence provide a unique historical path or could the present condition have been reached in various ways by the interaction of many forces which are now extinct?

In this discussion, I have indicated reasons why the attempt to construct a unique causal chain may fail. On the other hand, there have been surprising successes in palaeo-reconstruction. We need only note the apparent regularity and ridge symmetry of the magnetic reversals in the ocean floor, which opened up the paradigm of sea floor spreading in plate tectonics. The consequent explanations have been rewarding, at least in a broad sense and in local time, even if the exact designs of the engines causing the magnetic reversals and mantle convection are not known.

While the discussion has opened up some vexing problems, I have suggested a method of attack in which appropriate approximations of spectral turbulence be made at every scale in terms of spatio-temporally averaged parameters. There appears to be merit in shifting from deterministic treatments to statistical and fractal representations that allow quasi-periodicities to emerge from fundamentally chaotic systems. It may not be long before a transition to ensembles is forced on geophysicists. Then, as Mr Micawber used to say, "be ready—in case of anything turning up."

References

Batchelor G K 1967 *An introduction to fluid dynamics* (Cambridge: Cambridge University Press)
Bolt B A and Uhrhammer R 1975 Resolution techniques for density and heterogeneity in the earth; *Geophys. J. R. Astron. Soc.* **42** 419–435
Bolt B A 1976 Abnormal seismology; *Bull. Seismol. Soc. Am.* **66** 617–623
Bullen K E and Bolt B A 1985 *An introduction to the theory of seismology* (Cambridge: Cambridge University Press)
Cox A (ed.) 1973 *Plate tectonics and geomagnetic reversals* (New York: W H Freeman)
Gould S J 1987 *Time's arrow–Time's cycle* (Cambridge: Harvard University Press)
Hao Bai-Lin 1984 *Chaos* (Singapore: World Scientific)
Jeffreys H 1962 *The earth*, 4th ed. (Cambridge: Cambridge University Press)
Keilis-Borok V 1987 Address to the XIX General Assembly of IUGG, Vancouver
Lyell C 1830–1833 *Principles of geology*, being an attempt to explain the former changes of the earth's surface by reference to courses now in operation (London: John Murray)
Mandelbrot B B 1982 *The fractal geometry of nature* (New York: W H Freeman)
Munk W H and MacDonald G J F 1960 *The rotation of the earth* (Cambridge: Cambridge University Press)
Platzman G W 1971 Ocean tides and related waves; in *Mathematical problems in the geophysical sciences* (ed.) W H Reid (Providence: AMS)
Press F 1970 Earth models consistent with geophysical data; *Phys. Earth Planet. Inter.* **3** 3–22
Roberts P H 1971 Dynamo theory; in *Mathematical problems in the geophysical sciences* (ed) W H Reid (Providence: AMS)
Sieh K E 1984 Lateral offsets and revised dates of large prehistoric earthquakes at Pallett Creek, Southern California; *J. Geophys. Res.* **89** 7641–7670
Tarantola A 1987 *Inverse problems theory* (Amsterdam: Elsevier)
Vogel C R 1987 An overview of numerical methods for nonlinear ill-posed problems; in *Inverse and ill-posed problems* (eds) H W Engl and C W Groetsch (Orlando: Academic Press)
Wahba G 1987 Three topics in ill-posed problems; in *Inverse and ill-posed problems* (eds) H W Engl and C W Groetsch (Orlando: Academic Press)

The parameters controlling planetary degassing based on ^{40}Ar systematics

KARL K TUREKIAN
Department of Geology and Geophysics, Yale University P.O. Box 6666, New Haven, Connecticut 06511, USA

Abstract. The ^{40}Ar degassing process is used to define general degassing constants consistent with the present-day ^{40}Ar degassing rate and terrestrial potassium abundance. The application of the constants to ^{36}Ar and carbon indicates that carbon released at oceanic volcanic sites is dominated by primordial carbon.

Keywords. Planetary degassing; ^{40}Ar degassing equation; carbon flux.

1. Introduction

Degassing models for the Earth are constrained by the atmospheric growth history of radiogenic ^{40}Ar. The precursor material forming the Earth 4.55×10^9 years ago is presumed to have been virtually free of ^{40}Ar. Although the Earth has become zoned at some time in its history into crust, mantle and core, the degassing history of the planet can be constructed without introducing these complications into the discussion by choosing generalized parameters.

2. The ^{40}Ar degassing equation

The growth of atmospheric radiogenic ^{40}Ar by degassing from the solid Earth, in which it is formed, is a function of the amount of potassium present, the age of the Earth and the time-dependent pattern of release of ^{40}Ar to the atmosphere. The amount of ^{40}Ar produced in the Earth system as a function of time is

$$X^{40}(t) = K_0(1 - e^{\lambda t}), \tag{1}$$

where t is measured from the formation of the planet, K_0 is the amount of ^{40}K in the planet 4.55×10^9 years ago that will decay to ^{40}Ar by electron capture, and λ is the decay constant of ^{40}Ar (5.30×10^{-10} y^{-1}).

The growth of ^{40}Ar in the solid Earth is a function of production rate and degassing rate. The simplest equation describing these two processes is

$$dX^{40}_{SE}(t)/dt = \lambda K_0 \exp(-\lambda t) - \alpha \exp(-\beta t) X^{40}_{SE}(t), \tag{2}$$

where $X^{40}_{SE}(t)$ is the amount of ^{40}Ar in the solid Earth and α is a degassing constant which when multiplied by $\exp(-\beta t)$ becomes the time-dependent degassing coefficient. When $\beta = 0$ then the degassing coefficient is simply α and (2) becomes the original equation I used earlier (Turekian 1959). β was conceptually introduced by Sarda et al (1985) and Allègre et al (1987) as the index of attenuation of the capacity for mantle degassing over time mainly due to the rigidification of the mantle resulting from the decrease in internal heat flow.

The complete solution to (2) is

$$X^{40}_{SE}(t) = \frac{K_0 \lambda}{\beta} \exp\left(\frac{\alpha}{\beta} \exp(-\beta t)\right) \int_{\exp(-\beta t)}^{1} \tau^{(\lambda/\beta)-1} \exp[(-\alpha/\beta)\tau] \, d\tau. \quad (3)$$

The integral in (3) cannot be evaluated analytically, however, and it is therefore simpler to solve (2) numerically, using an accurate initial-value method such as the Runge-Kutta algorithm.

The growth of ^{40}Ar in the atmosphere is given by

$$X^{40}_{ATM}(t) = X^{40}(t) - X^{40}_{SE}(t). \quad (4)$$

The present-day value ($t = 4.55 \times 10^9$ y) of X^{40}_{ATM} is known ($= 9.85 \times 10^{41}$ atoms) and provides one limit to our degassing model.

The rate of growth of atmospheric ^{40}Ar as a function of time is given by

$$dX^{40}_{ATM}(t)/dt = \alpha \exp(-\beta t) X^{40}_{SE}(t). \quad (5)$$

We have two relationships describing the history of ^{40}Ar in the atmosphere. One gives us the constraint based on the present-day ^{40}Ar amount in the atmosphere. The other gives the rate of degassing of ^{40}Ar which can be related to present-day measurements. If we know X^{40}_{ATM} (present), K_0 and the present rate of degassing, dX^{40}_{ATM}/dt (present) then we are left with two unknowns (α, β) and two equations.

3. Solutions of the ^{40}Ar degassing equation

Figure 1 shows solutions to (2) using the Runge-Kutta algorithm and a "trial and error" process for a range of values of K_0 and dX^{40}_{ATM}/dt (present). Remembering that X^{40}_{ATM} (present) is set by "trial and error" we must seek independently determined values of K_0 and dX^{40}_{ATM}/dt (present) to arrive at acceptable values of α and β.

3.1 Present-day rate of ^{40}Ar degassing

The ^3He flux from the mantle in the oceans is our entry into the problem. This has been summarized by Craig and Lupton (1981). A minimal oceanic ^3He flux of 3 atoms cm^{-2} s^{-1} (Turekian 1984) combined with a ratio of MORB of $^{36}Ar/^3He$ of 5 (Atlantic) to 30 (Pacific) yields an ^{36}Ar flux of between 0.17×10^{28} atoms of ^{36}Ar per year and 1.0×10^{28} atoms ^{36}Ar per year, respectively. The $^{40}Ar/^{36}Ar$ ratio in MORB varies considerably. The $^3He/^4He$ ratio, however, for MORB is around 12.5×10^{-6} whereas "plume" ratios are between 21×10^{-6} and 28×10^{-6}. Pacific Ocean values for the $^3He/^4He$ ratio of the hydrothermally added helium average $15.4\ (\pm 3) \times 10^{-6}$. A simple inference from this is that roughly 80% of the helium supplied from the mantle to the ocean is from MORB sources and 20% from "plume" sources. The MORB source $^{40}Ar/^{36}Ar$ can be as high as 25,000 (Allègre et al

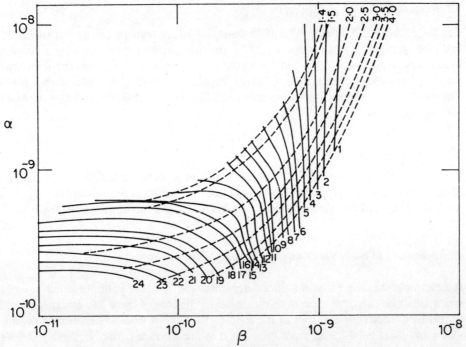

Figure 1. Solutions for (2) yielding $X_{ATM}^{40}(\text{present}) = 9.85 \times 10^{41}$ atoms. The solid lines represent values of present-day ^{40}Ar degassing rate in units of 10^{32} atoms per year. The dashed lines represent different values of K_0 in units of 10^{42} atoms. The most likely values of K_0 and present-day degassing rate yield $\alpha \cong 10-9$ and $\beta \cong 0.8 \times 10^{-9}$.

1987) but the average value is closer to 2500 (summary of Hart et al 1985), a value compatible with Allègre et al's analysis of the ^{3}He/^{4}He and ^{40}Ar/^{36}Ar data of oceanic basalts. Using this value the ^{40}Ar/^{3}He ratio venting to the atmosphere is between 12,500 and 75,000. This corresponds to a ^{40}Ar flux between 0.43×10^{31} and 2.5×10^{31} atoms per year.

A major flux of ^{40}Ar comes from continents. Using the excess ^{40}Ar in the Great Artesian System of Australia, Torgersen (1985) determines a total continental ^{40}Ar flux of about 2.5×10^{31} atoms/year. The total ^{40}Ar flux from all sources can be estimated from the ^{40}Ar/^{36}Ar of the Rhynie chert of Devonian age. Both Cadogan (1977) and Turner (1988) have found a value of 291 compared to a modern atmospheric value of 295.5. If we assume that the ^{36}Ar flux is negligibly small over the past 400 million years we arrive at a worldwide flux for the last 400 million years of about 4×10^{31} atoms per year. This is similar to the result obtained by Turner (1988) because ^{36}Ar flux does not markedly influence the ^{40}Ar/^{36}Ar ratio over the past 400 million years.

The worldwide flux is equal to the sum of the oceanic and the continental ^{40}Ar fluxes. Based on a total Earth and continental potassium concentration, we see that half the Earth's potassium is in the continents. This implies that degassing is controlled by the same α and β constants for all components of the solid Earth system. I will assume this to be a valid approximation throughout the history of the Earth.

3.2 Potassium content of the Earth

The lower limit of K_0 ($= 1\cdot082 \times 10^{42}$ atoms) is determined by the age of the Earth and the present-day quantity of ^{40}Ar in the atmosphere assuming complete outgassing. Values of K_0 between $1\cdot4 \times 10^{42}$ atoms and 4×10^{42} atoms encompass all suggested values for the Earth as a system (Zindler and Hart 1986; Allègre et al 1987). The most commonly accepted concentration for the whole Earth is 240 ppm which corresponds to $K_0 = 3\cdot2 \times 10^{42}$ atoms.

3.3 Determination of α and β

If we use a value of $K_0 = 3\cdot2 \times 10^{42}$ atoms and a present-day total ^{40}Ar degassing rate of between 4×10^{31} and 5×10^{31} atoms per year the limits on the values of α and β are set from figure 1. The value for α is about 1×10^{-9} and for β it is about $0\cdot8 \times 10^{-9}$.

4. Application to other gases and the history of the atmosphere

If we assume that the α and β values determined for ^{40}Ar are valid for other gaseous species then we can determine both the degassing history of these gaseous species and the initial burden of these gases in the solid earth. For the sake of simplicity I assume that α and β are valid for all gases, whether they are the rare gases, Ne, Ar, Kr and Xe or the chemically reactive volatiles CO_2 and H_2O. This is a gross oversimplification obviously, but it provides a basis for comparative histories of degassing and initial Earth composition.

Torgersen (1989) has shown that the flux of ^{3}He through the crust is immeasurably small. The ^{3}He flux from the oceans is clearly the dominant flux and I assume that all primary fluxes of the other volatiles also are at hot spots and spreading centers mainly in the ocean basin.

4.1 Argon-36 balance

Using α and β derived from the ^{40}Ar balance described above the present-day ^{36}Ar flux can be calculated as well as the initial bulk Earth abundance of ^{36}Ar since

$$dX^{36}_{SE}/dt = -\alpha \exp(-\beta t) X^{36}_{SE}. \tag{6}$$

Because of the nature of the curves and the estimated potassium content of the Earth, α cannot vary greatly around a value of 1×10^{-9} (figure 1). On the other hand the value of β is more strongly dependent on the value chosen for the present-day ^{40}Ar flux. Table 1 shows the variation of the parameters involving ^{36}Ar degassing as a function of β while holding α constant at $1\cdot0 \times 10^{-9}$ (the present-day atmospheric ^{36}Ar inventory is $3\cdot33 \times 10^{39}$ atoms).

The results are consistent with the upper limits of ^{36}Ar flux calculated from the ^{3}He flux and ^{36}Ar/^{3}He ratio in MORB rocks ($\sim 1 \times 10^{28}$ atoms per year). The uncertainties are so large in these balances that effective agreement is possible with appropriate adjustments. The aim is to obtain broad self-consistency and this is attained.

Table 1. Dependence of ^{36}Ar content of the solid earth and the present-day degassing rate on the value of β ($\alpha = 1 \times 10^{-9}$, present-day ^{36}Ar inventory = $3\cdot 33 \times 10^{39}$ atoms).

β (10^{-9})	(X_{SE}^{36}) initial $(10^{39}$ atoms$)$	Present-day ^{36}Ar flux $(10^{28}$ atoms/yr$)$
0·80	4·73	3·68
0·85	4·86	3·20
0·90	5·01	2·81
0·95	5·16	2·43
1·00	5·30	2·09

4.2 Carbon flux between crust and mantle

Des Marais and Moore (1984) determined the carbon flux from oceanic volcanism to be 5×10^{35} atoms per year. As we now have a measure of the ^{36}Ar flux (table 1) from the mantle, we gain an insight into the cyclic history of carbon between the crust and mantle by comparing the two components. Unlike argon, a rare gas which is found mainly in the atmosphere, carbon at the Earth's surface is found primarily in rocks (or sediments) and is subject to transport and subduction along with other components of the plates. This means that the carbon reservoir at the Earth's surface is not a unidirectionally accumulating reservoir but a cycling one. Indeed if we consider the observed reservoir abundance of carbon, $8\cdot5 \times 10^{44}$ atoms (Wedepohl 1969) and Des Marais and Moore's flux of 5×10^{35} atoms per year, the mean residence time of carbon in the crust relative to subduction and return to the mantle is $1\cdot7 \times 10^9$ years.

It is not possible from the carbon flux measurement and the crustal carbon reservoir alone to determine if the carbon flux at ocean ridges is primary bulk earth carbon or contains a significant component of recycled carbon as suggested by Javois et al (1982). It is evident, however, that a flux into the mantle must exist.

We can determine if the flux of carbon to the ocean basins is from primordial degassing or has a large recycled component by comparing the ^{36}Ar flux to the C flux. If the ratio of the two species resembles the ratio in carbonaceous chondrites and we assume that the "volatile" element carrier to the Earth is the same as that in carbonaceous chondrites (Turekian and Clark 1975; Turekian 1980), then we can be secure that we are seeing degassing of primordial material. If on the other hand the degassing ratio of C to ^{36}Ar is much larger than carbonaceous chondrites then it approaches the value expected from the subduction of crustal rocks rich in carbon and depleted in ^{36}Ar.

The C to ^{36}Ar ratio degassing from the mantle associated with oceanic volcanism is about $2\cdot5 \times 10^7$. The ratio in carbonaceous chondrites is $7\cdot9 \times 10^7$. Within the uncertainties of the assumptions these ratios are comparable. If anything, the degassing ratio is *lower* than the carbonaceous chondrite ratio. It would have to be *higher* to have a significant crustal recycled component. The C/^{3}He of C-1 carbonaceous chondrites is $3\cdot55 \times 10^9$ whereas in MORB it is about 2×10^9 (Marty and Jambon 1987). As both of these values have uncertainties to them, I do not think they are really different. Again, the difference, if present, is in the direction a lower C flux than predicted from the carbonaceous chondrite C/^{3}He and thereby not necessitating an additional cycled flux.

I conclude that although carbon and possibly other reactive volatiles such as water (protons) and nitrogen, is subducted into the mantle it is not resupplied to the Earth's surface by degassing on a short time scale but rather is lost to a less accessible mantle reservoir from whence recycling might occur on a time scale of the same order of magnitude as the age of the Earth.

Acknowledgements

Neil Ribe and Jordan F Clark helped me to integrate and solve the degassing equation numerically. My friends in our geochemistry coffee hour behaved in their usual tolerably intolerant fashion as these ideas developed. Although no grant has supported this specific work I am grateful to the National Science Foundation for funding other related enterprises that have nourished some of the concepts in this paper.

References

Allègre C J, Staudacher T and Sarda P 1987 Rare gas systematics: formation of the atmosphere, evolution and structure of the earth's mantle; *Earth Planet. Sci. Lett.* **81** 127–150

Cadogan P H 1987 Paleoatmospheric argon in Rhynie chert; *Nature (London)* **268** 38

Craig H and Lupton J E 1981 Helium-3 and mantle volatiles in the ocean and the oceanic crust; in *The sea, the oceanic lithosphere* (ed.) C Emiliani (New York: John Wiley) vol 7 391–428

Des Marais D J and Moore J G 1984 Carbon and its isotopes in mid-oceanic basaltic glasses; *Earth Planet. Sci. Lett.* **69** 43–57

Hart R, Hogan L and Dymond J 1985 The closed-system approximation for evolution of argon and helium in the mantle, crust and atmosphere; *Chem. Geol.* **52** 45–73

Javois M, Pineau F and Allègre C J 1982 Carbon geodynamics; *Nature (London)* **300** 171–173

Marty B and Jambon A 1987 C/^{3}He in volatile fluxes from the solid earth: implications for carbon geodynamics; *Earth Planet. Sci. Lett.* **83** 16–26

Sarda P, Staudacher T and Allègre C J 1985 ^{40}Ar/^{36}Ar in MORB glasses: constraints on atmosphere and mantle evolution; *Earth Planet. Sci. Lett.* **72** 357–375

Torgersen T 1985 ^{4}He and ^{40}Ar in natural gas reservoirs: implications for crustal degassing; *Eos* **66** 1119

Turekian K K 1959 The terrestrial economy of helium and argon; *Geochim. Cosmochim. Acta* **17** 37–43

Turekian K K 1980 Origin and evolution of the Archaean hydrosphere and atmosphere; in *The primitive earth revisited* (ed.) M H Hickman (Ohio: Miami University) pp. 86–93

Turekian K K 1984 Geochemical mass balances and cycles of the elements; in *Hydrothermal processes at seafloor spreading centers* (eds) P A Rona, K Bostrom, L Laubier and K L Smith Jr (New York: Plenum) 361–367

Turekian K K and Clark S P 1975 The non-homogeneous accumulation model for terrestrial planet formation and the consequences for the atmosphere of Venus; *J. Atmos. Sci.* **32** 1257–1261

Turner G 1988 Hydrothermal fluids and argon isotopes in quartz veins and cherts; *Geochim. Cosmochim. Acta* **52** 1443–1448

Wedepohl K H (ed.) 1969 Carbon; in *Handbook of geochemistry* 6-0-2 (Berlin: Springer)

Zindler A and Hart S R 1986 Chemical geodynamics; *Annu. Rev. Earth Planet. Sci.* **14** 493–571

Rare gases and hydrogen in native metals

R J POREDA[1,3], K MARTI[2] and H CRAIG[1]

[1] Isotope Laboratory, Scripps Institution of Oceanography, University of California at San Diego, La Jolla, California 92093-0220, USA
[2] Department of Chemistry, University of California at San Diego, La Jolla, California 92093-0317, USA
[3] Present Address: Department of Geological Sciences, University of Rochester, Rochester, New York 14627, USA

Abstract. Isotopic studies of He, Ne, Ar and H_2 in native iron from two sites in Greenland and in josephinite (native Ni-Fe) from southwest Oregon, are reported. The helium and neon isotopic ratios in the Greenland irons are completely consistent with crustal, atmospheric and cosmogenic origins for the gases. There is no evidence for a deep mantle-origin for the Greenland irons and the hypothesis that the iron formed during the intrusion of basalt into reducing sediments is consistent with rare-gas, hydrogen and carbon isotopic ratios.

The helium isotopic composition of unaltered josephinite ($R/R_A = 5\cdot6$) is consistent with formation of josephinite in either an upper mantle source or during serpentinization of the Josephine peridotite in the presence of "upper mantle" helium in the host rocks. Cosmic-ray produced ^{3}He may be a significant component in josephinite but is difficult to evaluate because of the lack of fluid inclusions as in olivines and because of the negligible production of ^{21}Ne in nickel-iron by cosmic rays. No evidence of the very large ^{3}He and ^{21}Ne anomalies reported by other laboratories has been found: as in the original "cold fusion" results of Paneth in 1926, these effects are shown to be artifacts due to scavenging in glass systems by H_2 extracted with rare gases from natural iron samples.

Keywords. Isotopic analyses; helium; neon; argon; hydrogen; josephinite; gas extraction procedures; rare gases; experimental artifacts.

1. Introduction

Noble gases, because of their lack of chemistry, are the most important tracers of the Earth's initial volatile inventory and its evolution through geologic time. Except for minor contributions from extraterrestrial sources and cosmic-ray production in the atmosphere and on the earth's surface, the present atmosphere has evolved from the mantle by degassing over geologic time. Two decades ago the discovery of the flux of "primordial" ^{3}He from the interior of the earth (Clarke *et al* 1969) showed that in 1967 A.D., when the first ^{3}He-enriched samples were collected, volatiles stored in the mantle during the accumulation of the solid earth were still being transported to the ocean and the atmosphere. Since then many studies have shown the ubiquitous presence of mantle ^{3}He and other rare gas components in Mid-Ocean Ridge and Ocean Island ("hotspot") basalts, volcanic gases, continental rifts, tectonic arcs and geothermal regions. Still, twenty years later, very little is yet known about the noble gas reservoirs in the deep interior of the earth and it is important to investigate each possible source of mantle-derived primordial gases. This paper describes the investigation of one such proposed reservoir, based on isotopic studies of helium, neon, argon and hydrogen.

Bird and Weathers (1975, 1977, 1979) proposed a deep-mantle origin for two occurrences of native iron and nickel-iron. The first of these native metals, *josephinite*, is found as stream placers in southwest Oregon. Conventional wisdom attributes the formation of this nickel-iron to reduction of Fe-silicates during serpentinization of the Josephine peridotite (Krishnarao 1964; Dick 1974; Botto and Morrison 1976; Dick and Gillete 1976), a serpentinized harzburgite tectonically emplaced into the Klamath Mountain Structural Arc during the Nevadan orogeny (Dick 1974). Josephinite is unique with respect to other occurrences of native nickel-iron (awaruite) associated with serpentinized ultramafic rocks, in that some specimens of josephinite contain oriented intergrowths of a two-phase nickel-iron, taenite (Fe, Ni) and awaruite ($FeNi_3$) and andradite garnet. The composition of the metal phases in josephinite is variable ranging from 62 to 71% Ni with average values of 68·2 wt% Ni and 28·9 wt% Fe. The remainder is Cu, As, Co plus minor silicate material. Botto and Morrison (1976) suggested that a higher temperature of serpentinization (450–500°C) stabilized the taenite phase and proposed a reaction for the reduction of the iron:

$$3\,CaMgSi_2O_6 + 7{\cdot}5\,Mg_2SiO_4 + 1{\cdot}5\,Fe_2SiO_4 + 12\,H_2O$$
$$\text{diopside} \quad + \quad \text{forsterite} \quad + \quad \text{fayalite} \quad + \quad \text{water}$$

$$\rightarrow 6\,Mg_3Si_2O_5(OH_4) + Ca_3Fe_3Si_3O_{12} + Fe$$
$$\text{serpentine} \quad + \quad \text{andradite} \quad + \text{iron}.$$

Nickel leached from olivines or sulphides and introduced by hydrothermal solutions results in the formation of taenite and awaruite instead of native iron (Botto and Morrison 1976).

Bird and Weathers (1975) argued that josephinite either originated from the "primitive mantle, possibly from the core-mantle boundary", or (Bird and Weathers 1979) that josephinite was "primitive material", incorporated in the mantle at the accretion of the earth, that has remained in a reduced state in the upper mantle ever since. They stated that the presence of Widmanstaetten figures (cf. Bird, Bassett and Weathers 1979) places the temperature of formation above 575°C, i.e. outside of the stability field for serpentinization. Thus the question at hand was simply whether josephinite itself, the native Fe-Ni metal, had been carried from the mantle to the crust *in reduced form* as $FeNi_3$, or whether the native metal had been formed in the crust as a product in the serpentinization.

Rare-gas isotopic components became part of the josephinite story with the publication, by Downing, Hennecke and Manuel (1977) and Bochsler, Stettler, Bird and Weathers (1978), of measurements of rare-gas isotopes, especially isotopes of He, Ne and Ar, in josephinite. The most extreme values were measured by Bochsler *et al* (1978): in some temperature fractions of the extracted gases, the $^3He/^4He$ ratio was found to be higher than the solar value (3×10^{-4}) and more than 10 times higher than the ratio in any mantle-derived sample. These high $^3He/^4He$ ratios were accompanied by similarly large and surprising ^{21}Ne enrichments. They thus proposed that these very large ^3He and ^{21}Ne enrichments showed that josephinite was not the product of serpentinization, but had indeed been derived in metallic form from the mantle. They also suggested that the large excess of ^{21}Ne might mark a "supernova-layer" somewhere in the mantle. Following this trail, Bird and Weathers (1979) suggested that josephinite could have accumulated with the solar nebula condensates forming the earth, and that the ^{21}Ne excess might have been produced in "original josephinite" in the solar nebula itself. In the same vein, Downing et al (1977) suggested that josephinite might actually be an iron meteorite, as old as the earth itself and developed a brief chronology of the early history of the earth and atmosphere based on the analysis of josephinite.

At the time of that work, high ^{3}He/^{4}He ratios were discussed solely in terms of mantle derivation. Recent work (Craig and Poreda 1986; Kurz 1986) has shown, however, that ^{3}He/^{4}He ratios well in excess of even the solar ratio can be produced *in situ* by exposure to cosmic radiation, which also produces excess ^{21}Ne (Marti and Craig 1987). These processes must also be considered in the context of the josephinite isotopic anomalies.

Other examples of native iron have been proposed as having a deep-mantle origin. The native iron from Disko Island, Greenland, and surrounding locales occurs as boulders up to 22 tons. These boulders appear to have weathered from late Mesozoic to early Tertiary basalts. Native iron is also found as disseminated grains within a basalt matrix at the nearby site of Kaersut. The Disko Island iron contains variable amounts of nickel (1–6%) and carbon (1–3%) (Smith 1879). The basalts which enclose the Greenland irons are linked to a volcanic "hot spot" track associated with the opening of the Labrador Sea. Early workers proposed that the Disko iron resulted from reduction of basalt magma by carbonaceous sediments (Smith 1879; Melson and Switzer 1966; Pederson 1969). Others have proposed that the basalt erupted through fissures containing large blocks of pyrrhotite that also exist as xenoliths in the basalt (Lofquist and Benedicks 1941). The Fe, Ni sulphides were altered to native iron by inclusions of carbonaceous sediments within the basalt.

Bird and Weathers (1977) rejected the basalt reduction mechanism because of the abundance of cohenite, a high-temperature iron carbide indicative of a long slow cooling history. They suggested that the iron was carried from the sub-lithospheric mantle as xenoliths within erupting basalts. Recently Bird, Goodrich and Weathers (1981) and Goodrich and Bird (1985) proposed that both high- and low-carbon metals crystallized at low pressure from iron-carbon liquids with approximately the bulk carbon contents of the present rocks and that Fe and Ni were reduced by carbon from assimilated sediments.

We address two aspects of the josephinite—rare gas story in this paper. First, the actual experimental data. We have published a brief discussion (Craig et al 1979) of our preliminary work which showed, on a single josephinite sample, 100 times less ^{3}He and a 66 times lower ^{3}He/^{4}He ratio, than the results of Bochsler et al (1978) as well as much lower excess ^{21}Ne. Similarly, our ^{3}He/^{4}He ratios were significantly smaller than those of Downing et al (1977) which were, in turn, about 10 × lower than those of Bochsler et al (1978). We had also shown that the use of RF induction heating to extract gases from these samples resulted in greatly enriched ^{3}He and ^{21}Ne contents suggesting that the high values so determined were procedural artifacts. The second question is what the rare gas isotopes can actually tell us about the origin of the host mineral josephinite: we address this problem in the light of presently known He and Ne components in rocks.

This study presents the results of rare gas, H_2 and D/H analyses of the Disko Island and Kaersut native irons of Western Greenland and for three individual cobbles of josephinite. We also discuss the cosmogenic contributions to ^{3}He and ^{21}Ne in these rocks and the necessity in future work of obtaining documented shielded samples which can have no cosmogenic components.

2. Experimental techniques

2.1 Sample description and preparation

The Greenland iron samples were part of an original set from the US National Museum, obtained long ago for measurements of the abundance and isotopic composition of the carbon in the irons (Craig 1953). The Disko Island iron, National Museum number R7696, was a 300 g block, 1·5 × 3·0 × 6·5 cm, cut from a larger piece of the iron. The sample is from Ofivak on the west coast of Disko Island, western Greenland. Chemical analysis showed that in addition to iron, this specimen contained 3·5 wt% nickel and 1·93 wt% carbon with a $\delta(^{13}C)$ value of $-24·3‰$, relative to the PDB carbonate standard (Craig 1953). Two surfaces were relatively free of basalt impurities (> 98% iron) and splits for rare-gas analyses were milled from these surfaces with a standard tungsten-carbide milling bit. At all times the temperature of the evolved chips was less than 100°C and microscopic inspection of the milling bit showed no evidence of chipping. After the outer 250 μm thick rusted surface was removed and discarded, subsequent 250 μm thick cuts were collected for analysis. Magnetic separation removed any basalt impurities. An attempt was made to sample from the side and top of this single block to determine if inhomogeneities existed; however the results did not indicate any systematic differences between fractions. Individual splits from this single block were designated DI-1A, DI-1B, DI-2B, DI-3B and DI-4B.

The Kaersut iron (KA-1), from the neighbouring peninsula of Nugsuaks, is part of the same set previously studied for carbon. In this sample (National Museum number 53479), iron occurs as discrete globules, several mm in diameter, in a basalt matrix. The metal is 3·4 wt% nickel. It was found to contain 0·13% C, with $\delta(^{13}C) = -20·9‰$; the enclosing basalt, also analysed, contained carbon with $\delta(^{13}C) = -21·8‰$ (Craig 1953). These very light carbon values compared to most mantle carbon ratios ($\delta(^{13}C) \sim -5‰$) are strongly indicative of reduction by carbonaceous sediments as the mechanism of iron formation. After crushing the basalt, individual iron grains were picked under a microscope, rinsed with 10% HNO_3, acetone and water and then dried at 60°C and stored in methanol. This process removed most of the visible surface alteration, most notably rust. While the Disko Island samples were free of any basalt impurities, the Kaersut iron contained small basalt inclusions (< 5%).

Josephinite rare-gas was studied on three specimens (J-77-1, J-3·1-D and J-1·1-C) kindly provided by Dr J Bird to K.M. All three samples, which weighted ~ 4 g each, were cut from larger josephinite pebbles. Each contained a rim of surface alteration with no part of any sample greater than 1 cm from this surface. After the surface rim was removed, the interior portions of J-77-1 and J-3·1-D were relatively free of alteration. J-77-1 was cut into three fractions: J-77-1A, J-77-1B, J-77-1C. J-77-1B and C and J-3·1-D were cut on a diamond saw to fit into 2·4 mm I.D. tantalum crucibles. For J-1·1-C the alteration was so extensive that the sample was easily broken into small pieces and the fractions (J-1·1-C-1 and J-1·1-C-2) were selected based on the degree of alteration, with J-1·1-C-2 containing abundant garnierite and other alteration products.

2.2 Gas extraction procedures

The extraction procedures were similar to those of Lupton and Craig (1975), Craig and Lupton (1976) and Poreda (1983) with minor modifications. All samples were fused in tantalum tubes in a Bieri *et al* (1966) type high-vacuum furnace in which the Ta tubes are the resistive heating elements: these tubes were outgassed at 1500°C to reduce the H_2 blank, after which the furnace was opened and ~ 600 mg of sample was loaded into each tube and degassed in vacuum for 24 hours at $\sim 100°C$. During the fusion (40 minutes at 1600°C) the gases were exposed to CuO at 600°C, to convert all H_2 to water: after collection of the non-condensible gases in two splits, the water, which had been frozen in U-traps on either side of the CuO, was collected in a break-seal tube for later measurement of the D/H ratio.

During the extractions a tantalum-iron-nickel alloy formed and melted through the wall of each crucible. This resulted in a loss of current to the crucible before the heating was complete. In one case inspection of the tantalum tube revealed that a 0·25 g piece of sample J-3·1-D did not melt during the initial heating. Re-extraction produced a significant quantity of helium, $\sim 11\%$ of the previous yield. All other samples were checked visually and appeared to have melted completely. The ranges of values for 4 single-tude extraction blanks were 1 to 5×10^{-9} cc of He, 1 to 4×10^{-10} cc of ^{21}Ne and 0·06 to 0·12 cc of H_2O.

For samples J-77-1A and DI-1A we used induction heating in a Pyrex line with a quartz extraction vessel. These samples were run in a molybdenum crucible which had been outgassed at 1600°C. For DI-1A the large amount of evolved gas produced a glow discharge in the extraction vessel. The RF power had to be turned off frequently to allow the Ti getter to purify the gas, increasing the extraction time by 30 minutes relative to the blank. A 125 ml flask was attached to the line and sealed off after the extraction was complete to provide an approximate 20% volumetric split of the evolved gas for ^{3}He/^{4}He analysis. The remaining gas was introduced into the mass spectrometer inlet line for further purification. Two temperature fractions of J-77-1A were obtained for neon and argon isotopes.

2.3 Isotopic analyses

Helium concentration and isotopic measurements were performed on a 25 cm radius, double-collecting mass spectrometer (GAD) using the procedures described by Lupton and Craig (1975) and Craig and Lupton (1976). Air helium served as the absolute standard. Instrumental precision estimates are based on the reproducibility of the air standards. Helium concentrations were determined by peak-height comparison to an air standard of known size.

The large errors associated with the ^{3}He/^{4}He ratios of the Disko Irons result from the low He concentrations and the low ^{3}He/^{4}He ratios. The typical ^{3}He signal represents ~ 20 ions per second with an error of $\sim 15\%$. The blank contributions were 20 to 40% of the total yield. These two factors resulted in the large uncertainties listed in table 1. The ^{3}He/^{4}He ratios in josephinite are subject to much lower uncertainties because of the higher He concentrations and ^{3}He/^{4}He ratios.

Neon and argon isotopes were analysed on a conventional Nier type, 15 cm radius, rare-gas spectrometer. During the neon analyses a second charcoal finger at $-195°C$ was included in the static volume to reduce the residual Ar, CO and H_2O peaks. The neon isotopes were measured on the electron multiplier for a period of 30 minutes. Peaks at masses 18, 40, 42 and 44 were scanned during the run to correct for the interferences of $H_2^{18}O^+ (\sim 1\%)$, $^{40}Ar^{++}$, $C_3H_6^{++}$ and $^{44}CO_2^{++}$. A correction of 1·0%

per amu was applied for mass discrimination. In two cases (a blank and J-3·1-D), the $C_3H_6^{++}$ correction was nearly 40% of the m/e = 21 peak. The corresponding $^{21}Ne/^{22}Ne$ ratio for J-3·1-D is much lower than the atmospheric value, indicating an overcorrection for 42^{++} ions. The hydrocarbon levels for josephinite samples were a factor of ten higher than the values normally observed. The neon results in table 2 are measured data, uncorrected for procedural blanks. Errors associated with the samples reflect a 30% uncertainty in the correction for doubly charged ions plus any statistical errors. Absolute amounts of ^{21}Ne were obtained by peak height comparison to a calibrated standard. This value is accurate to 10%.

Argon results are corrected for background (mainly HCl at m/e = 36) and mass discrimination (-0.5% per mass unit). The direct contribution of Hg^{5+} to m/e = 40 was negligible. Because of memory in the argon analysis, the isotope ratios were extrapolated to the time when the sample was admitted to the mass spectrometer. In two cases (J-3·1-D and J-1·1-C-1) the accuracy of the $^{40}Ar/^{36}Ar$ ratio is 10% due to the interferences at mass 36. Amounts of ^{36}Ar (accurate to 10%) were obtained by peak height comparison with a calibrated standard.

In a separate preparation line, the H_2O fraction was reduced to H_2 by reaction with metallic uranium at 790°C, measured in a calibrated manometer and analysed for D/H ratio in a 15 cm, split-tube double-collecting mass spectrometer (DELILAH: Craig and Gordon 1965). All but two of the samples were measured at a mass 2 ion beam intensity of 20 volts with an instrumental precision of 0·5‰. Corrections to the measured D/H ratio are principally due to the contribution of H^{3+} to the beam at mass 3. This H^{3+} value is monitored and the correction ranges from 1·019 to 1·033 of the D/H ratio. No valve mixing correction is necessary and the correction for the residual beam is 1·001 to 1·002. Isotope values for hydrogen are reported relative to SMOW (standard mean ocean water). Delta values represent per mil differences from this standard, defined by:

$$\delta D = (R/R_{SMOW} - 1) \times 1000,$$

where R is the D/H ratio.

For the sample measured at 13 volts (DI-2B), the intensity of the beam for the machine standard gas was reduced to match the sample voltage. To calculate the D/H ratio, the H^{3+} correction at 13 V was used. For the sample measured at 10 V (DI-3B), the machine standard could be reduced to only 11·3 V. To correct for the effect of this voltage difference between the sample and standard gas, a sample of SMOW was analysed at 10·0 V and compared with the machine standard at 11·3 V. A change in the apparent isotope ratio of -2.5‰ occurred for the SMOW run at 10 V, relative to the value at 20 V. The D/H ratio of the sample run at 10 V was adjusted by -2.5‰. The errors associated with this technique are reflected in the errors in the reported D/H ratios. A blank contribution of 0.16 cc H_2 per Ta tube with $\delta D = -122.4$‰ was measured and used in the calculations.

3. Experimental artifacts

Table 1 shows the comparison of the results from the Disko Island sample (DI-1A), heated in a Pyrex system with an induction furnace, with the same sample (DI-1B) extracted in the Bieri-type furnace. Sample DI-1A has 230 times (!) more ^{3}He per gram than DI-1B (2.8×10^{-12} cc/g vs. 0.012×10^{-12} cc/g), but 10 times *less* ^{4}He, with the ^{3}He/^{4}He ratios differing by a factor of 2000. The ^{21}Ne/^{22}Ne ratio clearly differs from the atmospheric value in DI-1A, with *excess* ^{21}Ne = $1.6 \pm 0.2 \times 10^{-12}$ cc/g and a ^{21}Ne (excess)/^{4}He ratio of 4×10^{-4}. DI-1B has an atmospheric neon isotopic ratio within the error bars. In DI-1A, the ^{3}He/^{4}He ratio is greater than 300 R_A with a ratio of ^{3}He to excess ^{21}Ne of 1.9. This rare gas composition observed in DI-1A is unlike any other terrestrial sample except the josephinite gases analysed by Bochsler *et al* (1978) but is comparable to the spallation gas composition of some extra-terrestrial samples.

The inductive-heating extraction system used for DI-1A had previously been used with meteorite and irradiated samples. Although the extraction blank for the induction furnace had atmospheric rare gas ratios, the blank procedure did not duplicate what occurred in the sample run. The large amount of gas, principally H_2 (table 1) and N_2 and CO, evolved from the metal sample, caused a glow discharge to occur in the extraction vessel. This glow discharge is responsible for the contamination of DI-1A with large excesses of ^{3}He and ^{21}Ne, scavenged from the system walls where it had been collected from previous runs of meteoritic material. (Glow discharges produced by Tesla coils are standard methods for cleaning glass tubing walls in vacuum systems. Additionally, atomic hydrogen is probably produced in the discharge and is well-known to be an outstanding "cleaner" for adsorbed gases in glass systems.) Later attempts to reproduce this effect in the same system were not successful. It is clear that analysis of terrestrial materials with an apparatus used extensively for meteorites can easily result in anomalous rare gas ratios, although details are not well understood.

This tremendous scavenging effect for helium in a glass system is strongly reminiscent of the so-called "cold fusion" experiments of Paneth and Peters in 1926 (Paneth 1927a, b) in which it was found that the liberation of helium from glass (and from Pd-asbestos), first thought to be a nuclear fusion effect, was found to depend on the presence of hydrogen. Just as in our experiments, glass tubes which gave off no He when heated in vacuum or in oxygen, were found to liberate *helium* whenever H_2 was present in the system. Paneth also showed that glass tubes, completely cleansed of He by heating in H_2, would, when exposed to air for a day, take up detectable amounts of *Ne-free* helium from the atmosphere. Thus, although we do not completely understand this effect, it seems clear that induction heating cannot be relied on to produce correct He isotope results when any H_2 is liberated during the extraction, especially when a glow discharge is formed. As shown in table 1, all these iron samples contain very large amounts of H_2, up to 3 cc(STP)/g, so that He isotope ratios measured on gases extracted by RF heating are not acceptable unless verified by resistance heating in a separate system.

4. Rare gas results

4.1 *Disko Island and Kaersut*

Table 1 shows that the native irons from Greenland (Disko and Kaersut) are relatively consistent in their ^{4}He concentrations (31 to 69×10^{-9} cc/g) and ^{3}He/^{4}He ratios of 0.15 to 0.39 R_A. Variability in these measurements could have resulted from the low

Table 1. Rare gas and H_2 measurements on native iron samples from Disko Island and Kaersut, Greenland. Gases from DI-1A (*) were extracted using RF heating in a Pyrex glass system: all others were heated with a Bieri-type furnace. δ^{21}Ne values are reported as percentage deviation from the ^{21}Ne/^{22}Ne ratio in air. Neon concentrations are not corrected for the procedural blank. Blank contributions are shown beneath the measured value as a percentage of total neon.

	Disko Island					Kaersut	
	DI-1B	DI-2B	DI-3B	DI-4B	DI-1A*	KA-1A	KA-1B
R/R_A	0.23 ±0.15	0.23 ±0.15	0.39 ±0.10	—	(500) ±300	—	0.15 ±0.08
^{4}He (10^{-6} cc/g)	0.037	0.031	0.064	—	(0.004)	—	0.069
^{20}Ne (10^{-9} cc/g)	2.3 85%	—	—	—	(0.91) 72%	—	0.94 95%
δ^{21}Ne (%)	6.2 ±10.1	—	—	—	(58) ±2.3	—	7.6 ±11.0
^{21}Ne (exc) (10^{-12} cc/g)	<0.4	—	—	—	(1.6) ±0.2	—	<0.2
H_2 (cc/g)	0.62 ±0.10	0.96 ±0.10	1.21 ±0.10	2.78 ±0.10	—	3.01 ±0.10	1.35 ±0.10
δD (‰) (SMOW)	−163 ±5	−154 ±5	−117 ±5	−178 ±5	—	−168 ±5	−160 ±5

^{4}He concentration, the presence of variable amounts of atmospheric helium, and the reliability of the blank correction, rather than from actual sample inhomogeneities. The ^{3}He/^{4}He ratios, 0.15 to 0.39 R_A, are less than the atmospheric ratio for both the Disko and Kaersut irons and are clearly not indicative of mantle ^{3}He/^{4}He ratios. If the Greenland irons had formed in the mantle they would reflect the ^{3}He enrichments relative to atmospheric that are typical of mantle-derived samples: i.e. $R/R_A = 8$ for Mid-Ocean Ridge Basalts, 6–8 in most volcanic arcs, and 5 to 30 in hotspot volcanic samples (Lupton 1983; Craig et al 1985; Poreda and Craig 1989).

The helium in these Greenland samples is primarily "crustal" or radiogenic in origin, and may have several different sources such as incorporation of crustal He during extrusion of the basalt, or *in situ* production of ^{4}He by decay of U and Th. The crust through which the basalt erupted is rich in radiogenic He, and as the iron formed in the near-surface environment it may have incorporated this crustal He. Radiogenic production of ^{4}He by *in situ* decay of U and Th in the ~ 50 My since the eruption of the basalt may also be significant. Less than 10 ppb of U within the iron is required to produce 3 to 7×10^{-8} cc/g of ^{4}He in 50 My.

Another effect which can be important for iron samples is inward diffusion of *tritium* from the atmosphere, which then decays to produce ^{3}He. It was long ago observed by Fireman (1958) that tritium could be found in recent nails (up to 0.02 dpm/g) and iron door-hinges with one-third that activity. In that case it was pointed out (Craig 1958) that metallurgists had established that iron samples in an atmosphere of D_2O quickly come to contain D_2, due to a catalytic production of hydrogen from water on iron surfaces, followed by subsequent diffusion of the H_2 (or D_2) into the iron object (Norton 1953). Thus these irons in Greenland have been exposed to the very high atmospheric tritium burdens due to nuclear weapons testing, unless it can be established that samples have been shielded by rock or stored in closed systems (in the present case the samples were simply in a laboratory drawer).

The isotopic composition of neon in the Disko Island and Kaersut irons is atmospheric within error limits (table 1). The low neon concentrations ($\sim 10^{-9}$ cc/g of ^{20}Ne) and large corrections made for contaminating ions caused large errors to be assigned to the neon values. An excess of $< 10^{-12}$ cc/g of ^{21}Ne was in the native irons from Greenland extracted in the Bieri-type furnace. Because of the low yields and high blank values, argon was not analysed in these samples. The amount of nucleogenic ^{21}Ne produced by *in situ* decay of U and Th can be estimated from the He concentration. If the source of the helium is the basalt impurities, then $< 10^{-14}$ cc/g of ^{21}Ne would be produced, based on a ^{4}He/^{21}Ne production ratio of 10^7 for silicates (Craig and Lupton 1976; Rison 1980). Lower ^{21}Ne production is expected for the iron phase because it contains almost no Mg and O, the principal target elements for nucleogenic production of ^{21}Ne.

4.2 Josephinite

Table 2 lists the rare gas results for josephinite. Our data do not confirm the high ^{3}He/^{4}He ratios ($R/R_A > 300$) observed by Bochsler *et al* (1978) or the somewhat lower values ($R/R_A = 17$ and 28) found by Downing *et al* (1977). The highest ^{3}He/^{4}He ratio for the three samples used in our study was $5\cdot6 R_A$, measured in the least altered sample (J-77-1B and J-77-1C) analyses. This value is only slightly lower than MORB ($R/R_A = 8 \pm 1$). The ^{3}He/^{4}He ratios are inversely correlated with the degree of alteration. The two most altered samples (J-1·1C-1, J-1·1C-2) have ^{3}He/^{4}He ratios lower than the atmospheric value ($R/R_A = 0\cdot84$ and $0\cdot33$). The alteration has removed some of the original He present in the metal phase and radiogenic production of ^{4}He in the serpentine and garnierite has lowered the ^{3}He/^{4}He ratio to its present value. Sample J-3·1-D is relatively free from alteration and its ^{3}He/^{4}He ratio of $2\cdot7 R_A$ is indicative of either a mantle or a cosmogenic component (or both) in the helium.

These josephinite samples typically contain ~ 10 times more helium than the Greenland irons, perhaps reflecting a higher He pressure in the environment of formation. No values exist for the concentrations of U and Th in the metal phase of josephinite, so that reduction of the ^{3}He/^{4}He ratio over the past 150 My by radiogenic production of ^{4}He cannot be ruled out. However, 5·4 ppb U with Th/U = 4 would be required to reduce a hypothetical original value of $10 R_A$ to the present ratio of $5\cdot6 R_A$ in 150 My. A uranium concentration of 5 ppb is certainly possible: Botto and Morrison measured a K concentration of 13 ppm and a K/U ratio of 3000 is a reasonable value for mantle-derived rocks.

Neon isotopic analyses of josephinite in our study do not show the tremendous enrichments in ^{21}Ne and ^{22}Ne measured by Bochsler *et al* (up to 1600% for ^{21}Ne). Our results show that josephinite has neon of approximately atmospheric composition. Although most of the samples had neon concentrations at the level of the blank, there was no discernable enrichment of ^{21}Ne that was greater than 10^{-12} cc/g. The ^{20}Ne/^{22}Ne ratios are atmospheric within very large error bars. The one sample which has a neon concentration significantly above background level was the 800–1600°C heating step for J-77-1A. The ^{21}Ne/^{22}Ne ratio in this sample is slightly enriched in ^{21}Ne relative to the atmosphere, with an excess ^{21}Ne of $0\cdot5 \pm 0\cdot3 \times 10^{-12}$ cc/g. It is not likely that the excess ^{21}Ne was produced by nucleogenic effects associated with *in situ* radioactive decay of U and Th. The principal nucleogenic

Table 2. Rare gas and H_2 measurements on josephinite samples. Gases from J-77-1A* were extracted in two temperature steps using RF heating in a Pyrex glass system: all other extractions were single-step, total-fusion runs using a Bieri-type furnace. Blank values are reported beneath the tabulated concentration as a percentage of total gas. Neon isotopes are reported as δ^{20}Ne and δ^{21}Ne values, the percentage deviations from the ^{20}Ne/^{22}Ne ratios in air.

Sample	J-77-1A* 800–1600°	J-77-1B	J-77-1C	J-3·1D	J-1·1-C-1	J-1·1-C-2
R/R_A	—	5·6 ±0·1	5·3 ±0·1	2·7 ±0·1	0·84 ±0·4	0·33 ±0·3
^{4}He (10^{-6} cc/g)	—	0·33	0·43	0·32	0·11	0·38
^{20}Ne (10^{-9} cc/g)	1·6 44%	1·5 60%	2·9 100%	1·8 60%	1·4 100%	—
^{36}Ar (10^{-9} cc/g)	9·2 1·6%	1·6 100%	4·1 72%	0·3 100%	1·0 100%	3·3 100%
δ^{20}Ne (%)	0 ±20	—	—	—	—	—
δ^{21}Ne (%)	11 ±6	—	—	—	—	—
^{21}Ne (exc) (10^{-12} cc/g)	0·5 ±0·3	<0·5	—	<0·5	—	—
^{40}Ar/^{36}Ar	338 ±3	305 ±31	321 ±32	294 ±29	287 ±29	321 ±32
H_2 (cc/g)	—	6·8	7·2	8·6	12·1	28
δD (‰) (Smow)		10·4	10·1	53·2	81·4	64·9

reactions for ^{21}Ne production (Wetherill 1954) are ^{18}O$(\alpha, n)^{21}$Ne and ^{24}Mg$(n, \alpha)^{21}$Ne and Mg and O are not abundant in josephinite. The excess ^{21}Ne is most probably the result of nucleogenic reactions within silicates perhaps introduced into the josephinite during formation in the upper mantle or crust. This small amount of excess ^{21}Ne is much lower than the results of Bochsler et al (1978) and it does not provide any support for a deep-mantle origin of josephinite.

For the argon in josephinite only one sample differed significantly from the atmospheric ratio. The 800–1600°C fraction of J-77-1A has a ^{40}Ar/^{36}Ar ratio of 338 ± 3 and an atmospheric ^{38}Ar/^{36}Ar ratio of 0·188 ± 0·003. This sample also had the highest ^{36}Ar concentration of $9·2 \times 10^{-9}$ cc/g. All other samples had atmospheric ^{40}Ar/^{36}Ar ratios within the error limits. The ^{40}Ar/^{36}Ar ratio of 338 is in the range of previously reported values for josephinite of 305 to 363 (Bernatowicz et al 1979; Bochsler et al 1978).

5. Cosmogenic production of ^{3}He and ^{21}Ne

Rocks at the earth's surface are subject to cosmic radiation and accumulate cosmic-ray-produced ("cosmogenic") ^{3}He (Craig and Poreda 1986; Kurz 1986) and ^{21}Ne (Marti and Craig 1987). The concentration of these isotopes, compared to the production rates, gives an "exposure age" or residence time at the surface of the earth if the erosion rate is small or can be neglected (Craig and Poreda 1986). In order to estimate the cosmogenic contribution in these native-iron samples one needs the production rates in Fe and Ni: these rates were kindly calculated for us by Professor Lal (cf. Lal 1988). The production rates at geomagnetic latitude (λ) > 50° and an altitude of 3·35 km (680 g/cm² atmospheric depth) are:

^{3}He(Fe or Ni): 236 atoms/g/y (8·8 $\mu\mu$cc/g/My),
^{21}Ne(in Fe): 0·61 atoms/g/y (0·023 $\mu\mu$cc/g/My),
^{21}Ne(in Ni): 0·47 atoms/g/y (0·018 $\mu\mu$cc/g/My).

Given these values estimated for a specific altitude and geomagnetic latitude, we scale the production rates to other latitudes and altitudes using the nomograph given for this purpose by Lal (1987).

The production rates for the Disko Island and Kaersut irons ($\lambda = 70°$, altitude = sealevel, pure Fe) are: 20·2 atoms/g/y (0·75 $\mu\mu$cc/g/My) for ^{3}He. The corresponding rates for ^{21}Ne are lower by a factor of 380 in pure Fe. At an elevation of 1000 m both rates will be higher by a factor of 2·1. The maximum ^{3}He concentration, calculated from the data in table 1, is 0·035 $\mu\mu$cc/g, corresponding to a cosmogenic exposure age of 47,000 years at sea level, or only 21,000 years at an elevation of 1 km. These ages assume no shielding of the samples. The Disko Island sample comes from the interior of a larger piece of iron occurring in the basalt (Craig 1953) and nothing is known about the original shielding, so that the calculated age is a minimum value and the actual exposure ages could be much longer. The Kaersut sample consisted of small irregular globules of iron, several mm in diameter, disseminated through basalt (Craig, op. cit.) and may also have been shielded by rock or ice at various times. Thus all that can be said about a cosmogenic contribution to the ^{3}He in these samples is that it is entirely possible, given the low amounts of ^{3}He that are present.

The measured Ne isotopic compositions in these Greenland irons (table 1: samples DI-1B and KA-1B) are atmospheric within error limits ($< 10^{-12}$ cc/g of excess ^{21}Ne relative to the atmospheric composition). The expected amounts for a cosmogenic origin would be a factor of 380 times less than the cosmogenic ^{3}He component, so that the maximum amount of ^{21}Ne we could expect in our samples would be $\sim 10^{-16}$ cc(STP)/g, an amount which cannot be measured.

For josephinite ($\lambda = 50°$, elevation = 1000 m), the production rate for ^{3}He is found to be 45·4 atoms/g/y (= 1·7 $\mu\mu$cc/g/My), $\sim$ twice the rate in Greenland at sea level. The ^{3}He concentration (samples J-77-1B and 1C in table 2) is 3·0 $\mu\mu$cc(STP)/g, so that the ^{3}He cosmogenic exposure age (assuming all the ^{3}He is cosmogenic) is only 1·8 My if the samples have not been shielded. This "age" is very short in comparison with the sample age ($\sim$ 150 My), so that it is entirely possible for all the ^{3}He in these samples to be cosmogenic. The problem is a familiar one which has recently been found to be similarly important in diamonds, which have a strong cosmogenic ^{3}He component in alluvial samples (Lal et al 1989). That is, just as with diamonds, it is necessary to have documented, shielded samples of josephinite, from drill cores or obtained by blasting or sectioning boulders. A 50 kg boulder has been reported, as described by Bird and Weathers, (1979): if approximately spherical, such a boulder (radius = 12 cm) would show a decrease of 50% in cosmogenic components from surface to centre, so that it would be possible to establish the actual non-cosmogenic ^{3}He/^{4}He ratio in josephinite by radial measurements. Until this experiment is done, all that can be said about the ^{3}He/^{4}He ratios we have measured is that they are upper limits for *either* the cosmogenic ratio *or* the mantle-component ratio assuming no cosmic-ray component.

For ^{21}Ne the production rate is much lower: 0·10 atoms/g/y (= 0·0036 $\mu\mu$cc/g/y), i.e. lower than the ^{3}He rate by a factor of 460. Our value for excess ^{21}Ne (table 2) is very rough because the value is low: $\sim 0.5 \pm 0.3$ $\mu\mu$cc/g, but taken at face value, would require an exposure age of 140 My which is much too long, and probably indicates that either our ^{21}Ne value is still too high, or that this is a non-cosmogenic mantle Ne component, probably produced by nuclear reactions such as ^{18}O(α, n)^{21}Ne in the earth's crust and incorporated into josephinite at the time of formation during the serpentinization process. Alternatively the excess ^{21}Ne could be a mantle component.

The very high ^{3}He and excess ^{21}Ne concentrations measured by Bochsler *et al* (1978) cannot result from cosmogenic exposure. A He concentration of 340 $\mu\mu$cc/g requires a surface exposure of 200 My, while 21×10^9 years (!) are required to produce their value of 77 $\mu\mu$cc/g of excess ^{21}Ne. Both these times are far too long to explain the Bochsler *et al* josephinite data.

6. Josephinite—Interlaboratory comparison

Table 3 shows the results of rare gas measurements on josephinite specimens measured in three laboratories other than our two laboratories at UCSD. The data are those of Downing *et al* (1977, Rolla, Missouri), Bochsler *et al* (1978, Bern) and

Table 3. Rare gas results from previously published analyses of josephinite. The Bern values are from Bochsler *et al* (1978). St Louis data are from Bernatowicz *et al* (1979). Rolla data are from Downing *et al* (1977). Neon isotopes are reported as in table 2.

Sample	Bern					St. Louis	Rolla	
	J21·2-W4 370°	J21·2-W4 1490°	J21·2-W3 530°	J21·2-W3 1600°	J21·2-W2 Fusion	Fusion	J1 Fusion	J2 Fusion
R/R_A	395 ±32	5·9 ±2·9	396 ±37	4·8 ±3·2	140 —	—	28·6 ±8·6	17·2 ±13·2
^{4}He (10^{-6} cc/g)	0·2	0·022	0·0093	0·49	1·74	—	0·16	0·14
^{20}Ne (10^{-9} cc/g)	0·78	0·13	0·15	0·32	2·15	8·2	1·8	2·8
^{36}Ar (10^{-9} cc/g)	0·04	1·37	0·026	3·8	6·3	27·1	35·9	34·4
δ^{20}Ne (‰)	−4·2 ±0·5	5·0 ±3·0	−9·7 ±1·1	−0·5 ±1·0	−3·7 ±2·5	—	9·2 ±5·1	6·1 ±4·1
δ^{21}Ne (‰)	1638 ±20	−18 ±15	1542 ±77	5·8 ±6·4	1168 ±15	—	—	—
^{21}Ne (exc) (10^{-12} cc/g)	37 ±1	<0·1	7·6 ±0·9	1·0 ±0·2	77 ±10	<1·0	—	—
^{40}Ar/^{36}Ar	305 ±3	308 ±2	—	305 ±3	297 ±3	312 ±10	327 ±3	326 ±3

Bernatowicz *et al* (1979, St. Louis). Although different pieces of josephinite were analysed, the rare gas results from the four laboratories should be comparable if the individual specimens were produced in the same geologic environment. With the exception of our samples DI-1A and J-77-1A in tables 1 and 2, all the UCSD samples were extracted by resistance heating. The other three laboratories all used RF heating for their gas extractions.

Table 3 gives examples of the high-temperature ($>1500°C$) and low-temperature ($<800°C$) rare-gas fractions of Bochsler *et al* (1978) and the total fusion results from Rolla and Washington University. There appears to be a systematic difference between the results which seems to be temperature-related.

6.1 *Helium isotope ratios*

Bochsler *et al* (1978), Downing *et al* (1977) and our study report helium isotope results. On an unaltered specimen of josephinite, 77-1, we measured a total-fusion value of $R/R_A = 5·6$. Samples with a greater degree of alteration had lower $^3He/^4He$ ratios, presumably related to radiogenic production of ^4He. In the Bochsler *et al* (1978) step-heating experiments, at least five temperature steps less than 750°C produced $^3He/^4He$ ratios of greater than $100 R_A$. The values in table 3 show a particularly gas-rich fraction at 370°C which has a $^3He/^4He$ ratio of $\sim 400 R_A$. By comparison, the $^3He/^4He$ ratios in all high-temperature fractions ($>1400°C$) of 4·8 to 5·9 R_A are in remarkable agreement with our data (5·6 R_A). These high-temperature steps presumably represent the gas released upon melting of the sample: they contain a sizeable fraction of the total helium released (between 10 and 40%). Thus the discrepancy in $^3He/^4He$ ratios exists only for the low-temperature data of Bochsler *et al* (1978).

The He ratios of Downing *et al* (1977) are less than the Bern ratios, but also much higher than our values, with $R/R_A = 17$ to 28, with large errors up to ± 13. The gases were also extracted in temperature steps, at 900°, 1300°, 1500° and 1700°C, and similar decreases of ratio with increasing temperatures were observed. Maximum $^3He/^4He$ ratios ($R/R_A = 76$ and 54 in two separate extractions) were found in the 900° steps, with values decreasing to $R/R_A = 19$ and 9 in the highest temperature steps. All of these ratios are much higher than our measured values.

6.2 *Neon and Argon isotope ratios*

For the neon isotopic data there is basic agreement between UCSD and Washington University. Each found atmospheric $^{20}Ne/^{22}Ne$ ratios within the error limits. For all samples the excess ^{21}Ne above the atmospheric ratio is less than 2×10^{-12} cc/g. In the high-temperature fractions of Bochsler *et al* (1978), the amount of excess ^{21}Ne is also less than 2×10^{-12} cc/g. However, *in the same low-temperature fractions* which produced the $^3He/^4He$ ratios of $R/R_A > 800$, the amount of excess ^{21}Ne ranged from $12-77 \times 10^{-12}$ cc/g, with $^{21}Ne/^{20}Ne$ ratios as high as $8 \times$ air in some fractions. In addition, excess ^{22}Ne was observed in these samples, also with maximum values in the lower T extraction steps. Thus the low-temperature data of Bochsler *et al* are in direct conflict with our data, the Washington University data, and the high-temperature data of Bochsler *et al* (1978) themselves.

Neon in the data of Downing *et al* (1977) showed temperature correlations similar to their helium ratios discussed above. The ^{20}Ne/^{22}Ne ratios decreased with increasing temperature steps in exactly the same way, from 18 and 27 in the initial low T steps (vs. air = 9·8), to ratios of ~ 10.5 in the highest temperature steps.

The argon isotopic results are in approximate agreement for the four laboratories. All report atmospheric ^{38}Ar/^{36}Ar ratios. The ^{40}Ar/^{36}Ar ratios range from atmospheric (296) to 338 for total fusion analyses of josephinite with minor differences among the laboratories. The present study has the highest ^{40}Ar/^{36}Ar ratio at 338, whereas Bernatowicz et al (1979) report a value of 312 and Bochsler et al (1978) report a range from atmospheric to 314. These minor differences in the ^{40}Ar/^{36}Ar ratios may reflect slight inhomogeneities in K content among the specimens (Bernatowicz et al 1979).

6.3 Helium and neon effects

Bernatowicz et al (1979) have given a thoughtful discussion of the Ne, Ar, Kr and Xe isotopic data with which we generally agree: they concluded that the isotopic anomalies in these gases reported by Bochsler et al (1978) and Downing et al (1977) were not real effects and that only the ^{3}He and possibly the ^{21}Ne, reported anomalies were evidence of an exotic origin for josephinite. These are, in fact, the principal differences in the data from the four laboratories, and they are present only in the low-temperature fractions of Bochsler et al (1978) and Downing et al (1977). Bernatowicz et al (1979) did not observe any low-temperature neon anomalies in their step-heating experiments and we did not observe anomalous neon or helium in the two-step heating experiment (J-77-1A). All of the high-temperature data except those of Downing et al (1977) are in reasonable agreement. If the low-T ^{3}He and ^{21}Ne anomalies are derived from the mantle, it seems remarkable that this unique rare gas signature remains distinct from the surrounding material at mantle temperatures, yet releases its rare gas component at temperatures as low as 370°C.

Our extractions produced significant anomalies in ^{3}He and ^{21}Ne in a sample from Disko Island. Subsequent analyses of the same sample fused in another extraction system showed no such anomalies. The anomalies in DI-1A were simply the result of scavenging by the large quantities of gas, principally H_2 released during the sample fusion which used RF heating in a line used routinely for meteoritic samples. The same conditions (high gas pressure, RF heating in Pyrex and previous high ^{3}He and ^{21}Ne samples) existed for the josephinite analysed by Bochsler et al (1978). Bochsler and Stettler (personal communication to K.M.) stated that a new extraction vessel and molybdenum crucible were used. In their 1978 paper (p. 70) they state that neutron-irradiated silicates (with ^{3}He and ^{21}Ne concentrations of $\sim 10^{-7}$ cc/g: Stettler and Bochsler 1979) were analysed prior to the josephinite runs. In addition, their anomalous neon signature did not resemble the spallation, solar wind, or meteoritic composition, but rather that produced by fast neutrons on magnesium (see figure 1 of Bochsler et al 1978). This is consistent with contamination from previous neutron-irradiated samples. Thus the probability that these anomalies resulted from an experimental artifact origin appears very likely. The anomalous rare gas composition in the low-temperature fraction of the josephinite analyses was simply triggered by the release of large quantities of H_2 during these same steps (cf. Bochsler et al 1978, p. 72) with subsequent scavenging by atomic H produced by the RF system.

A similar effect is seen in the very similar correlations in the data of Downing et al

(1977) and must reflect the same procedural artifact. In this case however, it is the $^{20}Ne/^{22}Ne$ ratio which is enriched with the $^3He/^4He$ ratios in the low-temperature stages (^{21}Ne was not reported) because neutron-irradiated material had not been used in the Rolla system as it was in Bern. No such scavenging effects are observed in the data of Bernatowicz et al (1979), probably because these workers kept a Cu-CuO reactor at 470° on line during the extractions and carried out the low-T extraction stages very slowly, slowly enough that high hydrogen pressures were probably prevented.

7. Hydrogen results

7.1 H_2 in Disko Island and Kaersut irons

Hydrogen concentrations in the Disko Island and Kaersut irons range from 1·02 to 3·26 cc/g (table 1). Experimental solubilities of hydrogen in iron and Fe-Ni alloys indicate a one-atmosphere solubility of between 0·06 and 0·12 cc/g at 1000°C (Dushman 1962). The high hydrogen concentrations reflect formation of these irons under reducing conditions and a significant hydrogen partial pressure. The reason for the variability in concentration is unclear but may reflect diffusion of hydrogen out of the iron. The δD values also exhibit a range from −117‰ to −178‰. All but one of the samples is in the range of −154 to −178‰. The sample with the highest D/H ratio, $\delta D = -118$‰, was the first sample analysed and was closest to the surface of the sample block. The heavier D/H ratio relative to the other iron samples may reflect either diffusion of hydrogen out of the iron or alteration of the iron and formation of iron hydroxides ("rust").

The other samples have much lower δD values of −154 to −178‰, which do not represent the typical mantle values of −50 to −80‰ (Shepard and Epstein 1970; Craig and Lupton 1976). The low D/H ratios in these native irons are similar to the local meteoric waters in Greenland, with δD values of −140 to −180‰ (Dansgaard 1964). Thus the D/H ratios are not indicative of origin in the mantle but are consistent with derivation of hydrogen from local meteoric water and origin in a crustal environment. It is of course possible that the D/H ratios may represent fractionated mantle values or post-eruptive alteration by meteoric water and these possibilities cannot be excluded. However it is most likely that the hydrogen in these Greenland irons was derived from local meteoric water which penetrated the basalts after eruption.

7.2 H_2 in josephinite

Unaltered josephinite (samples J-77-1 and J-3·1-D) have much higher hydrogen concentrations than the Greenland irons: 6·8 to 8·7 cc(STP)/g (table 2). These higher values reflect higher hydrogen partial pressures or higher solubility in josephinite (∼0·7 Ni, 0·3 Fe) relative to the Greenland irons (< 5% Ni). The altered josephinite samples, J-1·1-C-1 and J-1·1-C-2 have even higher H_2 concentrations, 12·3 and 28·3 cc/g respectively, due to the presence of hydrous alteration phases (recall that our extraction procedure extracts *both* H_2 in the metal and any H_2O in included alteration minerals, producing a combined H_2O sample for the total sample). If the measured hydrogen were entirely present in the samples as water, 28·3 cc/g of H_2 is equivalent to 2·3 wt% H_2O. The δD value of -81‰ for J-1·1-C-2, although similar to the MORB D/H ratio of -80‰ (Craig and Lupton 1976), is also similar to the value for low-temperature alteration products (Magaritz and Taylor 1976). J-1·1-C-2, which has $\delta D = -65$‰ appears transitional, with a D/H ratio between unaltered josephinite (-10 to -53‰) and J-1·1-C-1.

The unaltered josephinite samples (J-77-1 and J-3·1-D) exhibit a wide range in δD values, -10 to -53‰, with a 30‰ difference between the two splits of J-77-1 (-10‰ and -40‰). This variability is not understood: it may be an intrinsic feature, or loss of H_2 by diffusion may have increased the D/H ratio. These ratios are somewhat higher than in present meteoric water in the region (-60 to -80‰) but could be similar to crustal waters 150 My ago.

8. Discussion

We have shown by replicate analyses in different extraction systems, that very large He and Ne isotopic anomalies can be produced by the use of radiofrequency induction heating to extract gases from native iron and nickel-iron samples. These samples contain large amounts of H_2 which is remarkably efficient at scavenging He from previous experiments in glass systems, probably through the formation of atomic hydrogen. Similar scavenging effects were in fact observed 50 years ago by F Paneth, one of the early pioneers of ^{3}He research in meteorites. Our results show that the large He and Ne isotopic anomalies observed in josephinite samples analysed in other laboratories must be considered as probable procedural artifacts unless they can be verified by extraction of gases by different methods such as resistance heating, or possibly by solution.

A second problem with He isotope studies of native iron and nickel iron samples is the possible presence of cosmogenic ^{3}He in samples which have been exposed for long periods at the earth's surface and are often found only in placers. In the case of olivine phenocrysts in basalts, the cosmogenic and mantle He components can be readily distinguished by the comparison of He extracted by crushing from fluid inclusions with He extracted by total fusion, which removes the cosmogenic component. Secondly, cosmogenic Ne can be measured in the same samples because the cosmic-ray production rate is reasonably high relative to ^{3}He (Marti and Craig 1987). In the case of iron samples, neither procedure can be used to separate the cosmogenic component, and shielded samples have ultimately to be obtained by drilling, blasting, or sectioning the odd large boulder. This has not yet been accomplished for the samples analysed so far, so that the cosmogenic component is not yet evaluated.

Nevertheless, our He, Ne and H_2 results place some constraints on the environment of formation of these native metals. The clearest results are for the Greenland irons, in which the low $^3He/^4He$ ratios (0·15–0·4 R_A) and the atmospheric values for ^{21}Ne show that the source of the rare gases is the earth's crust or at the surface, from radiogenic and/or cosmogenic components. There is no evidence for any significant amount of mantle-derived rare gases, and even the H_2 in the irons has D/H ratios resembling Greenland meteoric water. These data are consistent with previous studies of carbon isotopes in these irons (Craig 1953) and support the proposition that the irons formed by reduction of Fe-bearing basalts in a carbon-rich crustal environment. If the Disko

Island iron had originated in the lower mantle it should have retained ^{3}He/^{4}He ratio similar to other mantle-derived samples (6–30 R_A), especially as helium is not known to diffuse in iron. The helium in these samples may have been incorporated into the metal as it formed in a near-surface environment or may be the result of *in situ* production of He by U and Th, in addition to cosmogenic contributions and possible production from ^{3}H diffusing in after formation on the iron surface by catalytic decomposition of atmospheric water after nuclear weapons tests.

The postulates of an origin of josephinite, a simple Fe-Ni alloy, from such far-off places as the earth's core, a "supernova" layer in the deep mantle, or the solar nebula, etc are examples of the "exotic theorem" which attempts to associate remarkable findings with remarkable places. Curiously enough, the *only* terrestrial samples in which it is established that ^{3}He/^{4}He ratios as high as those claimed to exist in josephinite can be found, are simply *"garden variety" basalts* on the summit of Maui, which have been exposed to cosmic radiation for some tens of thousands of years. This simple result shows the difficulty in establishing the usefulness of the exotic theorem for geologic research: like most mysterious entities such as bluebirds, high ^{3}He ratios can be found in some of the most ordinary places on earth.

Rare gas studies do not solve the major problems of the formation of these native irons, i.e. the predominance of high-temperature phases which require a long, slow cooling history, and the tremendous Fe localization necessary to produce 10,000 kg boulders (Bird and Weathers 1977). However, the ^{3}He data do show that native irons from Greenland are not similar in ^{3}He/^{4}He ratio to josephinite: all native irons are not alike. Moreover, the Greenland helium data are consistent with D/H and ^{13}C/^{12}C ratios in establishing that the reduction process which produced the iron occurred in a crustal environment and that iron has not been transported as huge native pigs in basalt from the mantle.

The ^{3}He/^{4}He ratio of 5.6 R_A is consistent with derivation of josephinite from an upper mantle source, although the value is lower than the ^{3}He/^{4}He ratio typical for MORB (8 R_A) and is more similar to helium in volcanic arcs and "Atlantic" or "low-^{3}He" type hotspots. This helium could have been incorporated into josephinite in the mantle, *or* during reduction or serpentinization processes after eruption of the host rocks. The helium ratio characterizes the *rock system: josephinite plus peridotite*, not the individual rock or mineral in a complicated petrological chemical processing plant. Additionally, if josephinite incorporated its helium while forming in the crust, it almost certainly took up a mixture of mantle and crustal gases during the serpentinization reactions as hydrothermal fluids transported crustal helium (0.01 R_A) and meteoric gases into the system. Radiogenic ^{4}He production would also reduce the ^{3}He/^{4}He ratio in 150 My. Finally, the possible presence of a cosmogenic ^{3}He component has to be determined before more explicit meaning can be attached to the helium ratio. Despite these complexities and the rich spectrum of possible He components, what *can* be said from the He data at present is that josephinite may have been formed in the upper mantle or the crust, but there is no indication of any origin more exotic than either of these locales.

There are several possible interpretations for the ^{40}Ar/^{36}Ar ratios present in josephinite. Since ^{3}He is a mantle tracer, the ratio of ^{3}He to ^{36}Ar acts as a constraint on the mantle source. Typical MORB, derived from an upper mantle source, has a ^{36}Ar/^{3}He ratio of 0.1–1.0 (e.g. Allegre et al 1983) while in josephinite, this ratio is 2000 (for J-77-1). If josephinite formed in a mantle region with a ^{36}Ar/^{3}He ratio similar to

MORB, then any upper mantle ^{36}Ar which may exist is overwhelmed by the presence of ^{36}Ar from another source. The nearly atmospheric ^{40}Ar/^{36}Ar ratio indicates that most of the Ar and all of the ^{36}Ar were almost certainly derived from a hydrothermal fluid, meteoric in origin, which serpentinized the peridotite: MORB ^{40}Ar/^{36}Ar ratios are much higher ($\sim 10{,}000$) than the atmospheric ratio and should be resident in the metal if it formed in the mantle.

All rare gas and hydrogen isotope data from this study are consistent with the interpretation that josephinite formed as a result of metamorphism of a section of oceanic crust by hydrothermal fluids. The Josephine peridotite represents an abducted section of typical oceanic upper mantle. The ^{3}He/^{4}He ratios, although slightly lower than MORB, are also consistent with an oceanic upper mantle source: the ^{40}Ar/^{36}Ar ratios probably are not. Until the possible cosmogenic ^{3}He component is evaluated, the helium data cannot be clearly labelled as indicative of a crustal or mantle isotopic signature, but a mantle helium signature would not preclude a crustal origin of the josephinite. On the other hand, a cosmogenic signature combined with an atmospheric Ar signature in the Josephinite gases would be difficult to ascribe to a mantle origin.

9. Conclusions

Rare gas and hydrogen isotopes are indicative of the environment of formation for native metals. The ^{3}He/^{4}He ratio of 0.15 to 0.40 R_A for the Disko Island and Kaersut irons from Greenland reflects crustal helium, in contrast to the presumed mantle signature of the enclosing basalts. The ^{3}He/^{4}He ratios in josephinite (up to 5.6 R_A) are approximately equal to normal upper-mantle values and are consistent with a serpentinization origin for the metal. There is no evidence of anomalous ^{3}He and ^{21}Ne enrichments that would indicate an exotic origin in the deep-mantle or core, or from 4.5 billion-year old meteorites, supernovae, etc.: reported data for these isotopic anomalies are due to experimental artifacts.

Acknowledgements

We have benefitted greatly during this work (and in other work as well) from penetrating discussions with Devendra Lal, to whom this paper is dedicated. The calculated baseline ^{3}He and ^{21}Ne cosmic-ray production rates in Fe and Ni were kindly supplied by him along with considerable wisdom. We are grateful to J Lupton and S Regnier who participated in the original stages of the work, and to John Bird for graciously supplying the josephinite samples. This research was supported by NSF-EAR grants in Mantle Geochemistry to the Isotope Laboratory (H.C.) and by NASA (grant NAG 9-41 to K.M.).

References

Allegre C J, Staudacher T, Sarda P and Kurz M 1983 Constraints on evolution of earth's mantle from rare gas systematics; *Nature (London)* **303** 762–766

Bernatowicz T J, Gottel K A, Hohenberg C M and Podosek F A 1979 Anomalous noble gases in josephinite and associated rocks? *Earth Planet Sci. Lett.* **43** 368–384

Bieri R H, Dymond J D and Koide M 1966 A high-temperature ultra-high vacuum furnace for gas extractions from geologic samples; *Earth Planet. Sci. Lett.* **1** 395

Bird J M and Weathers M S 1975 Josephinite: specimens from the earth's core? *Earth Planet. Sci. Lett.* **28** 51–64

Bird J M and Weathers M S 1977 Native iron occurrences of Disko Island, Greenland; *J. Geol.* **85** 359–371

Bird J M and Weathers M S 1979 Origin of josephinite; *Geochem. J.* **13** 41–55

Bird J M, Bassett W A and Weathers M S 1979 Widmanstaetten patterns in josephinite, a metal-bearing terrestrial rock; *Science* **206** 832–834

Bird J M, Goodrich C A and Weathers M S 1981 Petrogenesis of Uivfaq iron, Disko Island, Greenland; *J. Geophys. Res.* **86** 11787–11805

Bochsler P, Stettler A, Bird J M and Weathers M S 1978 Excess ^{3}He and ^{21}Ne in josephinite; *Earth Planet Sci. Lett.* **39** 67–74

Botto R I and Morrison G H 1976 Josephinite: a unique nickel-iron; *Am. J. Sci.* **276** 241–274

Clarke W B, Beg M A and Craig H 1969 Excess ^{3}He in the sea: evidence for terrestrial primordial helium; *Earth Planet. Sci. Lett.* **6** 213–220

Craig H 1953 The geochemistry of the stable carbon isotopes; *Geochim. Cosmochim. Acta* **3** 53–92

Craig H 1958 in *Cosmological and geological implications of isotope ratio variations*; Nat. Acad. Sci. Pub. No. 572 pp. 53–54

Craig H and Gordon L I 1965 Deuterium and oxygen 18 variations in the ocean and marine atmosphere; in *Stable isotopes in oceanographic studies and paleo-temperatures*, (ed.) E Tongiorgi (Pisa: V Lischi and Figli) pp. 9–130

Craig H and Lupton J E 1976 Primordial neon, helium and hydrogen in oceanic basalts; *Earth Planet. Sci. Lett.* **31** 369–385

Craig H, Poreda R, Lupton J E, Marti K and Regnier S 1979 Rare gases and hydrogen in josephinite; *Eos* **60** 970

Craig H and Lupton J E 1981 Helium-3 and mantle volatiles in the ocean and the oceanic crust, Chapter 11; in *The sea, Volume 7: The oceanic lithosphere*, (ed.) C Emiliani (New York: John Wiley) pp. 391–428

Craig H and Poreda R J 1986 Cosmogenic ^{3}He in terrestrial rocks: The summit lavas of Maui; *Proc. Natl. Acad. Sci.* **83** 1970–1974

Craig H, Rison W and Poreda R J 1985 Helium isotope variations in ocean island basalts; *Eos* **66** 1079

Dansgaard W 1964 Stable isotopes in precipitation; *Tellus* **16** 436–468

Dick H J B 1974 Terrestrial nickel iron from the Josephine peridotite, its geological occurrence, associations and origin; *Earth Planet. Sci. Lett.* **31** 291–298

Dick H J B and Gillete H 1976 Josephinite: specimens from the earth's core? A discussion; *Earth Planet. Sci. Lett.* **31** 308–311

Downing R G, Hennecke E W and Manuel O K 1977 Josephinite: A terrestrial alloy with radiogenic xenon-129 and the noble gas imprint of iron meteorites; *Geochem. J.* **11** 219–229

Dushman W 1962 *Foundation of scientific vacuum techniques* (New York: John Wiley) pp. 435

Fireman E L 1958 in *Cosmological and geological implications of isotope ratio variations*; Natl. Acad. Sci. Pub. No. 572 p. 53

Goodrich C A and Bird J M 1985 Formation of iron-carbon alloys in basaltic magma at Uivfaq, Disko Island: The role of carbon in mafic magmas; *J. Geol.* **93** 474–492

Krishnarao J S R 1964 Native nickel-iron alloy, its mode of occurrence, distribution and origin; *Econ. Geol.* **59** 443–448

Kurz M D 1986 Cosmogenic helium in a terrestrial igneous rock; *Nature (London)* **320** 435–439

Lal D 1987 Production of ^{3}He in terrestrial rocks; *Chem. Geol.* **66** 89–98

Lal D 1988 *In situ* produced cosmogenic isotopes in terrestrial rocks; *Annu. Rev. Earth Planet. Sci.* **16** 355–388

Lal D, Craig H, Wacker J F and Poreda R 1989 ^{3}He in diamonds: the cosmogenic component; *Geochim. Cosmochim. Acta* **53** 569–574

Lofquist H and Benedicks C 1941 Det stora Nordenskiolda jarn-blocket fran Ovifak-mikrostruktur och bildningssatt; *K. Sven. Vetenskapsakad. Handl. Tredjie Ser.* **19**

Lupton J E 1983 Terrestrial inert gases: isotope tracer studies and clues to primordial components in the mantle; *Annu. Rev. Earth Planet. Sci.* **11** 371–414

Lupton J E and Craig H 1975 Excess ^{3}He in oceanic basalts: evidence for terrestrial primordial helium; *Earth Planet. Sci. Lett.* **26** 133–139

Magaritz M and Taylor H P 1976 Oxygen, hydrogen and carbon isotope studies of the Franciscan formation, Coast Range, California; *Geochim. Cosmochim. Acta* **40** 215–237

Marti W and Switzer G 1966 Plagioclase-spinel-graphite xenoliths in metallic iron-bearing basalts, Disko Island, Greenland; *Am. Mineral.* **51** 664–676

Norton F J 1953 Diffusion of D_2 from D_2O through steel; *J. Appl. Phys.* **24** 499

Paneth F 1927a The transmutation of hydrogen into helium; *Nature (London)* **119** 706–707

Paneth F 1927b Neuere versuche uber die verwandlung von wasserstoff in helium; *Naturwissenschaften* **16** 379

Pedersen A K 1969 Preliminary notes on the tertiary lavas of northern Disko; *Rapp. Groenland Geol. Unders.* **19** 21–24

Poreda R J 1983 *Helium, neon, water and carbon in volcanic rocks and gases*, Ph.D. Dissertation, University of California, San Diego, pp. 215

Poreda R J and Craig H 1989 Helium isotope ratios in circum-Pacific volcanic arcs; *Nature (London)* **338** 473–478

Rison W 1980 *Isotopic studies of the rare gases in igneous rocks: implications for the mantle and atmosphere*, Ph.D. Dissertation, University of California, Berkeley, pp. 189

Sheppard S M and Epstein S 1970 D/H and $^{18}O/^{16}O$ ratios of minerals of possible mantle or lower crustal origin; *Earth Planet. Sci.* **9** 232–239

Smith J L 1879 Memoire sur le fer natif du Groenland et sur la dolerité qui le renferme; *Ann. Chem. Phys.* **16** 452–505

Stettler A and Bochsler P 1979 He, Ne and Ar composition in a neutron activated sea-floor basalt glass; *Geochim. Cosmochim. Acta* **43** 157–169

Wetherill G W 1954 Variations in the isotopic abundances of neon and argon extracted from radioactive minerals; *Phys. Rev.* **96** 679–683

Emanation of radon from rock minerals

RAMA

Tata Institute of Fundamental Research, Bombay 400 005, India

Abstract. The mechanism of emanation of radon from terrestrial minerals and rocks is briefly discussed.

Keywords. Radon emanation; terrestrial minerals; internal area.

1. Introduction

Most of the natural minerals contain small amounts of uranium and thorium, and their decay products. Amongst these, ^{222}Rn, ^{220}Rn and ^{219}Rn arise from decay of ^{226}Ra (^{238}U series) ^{224}Ra (^{228}Th series) and ^{223}Ra (^{235}U series) respectively. Their half-lives are 3·8 d (^{222}Rn), 55 s (^{220}Rn) and 4 s (^{219}Rn). It has been known for long that somehow a sizable fraction of these three gaseous isotopes produced inside the terrestrial minerals is able to emerge out of the minerals into the pores from where it diffuses out into the atmosphere.

The inert gaseous isotopes of Rn have found numerous applications as tracers for investigating geophysical and geochemical phenomena. In some of these applications, a knowledge of the exact nature of the mechanism by which Rn atoms emerge out of crystalline minerals is important. Attempts have therefore been made to understand this mechanism.

The basic mechanism was worked out by Zimens in late thirties and early forties. But certain aspects of the phenomenon still require unravelling and details remain to be worked out. Review articles by Tanner (1978) give a comprehensive treatment of the subject. In this article, I propose to bring out a couple of problems which still need to be worked out.

2. Recoil-out

On decay, a nucleus of ^{226}Ra, ^{224}Ra or ^{223}Ra gives out an alpha particle and the residual nucleus of ^{222}Rn, ^{220}Rn or ^{219}Rn recoils with equal momentum in the opposite direction. The kinetic energy of the recoiling nucleus is in the 100 keV range, and differs slightly for the three isotopes; the respective ranges of the three recoiling isotopes in common rock minerals are about 60, 80 and 90 nanometers.

The recoiled Rn atom loses its energy by atomic collisions inside the mineral, and then comes to rest within the mineral (figure 1). There is no possibility of its coming out of the mineral since the diffusion coefficient of Rn in crystalline mineral is extremely small. However, there is a good chance for Rn atoms to be directly recoiled out of the mineral if they are generated in the skin of the mineral i.e. within a depth

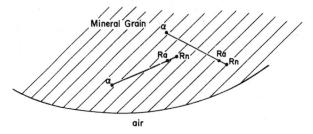

Figure 1. Rn generated inside a mineral grain is unable to emanate out.

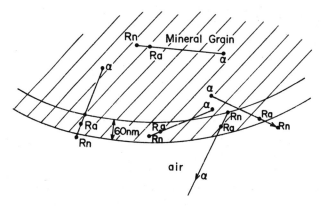

Figure 2. Some of the Rn generated in the skin of a mineral grain recoils out into air.

equal to the recoil range from the surface (figure 2). Considering the geometrical factors involved, one out of four such Rn atoms is expected to recoil out of the skin, while those generated deeper than the recoil range cannot come out at all. Thus only a small fraction, depending upon the size of the mineral grain, of the Rn atoms generated inside the grain can recoil out. If we assume that the Ra atoms are distributed uniformly within the grain, the fraction of Rn atoms recoiling out of the grain should be about 0·0005% for 1 cm size grain, 0·005% for 1 mm size, 0·05% for 100 μm, 0·5% for 10 μm and 5% for 1 μm size grain. This fraction, also termed emanation coefficient, increases sharply for submicronic grains. This is the expectation for the single grain lying in air. But if the grain is surrounded by other grains (as in soil or in rocks), some of the recoiled Rn atoms may get embedded in the neighbouring grain (figure 3). The recoil range in air is long ($\sim$ 150 μm). If the air gap between the grains is much less than 150 μm, (let us say, $\sim$ 1 μm), then there is essentially no stopping by air in the gap, and most of the recoiled Rn atoms should be reimplanted in the opposite grain. Thus the emanation coefficients should be far less than those indicated above. When the pores are however filled with water, all of the recoiled Rn should stop in water (recoil range in water $\sim$ 100–200 nm), and the emanation coefficients should shoot up to the values indicated above. Actual observations however show that the emanation coefficients are rather large; further, the enhancement due to introduction of water in the pores is rather small ($\sim$ a factor 2). To explain these observations, a hypothesis is invoked which says that the implantation efficiency in the opposite surface must be poor i.e. Rn atom somehow gets out of the skin of the neighbouring grain and comes back into the pore.

from rock minerals 175

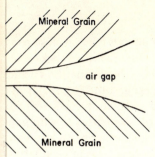

...out of the skin of a mineral grain gets embedded in the skin of a

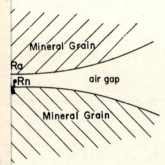

...into the opposite wall is able to diffuse back into the gap.

...atom deposits its kinetic energy very fast. It is further
...elt (perhaps even vaporize) the material, about 1 nm
...to diffuse out of the pit, back into the pore, before the
...appears to be a reasonable hypothesis, but needs to
...etails worked out. We have some experimental data
...dence that in granitic material only about 50% of the
...ted in the opposite surface and the other 50% are able
...leach out ^{224}Ra selectively out of a rock sample. This
...tion from the rock. But this reduction is only half as
...^{224}Ra in the leach. These experiments provide good
...lantation of ^{220}Rn in the rock is about 50%; neither
...hay differ with recoil energy, also with the nature of
...ould be possible to verify the hypothesis with direct
...tails. Indeed we are now in the process of attempting
...which recoils out ^{220}Rn atoms. This source when
...various minerals can be used to determine the
...rce is actually a dry electro-deposit of ^{224}Ra atoms,
...atmosphere of hydrogen.

4. Emanation from rocks

Rocks are assemblages of minerals lying in close contact with each other. The mineral grains can differ greatly in size in various rocks, but it is convenient to choose rocks with mm size mineral grains. The emanation coefficients for Rn in cm size pieces of such a rock often lie in the range 1 to 20%, i.e. far in excess of the value calculated for this size. The observed high emanation coefficient would indicate that the minerals in the rock are made of micron size particles, with gaps in between, i.e. with very large open area inside. Since the open volume inside a rock is very small, the gap width must be very small (a few nm). However, this interpretation rests on the assumption that Ra is distributed uniformly in each individual grain. An alternative explanation for the observed large emanation coefficients is more popular. It invokes two hypotheses: existence of high concentration of Ra on grain boundaries and existence of gaps at the grain boundaries. Accordingly, Ra-rich surfaces at the grain boundaries yield enough Rn in the intergranular gaps from where it diffuses out to the surface of 1 cm or even 100 cm size rock boulder. This explanation however does not appear to be valid in the case of several samples of granite we studied. Alpha radiographs of these granites do not show presence of excessive amount of Ra on grain boundaries, nor do we see gaps along the boundaries. On the other hand, direct and reliable evidence for the existence of large open areas inside the rocks is also not forthcoming. The problem is therefore an open one.

References

Amin B S and Rama 1986 Using radon as probe for investigating characteristics of fractures in crystalline minerals; *Nucl. Instrum. Meth. Phys. Res.* **B17** 527–529

Krishnaswamy S and Seidemann D E 1988 Comparative study of ^{222}Rn, ^{40}Ar, ^{39}Ar and ^{37}Ar leakage from rocks and minerals: Implications for the role of nanopores in gas transport through natural silicates; *Geochim. Cosmochim. Acta* **52** 655–658

Rama and Moore W S 1984 Mechanism of transport of U-Th series radioisotopes from solids into groundwater; *Geochim. Cosmochim. Acta* **48** 395–399

Tanner A B 1978 Radon migration in the ground; *Nat. Radiat. Environ.* 5–56

Several considerations on the early history of the Earth

E M GALIMOV

V I Vernadsky Institute of Geochemistry and Analytical Chemistry, Academy of Sciences of the USSR, Kosygin Street 19, 117334 Moscow, USSR

1. Introduction

Since the classical works by Urey, Ruby and Vinogradov in the early fifties it is commonly accepted that formation of the Earth proceeded by accretion of solid bodies, that neither the planet as a whole nor its upper mantle had ever been completely melted, that the rise of the ocean and the atmosphere was due to degassing of the mantle, and that the Earth crust was made up of derivatives of partial melting of the mantle rocks and therefore enriched in incompatable elements and large ion lithophiles.

These principles are well established physico-chemically and appear to be in agreement with geological observations connected with the relatively late geological history of the Earth. However, it is difficult to reconstruct the events of the early Earth from the accepted theory. This paper devoted to Professor D Lal on his 60th birthday is an attempt to look at this problem from another angle.

2. Problem of the ancient Earth crust

One would think that the early formation of the hydrosphere and the atmosphere and hence the development of the sedimentation process must have given rise to early stabilization of lithospheric blocks. However, as is well known, there is no geological formation on the Earth whose age exceeds 3·8–3·9 billion years. One might suggest that absence of any traces of the ancient sialic lithosphere is due to its subsequent reworking. However, when the Sm/Nd technique was applied to the study of Precambrian rocks, it had been discovered that ε^i_{Nd}-values for Archean rocks fell on the line of the Nd-isotopic evolution of chondritic reservoir. This meant that granites, gneisses and gabbros which composed the oldest nuclei of the Earth crust, derived directly from the mantle, and not from any differentiated precursor. Since petrological considerations argue against significant generation of acid magmas from a substance of mantle composition, a model was proposed that preliminary differentiation in fact existed. But the lifetime of the precursors was too short to show any appreciable change in their Nd-isotopic composition, so that the whole cycle including magmatism, weathering, sedimentation and remelting would not have exceeded 50 million years (e.g. De Paolo 1981). In terms of Sm/Nd systematics, this is equivalent to

177

immediate origin from the mantle. This idea was related to the concept of intensive mantle convection in the early Earth.

However as more data appeared, it became evident that ε_{Nd}^i-values for a majority of the Archean rocks deviated slightly from the line of chondritic evolution. Compilation of data shows that ε_{Nd}^i-values range between $+0.5$ and $+3$. Positive ε_{Nd}^i-values characterize rocks which are derived from a source previously depleted in light rare earths, in particular Nd relative to Sm ($f_{Sm/Nd} > 0$). It is difficult to reconcile this fact with the accepted model, as the depletion of the mantle has been suggested to be a consequence of the chemical differentiation which occurred due to the extraction of the crust. The positive ε_{Nd} values indicate that mantle was depleted even before formation of the earth's crust.

Geochemistry of the Early Archaean in many aspects is unique. The Archaean granites (tonalites and trondhjemites) are different in composition from the later ones. They are impoverished in K and enriched in Na. Sharp increase of K content in the crust occurred before 2·6–2·9 billion years. This boundary is also marked by a strong change of $^{87}Sr/^{86}Sr$ in the oceans. This suggests that before this time different mechanisms and sources were involved in the formation of the earth's crust.

3. Suggested model of formation of planets and meteorites

As mentioned earlier the classical view relates formation of the earth's crust to partial melting processes leading to the enrichment of incompatable elements in the liquid phase and eventually in the crust. However does a process exist which would produce crustal material depleted rather than enriched with incompatable elements? Yes, it does. Such a process apparently occurred on the moon. Many investigators agree that the upper mantle of the early moon was completely melted. As it cooled and crystallization began, heavy oxides containing Fe, Ti etc. and minerals of relatively high density like olivine formed the lower peridotitic layer while lighter minerals like plagioclase floated to the surface leading to the formation of the gabbro-anorthositic crust.

As this plagioclase crust is a cumulate and not a product of partial melting, it has to be depleted in incompatible elements. The same process could have occurred during formation of the Earth. However the belief exists that the Earth and other inner planets of the solar system have grown by a relatively low temperature accretion of solid bodies. And the "hot" model is not consistent with the presence of hydrosphere and other volatiles in the Earth.

The theory of collisional evolution of solid bodies developed by Safronov (1969) is still most popular. There exists an alternative model suggested by Gurevich and Lebedinskij (1950), the computer simulation of which was later developed by Kozlov and Eneyev (1977). This model envisages accumulation of a diffuse body by amalgamation of wisps of gas and dust. Proceeding from this model one can suggest that planets grew by compaction of diffuse disk-like protoplanetary bodies consisting of small particles.

What was the chemical composition of those particles? The protoplanetary material was earlier considered a product of condensation of high temperature gas of solar composition. However the discovery of the isotope anomalies in meteorites shows that grains of interstellar matter in the initial protosolar cloud were not completely evaporated during the formation of the Sun. They survived and might serve as initial material during creation of the planets. Therefore it is possible that the chemical composition of the initial particles was close to that of interstellar dust or roughly to the chemical composition of the most primitive carbonaceous chondrites.

Heat radiated by the condensed central body, due to adiabatic increase of temperature during compaction, must have led to melting of the particles in the surroundings. Volatiles evolved from the melted particles were swept away by the solar wind and degassed droplets (chondrules) fell to the surface of the growing planet. A gradient of chemical and physical properties was established along the profile of the compacting diffuse body from the melted and partly evaporated silicates in the central body through the shell consisting of melted and exhaustively degassed droplets in the inner part and less altered material on the periphery. I think that the different types of meteorites represent different parts of the former diffuse body which on compaction was catastrophically disrupted at a relatively early stage. I believe that it was an unrealized planet which presumably grew between Mars and Jupiter and was destroyed by the tidal forces from the side of Jupiter. The abbreviation UP (unrealized planet) is used hereafter to refer to this object.

After destruction of the UP, its material was scattered over the solar system. Part of it remained in the orbit and gave rise to a variety of asteroids and meteorites. The former droplets provided the main source of chondrites whose eventual composition (from H to L type of ordinary chondrites or from C_3 to C_1 type of carbonaceous chondrites) was determined by the position of the starting material within the original diffuse body. Achondrites represent silicate material of the central condensed body and iron meteorites—its metallic core.

The suggested model implies that the metallic core of planets formed immediately after the accretion process. It has been inferred from geophysical data that the core of the Earth contains, apart from Fe and Ni, some light elements like H, C, O, Si and S. But our model rules out H, C and S in the core as these elements were lost during accretion. The same applies to Si as its incorporation in the core requires the presence of reducing agents like H or C. The model is consistent with the presence of FeO as an additional component in the core. This idea was suggested by Dubrovskij and Pan'kov (1972). It is essential that solubility of FeO in molten iron depends on pressure. It amounts to 4 mol% under 100 kbar, 11 mol% under 300 kbar and 52 mol% under 600 kbar (Ringwood 1979). In this connection it is noteworthy that the Moon, where even in the central part the pressure does not exceed 50 kbar, contains only a small core ($\sim 2\%$ of the whole mass of the Moon). Apparently it consists only of metallic Fe and Ni. This explains the exceptionally low concentration of siderophyles and at the same time relatively high concentration of FeO (15–25%) in the lunar mantle. One can suggest that in general at the early stage of accretion only relatively small metallic Fe-Ni core can form. As the growing planet exceeds a certain size further increase of core becomes possible by dissolution of FeO in metallic iron. It is of interest to note that the inner part of the Earth core, which is believed to consist of metallic Fe + Ni, is also about 2% of the mass of the planet. Massive cores could be expected only in relatively large bodies. Indeed the Earth and the Venus have cores which are approximately 30% of their total mass. Mars has a much smaller core ($\sim 10\%$ of the mass of the planet). The exception is Mercury, in which the core constitutes 70% of its mass. However because of its proximity to the Sun it must have lost not only the volatiles but also a significant part of silicates.

Probably the Moon acquired only inner part of its initial nebula whereas the external part was captured by the Earth with which the Moon formed a binary system. Apparently the external part of the proto-Martian diffuse shell was also disturbed and partly lost during the catastrophic event which occurred when a protoplanetary body developed between Mars and Jupiter.

The fact that iron meteorites do not contain FeO shows that during the catastrophic destruction the inner condensed part of UP was approximately of lunar size.

It is well known that planets and meteorites are depleted in volatiles relative to solar abundances of the corresponding elements. The degree of depletion increases with increase of volatility of the elements: Si, Cr, Mn, Na, K, As, Ga, Rb, Tl, Cs, In, Ag, Zn etc (Ringwood 1979). This has been interpreted as a consequence of condensation of protoplanetary material from an initial high temperature gas and served as a cornerstone of the theory of heterogeneous accretion. However it may also be due to evaporation and loss of volatiles by melted droplets during accretion. The logic of the suggested model implies that the Moon should be depleted in volatiles to a greater extent than the Earth. In turn the Earth should be depleted relative to the ordinary chondrites and the latter relative to the carbonaceous chondrites.

The degree of depletion can be estimated by comparing elements which behave similarly in geochemical processes but show different volatility. For instance the K/U ratio is known to change only slightly in different terrestrial rocks. However, as one can see from table 1, this ratio is completely different for the planetary bodies with different accretion history. Depletion of K in the Earth relative to the ordinary chondrites is characterized by factor 0·15–0·18. For the Moon, this factor is 0·02–0·03.

It is noteworthy that the Moon and the achondrites have almost similar K/U ratios. Since achondrites in our model represent the silicate phase of the inner condensed body of UP, it is clear that this body was approximately of lunar size before destruction.

The Moon is depleted in Rb even to a greater extent than in K. Rb is heavier but a more volatile element. Consequently the difference in depletion is actually controlled by the difference in volatility. This is consistent with the idea that loss of volatiles occurred during compaction from the surface of small melted particles but not from the surface of a planetary-sized body.

The Moon is depleted not only in K and Rb but also in Na. Consequently, coming back to the problem of the origin of the earth crust, we can conclude that the uppermost layer of the primitive Earth, even if it was formed in the same way as the lunar gabbro-anortositic crust, could not be chemically identical to the latter. The earth's mantle contains 8–10 times more K and 5–6 times Na than the lunar mantle. The protocrust of the Earth has to be enriched in $K_2O(\sim 0.8\%)$ and $Na_2O(\sim 3.0\%)$.

Table 1. K/U and K/La ratios for the Earth, Moon and meteorites.

Ratio	Earth	Moon	Ordinary chondrites	Achondrites (eucrites)
$K/U \times 10^{-3}$	11·2	1·6	70·8	2·7
$K/La \times 10^{-5}$	4·5	0·7	24·2	1·2

In order to evaluate the chemical composition of the protocrust of the Earth one can modify the composition of the anorthositic lunar crust by a corresponding addition of potash feldspar and albite components and subtraction of an equivalent portion of anorthite component. Obviously it will be more acidic than the lunar crust. It would correspond to a pyroxene-feldspathic composition, with the normative plagioclase being close to 50, i.e. to andesine-labradorite.

The protocrust of this composition fits well with the Archaean tonalites and pyroxene granulites in a certain proportion. It could be a source of tonalitic magma as pyroxene and plagioclase are on the liquidus of tonalitic composition below 10 kbar. A high degree of melting along with the initial relatively low content of K in the protocrust explains the specific low K and high Na composition of the Archaean granites and gneisses. Besides, enrichment of the Archaean rocks in Sr and Ba, excess of Eu, positive ε_{Nd}^i-values and some other features can be explained.

4. Beginning of the ocean formation

As mentioned earlier ordinary chondrites are depleted in volatiles to a lesser extent than the Earth. Therefore even if ordinary chondrites contain $< 0.02\%$ of carbon and negligible amount of water, the Earth must have been depleted in carbon and water almost completely. It follows from the model that the surface of the Earth by the end of its accretion was low in water and other volatiles and could not gain them through degassing of the mantle. The question then arises when and how water comes to the surface of the Earth? More than 20 years ago, I suggested that the earth's crust received its volatiles from carbonaceous-chondrite-like-material (Galimov 1968). I based my theory upon the unique similarity in isotope composition between the volatile elements of the earth's crust and carbonaceous chondrites. Different versions of this idea have been proposed in the literature from time to time.

The origin of the hydrosphere of Earth can be investigated by oxygen isotope studies. Water plays an important role in the oxygen isotope balance and the oxygen isotope distribution between geospheres. In table 2, the oxygen isotope inventory is presented. It illustrates the well-known fact that the earth's crust-ocean system is enriched in ^{18}O relative to the upper mantle. The present calculation gives $\delta^{18}O = +8.8‰$.

If the model of the Earth suggested above had a primitive plagioclase crust, the difference between the isotope composition of the crust and the mantle could be partly due to isotope fractionation during crystallization. Plagioclases are enriched in ^{18}O relative to olivine. For andesine the isotope effect is about 1·5‰ below 1000°C. However it could give utmost $\delta^{18}O \simeq +7.5‰$ for the protocrust, which is not sufficient to explain the observed enrichment of the crust in ^{18}O. The earth crust is composed of primary magmatic rocks and secondary rocks through the sedimentary cycle. It is obvious that there are basic differences between the oxygen isotope composition of the secondary and the primary rocks. The sedimentary and metamorphic rocks have an average oxygen isotopic composition of $+13.8‰$. Part of the granites is also secondary. Even if we take half of all granites to be secondary, the $\delta^{18}O$-value for secondary rocks as a whole would be no less than $+12.8‰$. On the other hand, for primary rocks in total, $\delta^{18}O$ cannot be higher than $+7.0‰$. The oceanic water has $\delta^{18}O = 0‰$, and that of the juvenile water is about $+6‰$, since

Table 2. ^{18}O-budget in the Earth crust.

	Mass $\times 10^{24}$ g*	O_2-content $\times 10^{22}$ mol	δ^{18}O‰
Sedimentary rocks			
Sand and sandstones	0·43	0·69	+15
Clays and shales	1·14	0·81	+15
Carbonates	0·71	1·10	+25
Evaporites	0·02	0·03	—
Metamorphic rocks			
Metasandstones	0·47	0·75	+12
Paragneisses and shales	4·74	7·55	+12
Metamorph. carbonates	0·18	0·28	+15
Iron formation rocks	0·06	0·09	+6
Magmatic rocks			
Granites	5·68	8·69	+9
Alkalines	0·01	0·01	+7
Basalts	15·0	22·0	+6
Ultrabasic	0·02	0·03	+5
Earth crust:	28·46	43·06	+8·8
Ocean:	1·4	3·89	0
Total:		46·95	+8·1
Upper mantle rocks			+6·0

* Taken from Ronov (1980)

water in contact with molten silicates acquires isotope composition very close to the oxygen isotope composition of minerals.

Hence one can obtain an estimate of the minimum water flow through the mantle-crust boundary, ΔM, by considering the isotope balance as

$$M_{SR}\delta^{18}O_{SR} + (M_{OW} + \Delta M)\delta^{18}O_{OW} = M_{PR}\delta^{18}O_{PR} + (M_{JW} + \Delta M)\delta^{18}O_{JW}$$

where M is the mass of the corresponding oxygen reservoir (SR, secondary rocks; OW, oceanic water; PR, primary rocks; JW, juvenile or primary water). The calculation gives ΔM equal to $11·9 \times 10^{22}$ moles. This means that the amount of water passing through the mantle-crust boundary during geological history exceeds three times the present volume of the world oceans. This value gives an idea of the scale of lithospheric circulation of water and in particular about the scale of hydrothermal deposits related to this cycle.

Thus owing to the global circulation of water ^{18}O is transferred from the mantle to the ocean. On the contrary, as isotope fractionation under low temperature results in enrichment of ^{18}O in sedimentary minerals (e.g. +28‰ in calcite and +34‰ in silica relative to water under 25°C), sedimentary process takes out ^{18}O from the ocean. The present isotope composition of the ocean is under the control of these two competing processes. The oxygen isotopic composition of sedimentary rocks reflects the isotopic composition of contemporary oceanic water. Perry (1967) was the first to draw attention to the enigmatic trend of oxygen isotopic composition of cherts and carbonates with geological time. Figure 1 illustrates such a trend for carbonates. The data are taken from different literature sources and supplemented by measurements made recently in our laboratory. The latter, labelled with asterisk cover a 600 million year interval of the Late Proterozoic which was poorly covered by $\delta^{18}O$ data before. One can recognize some details of the background of the general trend. The most pronounced shifts towards positive $\delta^{18}O$ values occur during periods of tectonic activities, Caledonian, Hercynian and Alpinian. Apparently during these periods the global circulation of water increased. On the contrary during periods of steady sedimentation the trend weakened or was even reversed. The ^{18}O-trend indicates unsteady state of oxygen isotope distribution between the ocean and the earth's crust which might be caused by initial depletion of the Archaean ocean in ^{18}O as suggested by Perry (1967). Some investigators, however, interpret the observed change in oxygen isotope composition of carbonates and cherts as a secondary effect which is not related to change of the isotope composition of the ocean. The matter needs further study. However the suggested depletion of the Archaean ocean in ^{18}O is most likely. I believe that this phenomenon is due to change with geological time of the relation

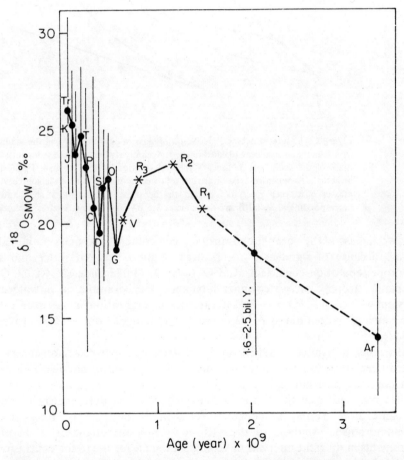

Figure 1. Variations of oxygen isotope composition of carbonates during geological time. The horizontal lines are ranges of variation of $\delta^{18}O$ (relative to SMOW) of carbonates of a given age. The dots are the averaged values. The points marked by an asterisk are the data obtained recently in our laboratory for representative samples of the Late Proterozoic (V, R_1-R_3) carbonates of the Russian platform prepared by G A Kazakov. The data for the Phanerozoic rocks are compilation from different literature sources. Archaean and the Early Proterozoic data have been taken from Veizer and Hoefs (1976).

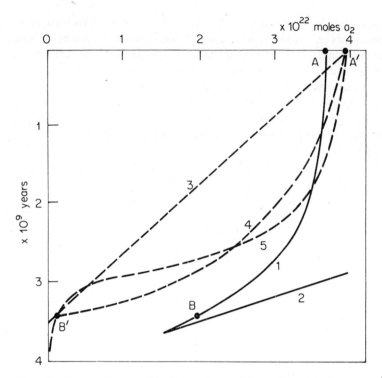

Figure 2. Suggested curves of evolution of mass of oxygen bound in the sedimentary shell (solid lines) and the ocean (dashed lines); 1. Accumulated mass of the sediments; 2. Deposited mass of the sediments; 3. Linear type of evolution of the ocean; 4. Exhaustive type of growth; 5. Exponential type of growth followed by exhaustion of the source. A and A′ are masses of oxygen reservoirs related to the present sedimentary shell and the ocean correspondingly, B and B′ are the same at 3,5 b.y. ago as suggested by the model.

between mass of the ocean (in terms of oxygen content) and mass of the sedimentary shell. Because of significant isotope fractionation between water and sediments (isotope separation coefficient $\alpha CaCO_3 - H_2O = 1\cdot028$ under 25°C), $\delta^{18}O$ of water depends strongly on water-mineral balance. For example, in a water-carbonate system with total $\delta^{18}O = +10‰$ if water-to-mineral ratio changes from 1:10 to 10:1 (in terms of oxygen mass), $\delta^{18}O$ of the water changes from $-15\cdot5‰$ to $+7\cdot5‰$ (at 25°C).

In order to evaluate oxygen isotope distribution between sedimentary shell and ocean one needs to know how their masses evolved during geological history. This is an unsolved problem and only a very approximate evaluation is attempted in the following. The growth of the sedimentary shell is determined by the rate of accumulation of the sedimentary rocks and the rate of their erosion and metamorphism. Analysis of data (Ronov 1980) on quantitative distribution of sediments in different intervals of geological time leads me to the conclusion that the decrease in amount of sediments of a given age due to erosion and metamorphism obeys the following relationship: $M_t = M_o \exp(-1\cdot35 \times 10^{-3} t)$, where M_o is the initial mass, t is time (in 10^6 years) elapsed since sedimentation and M_t is the preserved mass at t. The rate of accumulation of sediments, averaged over large intervals of time, appears to be more or less constant, given by 3×10^{23} g/100 million years. The integration then gives a generalized curve of growth of sedimentary shell as shown in figure 2. The present mass of sedimentary shell is 2.3×10^{24} g (3.63×10^{22} mol of oxygen) although over 4 billion years the sedimentation cycle produced about 12×10^{24} g of sediments. More than half of this amount was subjected to metamorphism and granitization and the rest recycled.

Since the ocean-crust system is an open system in terms of the oxygen balance, the rate of accumulation of the ocean cannot be inferred directly from the curve of growth of sedimentary shell and the known values of oxygen isotope distribution. However at the beginning of the process, a closed system model is an acceptable assumption. Calculation on this basis gives a value of about 0·05–0·08 for the ocean-sedimentary shell ratio 3·5 billion years ago. This means that the mass of the Early Archaean ocean was two orders smaller than its present value. As seen from figure 2 any conceivable curve of evolution of the mass of the ocean must begin from 3·9–4·1 billion years ago.

Thus one comes to the conclusion from the above considerations that during the first 500 million years, hydrosphere on the Earth was not significant, even if it existed at all.

5. Beginning of geological processes

After the accretion and formation of the core accompanied by evolution of great amounts of heat, the Earth became cool owing to convection and thermal radiation from the surface. Apparently during the growth of the Earth only the upper hundred kilometres of its mantle could be in the molten state. The melted layer concentrated incompatable elements. Therefore the upper mantle of the earth is enriched in such elements relative to ordinary chondrites. Solidification of the upper mantle proceeded simultaneously from below and above. The surficial plagioclases led to the formation of a primitive sialic protocrust—analogous to the lunar anorthositic crust but much more enriched in albitic component.

The residual liquid, enriched in volatiles, incompatable elements and large ion lithophyles (LIL) like K, REE, U, Th, P, Rb etc, were concentrated at a certain depth, possibly 100–200 km. This sort of matter has been recognized as a source of KREEP-material on the Moon. With the Earth enriched in volatiles relative to the Moon, the reservoir of KREEP-like material must have played a much more significant role.

The surface of the Earth was almost free of water and other volatiles with the exception of some noble gases. The landscape must have been similar to that of the Moon at present. Appearance of water completely changed the situation. As is well known, water drastically decreases the melting point of silicates. For example, dry andesite is stable up to 1100°C. However in the presence of water andesite is melted at 650–700°C and its solidus intersects the Archaean geotherm at a depth 15–20 km. Magmatic differentiation of the primitive pyroxene-plagioclase protocrust must have given rise to formation of tonalites, grey gneisses, gabbro and granulites. Increased temperature which was characteristic of the Early Archaean upper mantle led to the high degree of partial melting and formation of komatiitic lavas.

Weathering of igneous rocks by water and chemically active gases in the atmosphere supplied sedimentary material to the newly formed water basins. Such basins were obviously limited in volume and isolated from each other. The mass of sediments accumulated in such basins could exceed many-fold the mass of water contained in them. Accumulation of sediments alternating with effusive rocks resulted in formation of thick sedimentary-effusive complexes which now are known greenstone belts. At this early stage of geological history thermal conditions in the Earth were determined by heat evolved during accretion. The processes described occurred predominantly within the lithosphere possibly down to a depth of 80–150 km. Underlain layers including that enriched in KREEP-material were not touched at this stage. This layer was involved in active magmatism later when new increase of temperature occurred probably due to decay of radioactive elements. I believe that dramatic increase of the K content in rocks of the earth's crust from 2·5–2·9 billion years ago was related to volcanism which was supplied from this source.

6. Origin of the biosphere

Carbonaceous chondrites contain a great variety of organic molecules including amino acids, porphyrines, nucleic acids etc. These molecules being highly concentrated in incipient water basins must have given rise to fast prebiological evolution.

In an attempt to trace the earliest signs of life our attention was attracted by banded iron formations (BIF), firstly because rocks of this formation occur among the oldest rocks (3·8 billion years in Isua, West Greenland) and secondly because sedimentation of iron deposits calls for a source of molecular oxygen. Cloud (1976) was the first to put forward the idea that formation of the giant masses of iron ores in the Precambrian was due to release of oxygen by photosynthetic organisms. This idea has been disputed by others since molecular oxygen could be produced not necessarily by organisms but simply through photolysis of water. However, carbon isotope data support the idea that BIF is actually a product of the biosphere. Graphite is widely present in BIF. Its carbon isotope composition varies from $\delta^{13}C = -20‰$ to $-28‰$. These values are typical of photosynthetically-produced organic carbon and indicate the biological source of graphites. BIF terranes often include layers of siderites. Figure 3 presents the reactions resulting in the deposition of oxidized iron. According to this scheme the carbon of siderite is derived both from bicarbonate and biogenic source. Carbon isotope composition of siderites varies mainly from $\delta^{13}C = -6‰$ to $-13‰$. These are intermediate between $\delta^{13}C$ of biogenic and bicarbonate carbon.

Withdrawal of significant amounts of biogenic carbon from biospheric cycle during formation of iron deposits affects the carbon isotope balance and is eventually reflected in the isotope composition of contemporary marine carbonates. Depletion of the Precambrian carbonates in ^{12}C has been actually recorded. A pronounced depletion of carbonates in ^{12}C ($\delta^{13}C$ from $+8$ to $+13‰$) appears to have taken place during the Early Proterozoic (2·0–2·6 billion years ago) when 90% of the Precambrian iron ores was deposited.

Occurrences of phosphorites and apatites related to BIF are often reported. Phosphorus plays a crucial role in the biosphere. Availability of phosphorus determines the mass of biosphere. Hence the relationship between development of biosphere and volcanism. It is of interest to note that explosive proliferation of life manifesting in exceptional development of BIF occurred just after activization of the layer enriched in KREEP-material. The correlation between occurrence of BIF and

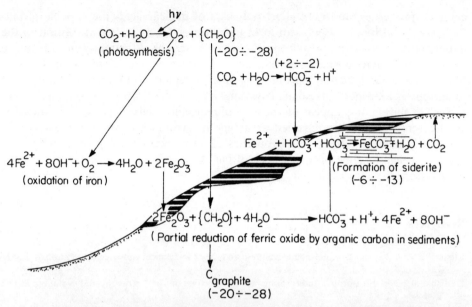

Figure 3. Isotopic evidences of involvement of biological carbon $\{CH_2O\}$ in the process of deposition of banded iron formation. Approximate carbon isotope composition ($\delta^{13}C$ in ‰, relative to PDB) of a compound is pointed out in brackets.

concentration of phosphorus is characteristic of deposits as old as 3·4–3·5 billion years.

Thus it appears that BIF is actually a phenomenon of the biosphere. Is it possible to extrapolate this conclusion to the Isua case? If it is, it would mean that photosynthetic organisms already existed 3·8–3·9 billion years ago. On the other hand as was just suggested, water and carbonaceous matter appeared on the surface of the Earth about 4·2–3·9 billion years ago. Then the time interval between the beginnings of the prebiological and biological evolution right up to photosynthesis is too short.

In this connection I suggested that photosynthesis, more precisely photochemical release of oxygen on the basis of organic compounds, appeared before life. In other words photosynthesis was suggested to be a prerequisite for the origin of life rather than its consequence. It seems logical that development of energy consuming and low entropy biological systems was preceded by the appearance of an effective and universal mechanism of conversion of external energy (light) into chemical energy. Besides experiments are known which demonstrate the release of oxygen by a simple system containing chlorophyll molecules and lipids. Porphyrins and bipolar compounds of lipid character have been found in carbonaceous chondrites and could be easily synthesized abiotically. Chlorophyll molecules on a lipid membrane are able to produce a photoeffect. The next step was the oxygen-release by the splitting of water molecules in the presence of an acceptor of hydrogen. Further development of this energy cell could result in realization of photoinduced chemical potential in the form of phosphorylation.

This could be one of the ways of prebiological evolution. In general prebiological evolution is conceivable as a separate development of functions inherent in different types of carbon compounds. In fact several combinations of amino acids may reveal catalytic properties, and even short polymers of nuclear acids are capable of self-replication and bipolar lipids can form membranes. I believe that transition to the living matter consists in putting together these functions. Apparently the advantage achieved by interconnection of different components and coordination of their functions is as fundamental a property of open system as disintegration and increase of entropy is a property of isolated systems.

Such evolution may appear on the level of molecules, cells, organisms, or societies. In this sense life can be regarded as a type of evolution of matter. It cannot be adequately defined through the properties of an object. Therefore it is difficult to indicate a definite time of appearance of life. As a type of evolution of material it existed from the very beginning.

References

Cloud P E 1976 Beginnings of biosphere evolution and their biochemical consequences; *Paleobiol.* **2** 357–387

De Paolo D J 1981 Nd isotopic studies: Some new perspectives on earth structure and evolution; *Eos* **62** 137–140

Dubrovskij B A and Pan'kov V L 1972 On the composition of the earth core; *Izv. Akad. Nauk. SSSR Fiz. Zemli* **7** 48–54 (in Russian)

Eneyev T M and Kozlov N N 1977 Numerical modelling of the process of formation of planets from protoplanetary nebula; *Preprint of the Institute of Applied Mathematics Acad. Sci. USSR* p. 80 (in Russian)

Galimov E M 1968 Geochemistry of stable carbon isotopes (Moscow: Nedra) p. 224 (in Russian)

Gurevich L E and Lebedinskij A K 1950 *Izv. Akad. Nauk SSSR Ser. Phyzicheskaya* **14** 765 (in Russian)

Perry E C 1967 The oxygen isotope chemistry of ancient charts; *Earth Planet. Sci. Lett.* **3** 62–66

Ringwood A E 1979 Origin of the earth and moon, (New York: Springer Verlag) p. 293

Ronov A B 1980 Sedimentary shell of the earth; *Nauka* (*Moscow*) p. 80 (in Russian)

Safronov V S 1969 Evolution of protoplanetary nebula and formation of the earth and planets; *Nauka* (*Moscow*) p. 244 (in Russian)

Veizer J and Hoefs J 1976 The nature of $^{18}O/^{16}O$ and $^{13}C/^{12}C$ secular trends in sedimentary carbonate rocks; *Geochim. Cosmochim. Acta* **40** 1387–1395

Life on the early Earth: Bridgehead from Cosmos or autochthonous phenomenon?

MANFRED SCHIDLOWSKI

Max-Planck-Institut für Chemie (Otto-Hahn-Institut), D-6500 Mainz, W. Germany

Abstract. There is by now ample paleontological and biogeochemical evidence indicating that the Earth was inhabited by life over almost 4 Gyr of geological history. Specifically, there is little doubt that photoautotrophy as the quantitatively most important process of biological carbon fixation had been operative as a biochemical process and as a geochemical agent since at least 3·8 Gyr ago. With an age of the Earth close to 4·5 Gyr, the question seems permissible whether the remaining time interval could have satisfied the temporal requirements for both early chemical evolution and the rise of the oldest life forms to the level of the prokaryotic cell that had abounded in microbial communities as from 3·5 Gyr ago (and probably before). These difficulties would be largely overcome if the juvenile planet had been vaccinated by either an extraterrestrial organic molecular seeding or by protobionts of lower organizational levels that found the terrestrial surface a hospitable setting for subsequent evolution. Recent astrophysical data and the latest findings of space life sciences seem to suggest that a tentative reappraisal of older panspermistic concepts might be in order.

Keywords. Life on Earth; chemical evolution; organic evolution; temporal constraints; panspermia; paleontological evidence; biogeochemical evidence; autochthonous phenomenon.

1. Introduction

During the last decade, several disparate lines of evidence have merged to indicate that the early Earth was inhabited by life as from almost 4 Gyr ago. Respective inferences are based on an impressive set of both paleontological and biogeochemical data which, in concert, leave virtually no doubt that the planet had hosted prolific microbial (prokaryotic) ecosystems not long after its formation (cf. Schopf 1983; Schidlowski 1984, 1987).

Accepting an age of the Earth between 4·55 and 4·57 Gyr based on latest lead isotope systematics (cf. Faure 1986; p. 311 ff.), and 3·8 Gyr as the time by which a quasi-modern biogeochemical carbon cycle was in operation (Schidlowski *et al* 1979) and from which the oldest putative (though not undebated) morphological evidence of life has been reported (Pflug and Jaeschke-Boyer 1979; Robbins 1987), we would end up with an uncomfortably short time interval of 0·7 Gyr for early chemical and organic evolution that must have preceded the appearance of the oldest life on Earth. Since an appreciable time span was required for the cooling of the planetary surface, and since the ancestral lines of the oldest prokaryotes had probably extended well beyond the benchmark of 3·8 Gyr, the estimate of 0·7 Gyr for this interval is certainly a conservative one, the period actually available probably bracketing not more than 0·3–0·4 Gyr. Hence, the question may be raised whether a time segment of such brevity could have satisfied the temporal requirements for an Earth-based evolution of life from the first simple organic precursor molecules to the level of the prokaryotic cell. These difficulties would be proportionately diminished if the ancient Earth had been inoculated by some organic "molecular seeding" from space, or by extraterrestrial protobionts of lower organizational levels that found the terrestrial surface a hospitable environment for their subsequent evolution. Although the empirical evidence bearing on these questions is still deplorably scanty, it seems nevertheless worthwhile to attempt a reappraisal of older panspermistic concepts in the light of recent astrophysical data and the latest findings of space life sciences.

2. Antiquity of life on Earth: summary of empirical evidence

In order to properly assess the time interval between the formation of the Earth (4·55 Gyr) and the appearance on the terrestrial surface of the first prolific microbial ecosystems with the concomitant establishment of a biogeochemical carbon cycle, a brief review of the oldest paleontological and biogeochemical evidence would seem in order.

Relatively easy to comprehend is part of the paleontological evidence made up of *biosedimentary structures* that mainly consist of aggregates of bun-shaped, interfering laminae conventionally termed "stromatolites" (figure 1). Such structures are known to be generated within present-day aquatic environments at the sediment-water interface as a result of either trapping or precipitation of detrital or chemical sediments, respectively, by mat-forming microbenthos of preferentially prokaryotic affinity. The interpretation of these stromatolites as lithified microbial mats or "microbialites" (cf. Burne and Moore 1987) rests on a straightforward morphological analogy to benthic or "surface dwelling" microbial colonies from contemporary aquatic habitats. Hence, whenever we encounter in geological record laminated and domical structures of the type represented in figure 1, they convey a piece of crucial paleontological information, namely, that communities of benthic prokaryotes (usually dominated by cyanobacteria) had been extant at the time the host sediments were formed.

It is by now firmly established that lithified microbial mats and related structures figure among the most conservative and persistent features of the paleontological record, covering a time span of > 3·5 Gyr from their first appearance in Early Archaean rocks until their manifestations in the present world (Walter 1983). It has also been demonstrated that the most ancient stromatolites are basically indistinguishable in their morphology from those occurring in geologically younger sediments, and that they can be well interpreted in terms of modern (Holocene) analogs. Moreover, there is reasonable evidence that the principal microbial stromatolite builders of Archaean times were filamentous and unicellular prokaryotes (often sheath-enclosed) entertaining a photoautotrophic metabolism and being capable of phototactic responses. Accordingly, the first occurrence of stromatolitic structures within the 3·3–3·5 Gyr-old Warrawoona Group of Western Australia (Dunlop *et al* 1978; Lowe 1980; Walter *et al* 1980) constitutes conclusive proof that prokaryotic communities were already widespread on the Archaean Earth, and that the impact of the oldest microbial biosphere on sediment-forming processes and on the terrestrial carbon cycle must necessarily date back to *at least* this time.

These conclusions are decidedly corroborated by reports of *cellularly preserved microfossils* from several Early Archaean formations including stromatolite-hosting

Figure 1. Aggregate of superimposed lithified microbial mats ("stromatolite") from the Precambrian Transvaal Dolomite Series, South Africa ($\sim 2 \cdot 3$ Gyr). The bun-shaped, partially interfering laminae represent successive growth stages of the original microbial community. The mat-building microbenthos was mainly made up of cyanobacteria (formerly known as "blue-green algae"). The oldest organosedimentary structures of this type occur in rocks $\sim 3 \cdot 5$ Gyr old, indicating that microbial ecosystems were prolific already on the ancient Earth.

ones. Although the biogenicity of the oldest putatively cellular morphologies from the 3·8 Gyr-old Isua supracrustals (Pflug and Jaeschke-Boyer 1979) has been questioned (Bridgwater *et al* 1981), there seems to be little doubt in view of other relevant evidence accrued during recent years that the ancestral lines of prokaryotes surely must extend to Isua times and possibly beyond. Moreover, additional categories of morphological evidence suggestive of microbiological affinities have been recently described from the Isua banded iron formation (Robbins 1987).

Apparently *bona fide* microbial morphologies principally comprising filamentous forms and sheath-enclosed colonial unicells that can be reasonably interpreted as fossil cyanobacteria have been documented in excellent preservation in cherty formations of the Australian Warrawoona Group (Awramik et al 1983; Schopf and Packer 1987), i.e. the same that hosts the oldest stromatolite-bearing strata (see above). With such and similar (e.g. Walsh and Lowe 1985) findings at hand, there is broad consensus today that morphologically diverse microbial ecosystems dominated by prokaryotic photoautotrophs inclusive of cyanobacteria were prolific as from at least 3·5 Gyr ago, if not appreciably earlier.

Complementary support for the inferences drawn from these paleontological data comes from *biogeochemical evidence*, notably the sedimentary carbon isotope record. It is well established today that sedimentary organic carbon or "kerogen" represents the highly polymerized (polycondensed) end-product of the diagenetic alteration of primary biogenic substances ultimately stemming from autotrophic carbon fixation (mostly photosynthesis). Moreover, it has been confirmed that all common photosynthetic pathways discriminate against the heavy carbon isotope (^{13}C), principally as a result of a kinetic isotope effect inherent in the first CO_2-fixing carboxylation reaction. This consequently leads to a preferential accumulation of the light C-isotope (^{12}C) in all forms of biogenic (reduced) carbon, with the heavy complement relegated to the residual pool of inorganic (oxidized) carbon mainly consisting of marine bicarbonate ion (HCO_3^-) and atmospheric carbon dioxide (CO_2).

The geochemical consequences of such biologically induced isotopic fractionation between organic and inorganic carbon are straightforward. Since these isotopic differences are basically "frozen" when biogenic materials and carbonate enter the sedimentary record, the kinetic isotope effect associated with autotrophic carbon fixation is propagated from the surficial exchange reservoir into the rock section of the carbon cycle. With both kerogenous substances and carbonates (limestones, dolomites) abundantly preserved in sediments, the isotopic evidence has come to be encoded in the sedimentary record as from its very onset about 3·8 Gyr ago. In the fullness of time, biologically-mediated ^{13}C/^{12}C fractionations were apt to bring about a conspicuous isotopic disproportionation of the Earth's original endowment of primordial carbon into a "light" organic and a "heavy" inorganic moiety such as stored in the forms of kerogen and sedimentary carbonate within the Earth's crust.

As had been pointed out previously (Broecker 1970; Schidlowski et al 1975), sedimentary carbonates formed in the absence of a biological carbon sink should have inherited the isotopic composition of the Earth's primordial carbon, with a δ^{13}C value close to $-5‰$ that reflects the average ^{13}C/^{12}C ratio of the principal forms of deep-seated (mantle) carbon such as carbonatites and diamonds. After the advent of life and the concomitant emergence of a biological sink, isotopically light carbon would have preferentially concentrated in organic matter, thus forcing the δ^{13}C values of carbonates from $-5‰$ in the direction of more positive (heavier) values, the magnitude of the shift depending on the percentage of total carbon ending up as organic matter. Hence, the most important single event in the evolution of the terrestrial carbon cycle was the establishment of a large-scale biological control exercised by an organic carbon sink that had emerged after the advent of photosynthetic carbon fixation. Since this time, the carbon cycle bears the mark of the Earth's biosphere and should be aptly referred to as a *biogeochemical cycle*.

As a result of the documentation of a continuous organic carbon record (in the form of kerogen and derivative graphite) over almost 4 Gyr of geological history, it could be shown that the bias in favour of the light carbon isotope (^{12}C) inherent in the principal carbon-fixing reaction of the photosynthetic pathway can be traced back over this whole time span (Schidlowski *et al* 1979, 1983; Hayes *et al* 1983; Schidlowski 1987, 1988). As is demonstrated by figure 2, the average difference in δ^{13}C of some 20 to 30‰ between organic and carbonate carbon as observed in the present environment holds over 3·5, if not 3·8 Gyr of the geological past. It might be added that the characteristic ^{12}C-enrichment in both living and fossil organic matter principally derives from the isotope-discriminating properties of one single enzyme responsible for the first CO_2-fixing carboxylation step, namely, ribulose-1, 5-bisphosphate carboxylase, the key enzyme of the Calvin cycle that channels most of the carbon transfer from the non-living to the living world. Since there is convincing evidence that the isotope shifts displayed by the 3·8 Gyr-old Isua supracrustals are due to the amphibolite-grade metamorphism to which this suite had been subjected (cf. Schidlowski 1987), the isotope age function of figure 2 can be most reasonably interpreted as an index line of biological carbon fixation over almost 4 Gyr. The uniformity through time of this isotopic signature would be consistent with the notion of an extreme degree of conservatism of the principal pathways of autotrophic carbon fixation, notably Calvin cycle or "C3" photosynthesis.

Summarizing the implications of both the paleontological and carbon isotopic evidence currently available, there can be little doubt that autotrophy was an early achievement in the evolution of life and that, in particular, communities of microbial (prokaryotic) photoautotrophs were probably prolific as from at least 3·8 Gyr ago. Accordingly, the dual nature of the carbon flux into the sedimentary shell (comprising an organic and an inorganic sink) must have been established well before that time, with all attendant corollaries and consequences for the terrestrial carbon cycle. Altogether, these findings would give eloquent testimony to the continuity of life processes on this planet over almost 4 billion years.

3. Early chemical and organic evolution on Earth: Temporal constraints

With the evidence set out above, we have to reconcile two obvious facts, namely, that the age of the Earth as a whole is about 4·55 Gyr, and that a fully evolved biogeochemical carbon cycle was ostensibly operating on its surface as from 3·8 Gyr ago suggesting the contemporaneous existence of a microbial veneer which is also documented by coeval or slightly younger paleontological evidence. It should be noted that the assumed temporal benchmark of 3·8 Gyr for the existence of this earliest terrestrial biosphere only indicates a *minimum age* as the record breaks off at this stage. In fact, with the currently available evidence at hand, a reasonable case can be made for the notion that biological processes had commenced on this planet not much later than 4·0 Gyr ago. Considering, furthermore, the requirement of approximately 0·2 Gyr for the establishment of hospitable surface conditions after the formation of the planet, the time available for the initiation of life processes on the early Earth may well shrink to as little as 0·3–0·4 Gyr. This seems, in fact, an

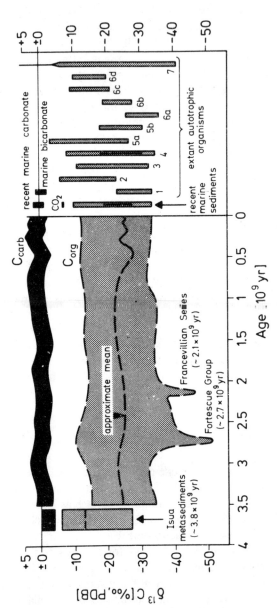

Figure 2. Isotope age functions of carbonate (C_{carb}) and organic carbon (C_{org}) over 3·8 Gyr of recorded geological history as compared with the isotopic compositions of their progenitor materials in the contemporary environment [marine bicarbonate and biogenic matter, see right box; 1 gigayear (Gyr) = 10^9 yr]. Isotopic compositions are given as $\delta^{13}C$ values indicating either an increase (+) or a decrease (−) in the $^{13}C/^{12}C$ ratio of the respective sample (in permil) relative to the PDB standard. Since organic carbon gives consistently negative readings of $\delta^{13}C$, it is correspondingly enriched in light carbon (^{12}C). Numbered groups of extant autotrophic organisms are (1) C3 plants, (2) C4 plants, (3) CAM plants, (4) eukaryotic algae, (5) cyanobacteria (natural and cultured), (6) photosynthetic bacteria other than cyanobacteria, (7) methanogenic bacteria. The envelope shown for fossil organic carbon covers an update of the data base originally presented by Schidlowski et al (1983) for some 150 Precambrian kerogen provinces; conspicuous negative offshoots such as at 2·7 and 2·1 Gyr indicate the involvement of methane-utilizing pathways in the formation of the respective kerogen precursors. Note that the isotope spreads of extant primary producers have been propagated into the record with just the extremes eliminated, the resulting isotope age function thus representing an index line of autotrophic carbon fixation over almost 4 Gyr of Earth history (both $\delta^{13}C_{carb}$ and $\delta^{13}C_{org}$ have been moderately reset by metamorphism in the 3·8 Gyr-old Isua suite). From Schidlowski (1987).

uncomfortably short time interval to accommodate the whole sequence of early chemical and organic evolution, i.e. the totality of processes that bridge the abyss between the formation of the first simple organic precursor molecules (known to exist already in the interstellar medium) and microbial life on the prokaryotic level.

Unfortunately, there is little empirical evidence available that would allow us to properly assess the temporal requirements of such an evolutionary process. With the tempo and mode notably of early chemical evolution largely depending on the *kinetics* of the principal organic reactions involved, it might be even argued that temporal constraints are rather loose, and that a probablistic approach to these problems based on a consideration of the time factor may not be relevant at all. After all, going back from the age of the Earth to that of the Universe (20 Gyr) would improve our postulated time allowance of 0·3–0·4 Gyr by two orders of magnitude only which may be still considered as small in probablistic terms. Nevertheless, the obvious fact that life—if it had ever originated on Earth—must have evolved within relatively narrow time bounds certainly deserves appropriate attention and should justify a reappraisal of alternative concepts that hold the promise of overcoming these difficulties.

4. Panspermia—a modern reappraisal

It is widely understood that the theory of *panspermia* introduced around the turn of the century (Arrhenius 1903) had lost much of its appeal during the subsequent decades when discussions on the origin of life were largely dominated (if not monopolized) by Earth-based or "autochthonous" concepts such as the Oparin-Haldane and Miller-Urey models, the hypothesis of the "prebiotic broth", and related scenarios (cf. Oparin 1938; Miller 1955). It is obvious, however, that pivotal aspects of these "terrestrial" scenarios have, of late, been subjected to severe criticism (cf. Chang *et al* 1983; King 1986; and others) whereas, on the other hand, panspermistic ideas were able to regain a moderate respectability, primarily due to the fact that the common dust particles from interstellar molecular clouds had been identified as sites of complex chemical reactions involving an appreciable number of organic compounds (Greenberg 1984). Specifically, it could be shown that among these were compounds such as hydrogen cyanide (HCN), acetaldehyde (CH_3CHO), cyanoacetylene (HC_2CN), formic acid (HCOOH), formaldehyde (HCHO) and several others (Irvine 1987) that consequently go into the making of more complex biological molecules. Moreover, there is increasing evidence (notably from the recent Giotto Mission to comet Halley) that cometary material largely resembles aggregates of interstellar dust in both structure and chemistry (Greenberg 1986), with its organic constituents continuously subjected to vigorous chemical transformations powered photochemistry. The set of molecules thereby generated could have provided a most efficient organic seeding material for prebiotic (chemical) evolution on *any* suitable planet.

Accordingly, there is little doubt today that simple organic molecules preferentially of the "prebiotic" type are widespread within, and dispersed throughout, major parts of the Cosmos, their largest concentrations occurring in interstellar dust clouds where they undergo steady recombination as a result of continuous photoprocessing. The complexity of these molecules may go well to the level of ethyl alcohol (ethanol) and probably of porphyrins, and an amusing side of the story is that the ethanol content of the interstellar medium has been estimated to surpass the order of magnitude of the mass of the Earth (10^{27} g). In sum, this and related evidence constitutes proof of a fairly complex organic chemistry operating throughout the vast stretches of the presently known Universe.

However, the panspermistic concept proper concerns itself primarily with the interplanetary or intergalactic transfer of *biological objects*. Here, gravitational considerations would confine the circle of potential space travellers to mainly microbes and spores; in fact, terrestrial airborne bacteria and spore-forming fungi have been identified at altitudes as high as 77 km (cf. Horneck 1981). With increasing upward travel, potential hazards (due to solar UV and corpuscular radiation) are proportionately increasing which, in conjunction with severe gravitational constraints, makes the escape to space of *viable* micro-organisms in the case of the Earth a rather unlikely event. Since the beginning of space travel, however, the accidental transport of microbial life remains a possibility. Alternatively, space missions offer an opportunity to deliberately expose micro-organisms to space conditions outside the Earth's gravitational and magnetic fields to study their response to such environment.

To survive in the inhospitable reaches of interplanetary and intergalactic space, life has to brave the hardships imposed by a number of environmental parameters typical of these regions, namely, (i) solar electromagnetic radiation (specifically the deleterious UV spectrum below 300 nm), (ii) solar and cosmic corpuscular radiation (mostly protons and He-nuclei), (iii) high vacuum with pressures down to 10^{-14} Pa, and (iv) extremely low temperatures approaching minima in the range of 2–4 K. Though the concerted action of these factors may result in both synergistic and antagonistic effects, their overall influence generally imposes clear-cut limits on life processes in space environments.

For an experimental approach to a viability assessment of terrestrial microbiota in outer space, the bacterial spore is a most suitable candidate as it represents the prototype of an extraordinarily resistant dormant ("cryptobiotic") cell that has proved capable of withstanding the harshest conditions ever encountered by terrestrial life in terms of heat, desiccation, radiation, etc. Accordingly, it had been singled out as an object for survival tests under genuine and simulated space conditions already during the earlier Apollo and Apollo-Soyuz Missions where respective investigations were conducted on spore assemblages of *Bacillus subtilis* (Bücker *et al* 1974; Horneck 1981). This work was followed up during subsequent Spacelab flights (notably Spacelab 1) as part of a programme to assess the response of microbial life to the potentially deleterious components of the space environment, notably the full spectrum (170–300 nm) of solar UV and high vacuum (Horneck *et al* 1984). Apart from broadening our understanding of the nature of the biochemical and histological damage caused by the principal space hazards, these programmes (along with complementary ground controls) have furnished a large set of inactivation spectra for bacterial spores giving the survival quotient N/N_0 of a population as a function of increasing UV irradiation and exposure time to space vacuum.

The results obtained by these experiments have clearly demonstrated that extreme dehydration of spores occurring under high vacuum condition causes a hypersensitivity to UV that, in turn, brings down viability counts to about 50% as compared to non-vacuum treated populations and increases mutation frequencies by a factor of 10 (Horneck *et al* 1984). The synergistic effect of these two factors is apt to substantially reduce the chances of microbial life to pass through space unharmed unless shielded against solar UV. The potentially most destructive wavelengths are those close to 260 nm as photons of this energy range are preferentially absorbed by the nucleic acids of the genetic apparatus. An ultimate limit for the survival of microbiota in space seems to be imposed by their bombardment with the heavy ions of the cosmic background radiation against which protection is virtually impossible. This has been calculated to confine the accident-free travel of a spore to a statistical maximum of 10^5 to 10^6 years (Horneck 1981). Such findings would be certainly compatible with panspermistic concepts in the *interplanetary* range, but would pose severe difficulties for respective communication on the *intergalactic* scale.

Recently, however, Greenberg and Weber (1985) have submitted a feasibility study of the survival chances of *Bacillus subtilis* spores under simulated interstellar space conditions that makes allowance for the extremely low temperatures of the interstellar medium. Viability tests were conducted in a vacuum chamber (equipped with UV irradiation facilities and a cold finger) originally designed to study the chemical evolution of interstellar dust grains. A surprising, but obviously predictable (cf. Ashwood-Smith *et al* 1968) result of these tests was that in the intergalactic temperature range close to absolute zero the sensitivity of spores to harsh UV was dramatically diminished as is testified by relatively "flat" inactivation curves at $T = 10$ Kelvin (figure 3). However, even with such ultra-deep-freeze protection, the ten-percent survival rate ($N/N_0 = 10^{-1}$) was calculated to only hold for exposure times on the order of 100 years in the diffuse (low-density) regions between interstellar clouds, offering a survival chance of some 1·000 years just for a mere 10^{-5} of an original spore population. On the other hand, the environment of common dense molecular clouds

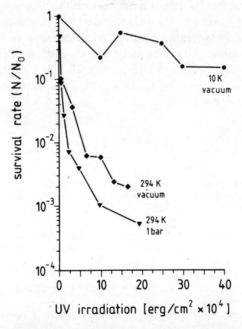

Figure 3. Survival function N/N_0 for spores of *Bacillus subtilis* (strain TKJ 6323) at 10 and 294 K after ultraviolet exposure. Note that at extremely low temperatures such as prevailing in the interstellar medium bacterial spores display a remarkable degree of resistance against UV-induced inactivation. After Greenberg and Weber (1985).

could be shown to not only substantially attenuate the incident ultraviolet, but to be also conducive to the accretion of UV-absorbing molecular coatings on vagabonding spore-like objects. Taken in conjunction, these two factors were estimated to push up permissible exposure times for ten-percent survival to some 10 million years. As this is the order of magnitude for the passage of a molecular cloud from one solar system to another (at velocities of $\sim 10 \text{ km s}^{-1}$), the random motion of such clouds might, in principle, provide a convenient vehicle for an intergalactic transfer of microbial life forms. The most hazardous parts of any long-distance travel would be the times spent outside the shelter of the interstellar medium, i.e. the departure from and the re-entry into, a host planet (Greenberg and Weber 1985). These segments involve an exposure to solar UV that surpasses the radiation in low-density clouds by a factor of 10^{10}. Here, safe passage will be contingent on the previous acquisition of UV-absorbing mantles or other protective strategies; micron-thick dust coatings have been considered sufficient for this purpose.

5. Summary and conclusions

With the astrophysical and exobiological evidence presently at hand, it can be stated with reasonable confidence (i) that a fairly complex organic chemistry is operating throughout the Universe (specifically in the interstellar medium), and (ii) that the space environment does not preclude the migration of simple life forms on the organizational level of bacterial spores over astronomical distances. This being granted, current difficulties with the establishment of life on the early Earth within an uncomfortably short time interval (probably not more than 400 million years) would be proportionately diminished if it were permissible to entertain the notion that the planet had been seeded with either organic material of interstellar or cometary provenance or even simple life forms during its early history. Although important facets of such scenario remain conjectural at this stage, it is felt that a panspermistic approach to the initiation of life processes on Earth would not appear outside the mainstream of contemporary scientific thought.

References

Arrhenius S 1903 Die Verbreitung des Lebens im Weltenraum; *Umschau* **7** 481–485
Ashwood-Smith M J, Copeland J and Wilcockson J 1968 Response of bacterial spores and *Micrococcus radiodurans* to ultraviolet irradiation at low temperatures; *Nature (London)* **217** 337–338
Awramik S M, Schopf J W and Walter M R 1983 Filamentous fossil bacteria from the Archaean of Western Australia. In *Developments and interactions of the Precambrian atmosphere, lithosphere and biosphere (Developments in Precambrian Geology 7)* (eds) B Nagy, R Weber, J C Guerrero and M Schidlowski (Amsterdam: Elsevier) 249–266
Bridgwater D, Allaart J H, Schopf J W, Klein C, Walter M R, Barghoorn E S, Strother P, Knoll A H and Gorman B E 1981 Microfossil-like objects from the Archaean of Greenland: A cautionary note; *Nature (London)* **289** 51–53
Broecker W S 1970 A boundary condition on the evolution of atmospheric oxygen; *J. Geophys. Res.* **75** 3553–3557
Bücker H, Horneck G, Wollenhaupt H, Schwager M and Taylor G R 1974 Viability of *Bacillus subtilis* spores exposed to space environment in the M-191 experiment system aboard Apollo 16. In *Life sciences and space research XII* (ed.) P H A Sneath (Berlin: Akademie-Verlag) 209–213

Burne R V and Moore L S 1987 Microbialites: Organosedimentary deposits of benthic microbial communities; *Palaios* **2** 241-254

Chang S, DesMarais D, Mack R, Miller S R and Strathearn G 1983 Prebiotic organic synthesis and the origin of life; in *Earth's earliest biosphere: Its origin and evolution* (ed.) J W Schopf (Princeton, N.J.: Princeton University Press) 53-92

Dunlop J S R, Muir M D, Milne V A and Groves D I 1978 A new microfossil assemblage from the Archaean of Western Australia; *Nature (London)* **274** 676-678

Faure G 1986 Principles of isotope geology (New York: Wiley) 2nd ed. XV + 589 pp

Greenberg J M 1984 The structure and evolution of interstellar grains; *Sci. Am.* **250** 124-135

Greenberg J M 1986 Evidence for the pristine nature of comet Halley. In *The comet nucleus sample return mission* (Brussels: European Space Agency) Special publication **249** 47-55

Greenberg J M and Weber P 1985 Panspermia—a modern astrophysical and biological approach. In *The search for extraterrestrial life: recent developments* (ed.) M D Papagiannis (Dordrecht: Reidel) 157-164

Hayes J M, Kaplan I R and Wedeking K W 1983 Precambrian organic geochemistry: Preservation of the record. In *Earth's earliest biosphere: Its origin and evolution* (ed.) J W Schopf (Princeton, N.J.: Princeton University Press) 93-134

Horneck G 1981 Survival of microorganisms in space: A review; *Adv. Space Res.* **1** 39-48

Horneck G, Bücker H, Reitz G, Requardt H, Dose K, Martens K D, Mennigmann H D and Weber P 1984 Microorganisms in the space environment; *Science* **225** 226-228

Irvine W M 1987 Chemistry between the stars; *The Planetary Report* **7(6)** 6-9

King G A M 1986 Was there a prebiotic soup? *J. Theor. Biol.* **123** 493-498

Lowe D R 1980 Stromatolites 3,400-Myr-old from the Archaean of Western Australia; *Nature (London)* **284** 441-443

Miller S L 1955 Production of some organic compounds under possible primitive earth conditions; *J. Am. Chem. Soc.* **77** 2351-2361

Oparin A I 1938 *The origin of life* (New York: Macmillan)

Pflug H D and Jaeschke-Boyer H 1979 Combined structural and chemical analysis of 3,800-Myr-old microfossils; *Nature (London)* **280** 483-486

Robbins E I 1987 *Appelella ferrifera*, a possible new iron-coated microfossil in the Isua iron-formation, southwestern Greenland. In *Precambrian iron-formations* (eds) P W U Appel and G L LaBerge (Athens: Theophrastus Publications) 141-154

Schidlowski M 1984 Biological modulation of the terrestrial carbon cycle: Isotope clues to early organic evolution; *Adv. Space Res.* **4(2)** 183-193

Schidlowski M 1987 Applications of stable carbon isotopes to early biochemical evolution on earth; *Annu. Rev. Earth Planet. Sci.* **15** 47-72

Schidlowski M 1988 A 3800-million-year isotopic record of life from carbon in sedimentary rocks; *Nature (London)* **333** 313-318

Schidlowski M, Appel P W U, Eichmann R and Junge C E 1979 Carbon isotope geochemistry of the 3.7×10^9 yr old Isua sediments, West Greenland: Implications for the Archaean carbon and oxygen cycles; *Geochim. Cosmochim. Acta* **43** 189-199

Schidlowski M, Eichmann R and Junge C E 1975 Precambrian sedimentary carbonates: Carbon and oxygen isotope geochemistry and implications for the terrestrial oxygen budget; *Precambrian Res.* **2** 1-69

Schidlowski M, Hayes J M and Kaplan I R 1983 Isotopic inferences of ancient biochemistries: Carbon, sulfur, hydrogen and nitrogen. In *Earth's earliest biosphere: Its origin and evolution* (ed.) J W Schopf (Princeton, N.J.: Princeton University Press) 149-186

Schopf J W (ed.) 1983 *Earth's earliest biosphere: Its origin and evolution* (Princeton, N.J.: Princeton University Press) XXV + 543 pp

Schopf J W and Packer B M 1987 Early Archaean (3·3 billion to 3·5 billion-year-old) microfossils from Warrawoona Group, Australia; *Science* **237** 70-73

Walsh M M and Lowe D R 1985 Filamentous microfossils from the 3,500-Myr-old Onverwacht group, Barberton Mountain Land, South Africa; *Nature (London)* **314** 530-532

Walter M R 1983 Archaean stromatolites: Evidence of the earth's earliest benthos. In *Earth's earliest biosphere: Its origin and evolution* (ed.) J W Schopf (Princeton, N.J.: Princeton University Press) 187-213

Walter M R, Buick R and Dunlop J S R 1980 Stromatolites 3,400-3,500-Myr old from the North Pole area, Western Australia; *Nature (London)* **284** 443-445

Factors controlling the distribution of ^{10}Be and ^{9}Be in the ocean

TSUNG-HUNG PENG[1], TEH-LUNG KU[2], JOHN SOUTHON[3], C MEASURES[4] and W S BROECKER[5]*

[1] Oak Ridge National Laboratory, Oak Ridge, TN 37830, USA
[2] University of Southern California, Los Angeles, CA 90089, USA
[3] McMaster University, Hamilton, Ontario L8S 4M1, Canada
[4] Massachusetts Institute of Technology, Cambridge, MA 02139, USA
[5] Lamont-Doherty Geological Observatory, Palisades, NY 10964, USA

Abstract. Observations show that while the concentration of ^{9}Be in the deep sea is geographically uniform, that of ^{10}Be increases from Atlantic to Antarctic to Pacific. Using a geochemical ocean box model calibrated by the radiocarbon distribution in the sea, we show that this situation requires (i) that ^{9}Be be added preferentially to the Atlantic Ocean, and (ii) that the residence time for beryllium in the sea be comparable to the oceanic mixing time. Both requirements are consistent with what is known regarding the geochemistry of beryllium.

Keywords. Beryllium concentration; beryllium distribution; geochemical ocean box model; radiocarbon distribution in sea.

1. Introduction

Measurements of the concentrations of ^{10}Be and ^{9}Be in the sea reveal an interesting global pattern (see table 1). While ^{10}Be shows a progressive increase from deep Atlantic to deep Antarctic to deep Pacific, ^{9}Be shows no significant trend. This difference in behaviour must have its roots in the fact that while the fallout of

Table 1. Concentrations of ^{9}Be and ^{10}Be measured in various parts of the ocean given as ratios to the concentration measured in the deep Pacific.

	^{9}Be	Reference*	^{10}Be	Reference*
Deep Atlantic	0·9 ± 0·2	1, 3, 4	0·4 ± 0·1	4, 7
Deep Antarctic	0·9 ± 0·2	5	0·7 ± 0·1	5, 8
Deep Pacific	1·0	1, 5, 6	1·0	4, 5
Surface Atlantic	0·9 ± 0·3	1–4	0·3 ± 0·1	4, 7
Surface Antarctic	0·4 ± 0·1	5	0·7 ± 0·1	5, 8
Surface Pacific	0·3 ± 0·1	5, 6	0·4 ± 0·1	4, 5

* References
1, Measures and Edmond (1983); 2, Measures *et al* (1984); 3, Measures (unpublished); 4, Kusakabe *et al* (1987a); 5, Kusakabe *et al* (1987b); 6, Measures and Edmond (1982); 7, Ku *et al* (1989); 8, Kusakabe *et al* (1982)

---

* To whom correspondence should be addressed

cosmogenic ^{10}Be is more or less uniform across the entire surface ocean, rock derived ^{9}Be is preferentially added by rivers entering the Atlantic. Intuition would lead one to expect that such an input pattern would yield enrichment of ^{9}Be in the Atlantic relative to the Pacific and a uniform ^{10}Be distribution. The question then is, why is this not the pattern which is observed?

The first point to be made in this regard is that intuition's pattern is to be expected only if the residence time of Be in the sea is short compared to the time constant for ocean mixing. If, for example, the residence time for Be were long compared to the ocean mixing time, the distribution of both Be isotopes should have a pattern akin to that for nutrient constituents (i.e. PO_4, NO_3, H_4SiO_4, Ba,...). Namely, both isotopes would be enriched in the deep Pacific relative to the deep Atlantic. Also, in this case the ^{10}Be/^{9}Be ratio should approach constancy throughout the sea. As the ^{10}Be/^{9}Be ratio is not constant, clearly the residence time for Be in the sea cannot be considerably greater than the ocean-mixing time. Thus, we are driven to the conclusion that the Be residence time must be neither much larger than, nor much smaller than the oceanic mixing time. Rather, the two times must be similar in magnitude. This conclusion is consistent with the radiocarbon based ocean mixing time of about 1400 years and the ^{10}Be inventory based residence time of several thousand years (Kusakabe *et al* 1987b).

If the Be residence time is similar to the oceanic mixing time, then the pacificward enrichment of ^{10}Be can be attributed to the nutrient effect, while the near uniformity in ^{9}Be can be attributed to a chance balance between the input effect on one hand and the nutrient effect on the other.

In an attempt to quantify this analysis, we have employed the 10-box Pandora, something-like-the-real-ocean geochemical model (Broecker and Peng 1986, 1987). The architecture of this model is shown in figure 1. In this model the Indian Ocean is

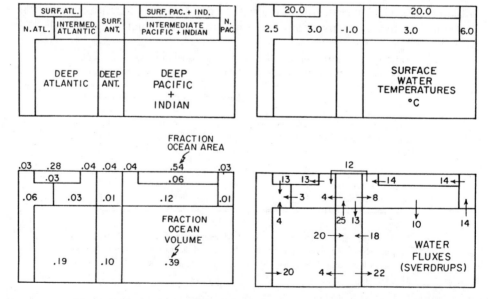

Figure 1. A ten box version of Pandora, something like the real ocean geochemical model (Broecker and Peng 1986, 1987).

taken to be a part of the Pacific Ocean. The water fluxes in this model are set to replicate the observed radiocarbon distribution in the sea.

We add ^{10}Be uniformly over the surface of the ocean. ^{9}Be is added to the surface Atlantic reservoir and to the surface Pacific reservoir. The fraction of ^{9}Be added to the Atlantic is a free parameter. Particulate cycling is handled by assigning removal times to Be in the three major surface reservoirs. All the Be removed from the surface is returned to solution in the underlying deep reservoir. This adds three more free parameters. Finally, a residence time for Be in the ocean as a whole is assigned and applied to all the model's reservoirs.

By trial and error we found the residence times with respect to particulate removal from surface water of 50 years for the surface Atlantic, Pacific and Antarctic yield satisfactory surface vs deep differences. We then made model runs for Be residence times of 1000, 2000 and 3000 years and for both 90–10 and 60–40 splits of the ^{9}Be input between the surface Atlantic and the surface Pacific. The results are shown in table 2. As in table 1, the concentrations are expressed as a ratio to the deep Pacific reservoir.

Reasonably good matches are achieved for the 90–10 input ratio and a Be residence time of 2000 years and for the 60–40 input ratio and a residence time of 1000 years. Hence our simplified ocean model confirms that the residence time of Be in the sea must be comparable to the oceanic mixing time and that more ^{9}Be must be added to the Atlantic than to the Pacific and Indian Oceans.

Our results are consistent with the residence time estimate of 1000 to 4000 years for ^{10}Be by Kusakabe et al (1987b).

Estimates for the residence time based on ^{9}Be fluxes have varied rather considerably (Measures and Edmond 1982, 1983) largely because new knowledge about the behaviour of Be in estuarine and hydrothermal systems has forced us to revise the estimates on the magnitude of Be entering the oceans. The most recent

Table 2. Model results for three Be residence times and for two choices of the fractions of the ^{9}Be input to the world ocean entering the surface Atlantic and surface Pacific reservoir.

90% Atlantic– 10% Pacific	Be 1000 yr ^{9}Be	^{10}Be	Residence 2000 yr ^{9}Be	^{10}Be	Time 3000 yr ^{9}Be	^{10}Be	Obs. ^{9}Be	^{10}Be
Surf. Atlantic	1.23	0.42	0.59	0.27	0.42	0.23	0.9	0.3
Deep Atlantic	1.85	0.76	1.16	0.65	0.93	0.60	0.9	0.4
Surf. Antarctic	1.02	0.67	0.82	0.64	0.74	0.62	0.4	0.7
Deep Antarctic	1.40	0.86	1.11	0.84	1.09	0.83	0.9	0.7
Surf. Pacific	0.27	0.48	0.23	0.34	0.22	0.29	0.3	0.4
Deep Pacific	1.00	1.00	1.00	1.00	1.00	1.00	1.0	1.0
60% Atlantic– 40% Pacific	1000 yr ^{9}Be	^{10}Be	2000 yr ^{9}Be	^{10}Be	3000 yr ^{9}Be	^{10}Be		
Surf. Atlantic	0.72	0.42	0.40	0.27	0.31	0.23		
Deep Atlantic	1.13	0.76	0.84	0.65	0.73	0.50		
Surf. Antarctic	0.75	0.67	0.69	0.64	0.66	0.63		
Deep Antarctic	1.03	0.86	0.94	0.84	0.90	0.83		
Surf. Pacific	0.42	0.48	0.30	0.34	0.27	0.29		
Deep Pacific	1.00	1.00	1.00	1.00	1.00	1.00		

estimate (Measures and Edmond 1983) based on river and hydrothermal sources puts the residence time at 3600 yrs. Some very recent unpublished data, however, show that an amount equivalent to 40% of the revised river flux is being passed to the Atlantic from the outflow of the Mediterranean. If the ultimate source of this extra Be is the partial dissolution of the eolian dust, the same as that believed to cause the elevated levels of Al, then we have to consider another important way of adding ^{9}Be to the oceans. At this time it is rather difficult to estimate the amount of ^{9}Be that might be added by this route, but we can take the estimates (Prospero 1981) of dust input to the Atlantic and Pacific (the Indian Ocean values are not considered very reliable) and make some assumptions about solubility and Be content. If we do this, then this new addition allows us to reduce our estimate of the residence time to something around 1700 years and also produces an input distribution of 60/40 (Atlantic/Indopac) similar to the observed residence time and distribution of one of the runs.

Obviously with rather a large number of barely known inputs this paper will not be the last word on residence times. However, the fact that we can get the model and the real world in some concordance is in itself significant.

Acknowledgements

We thank Lal for his many years of inspiration and leadership regarding the application of cosmogenic isotopes to the geochemical problems!

References

Broecker W S and Peng T-H 1986 Carbon cycle: 1985, Glacial to interglacial changes in the operation of the global carbon cycle; *Radiocarbon* **28** 309–327

Broecker W S and Peng T-H 1987 The role of $CaCO_3$ compensation in the glacial to interglacial atmospheric CO_2 change; *Global Biochem. Cycles* **1** 15–39

Ku T L, Kusakabe M, Measures C, Southon J R, Vogel J S, Nelson D E and Nakaya S 1989 Be isotope distribution in the Western North Atlantic: A comparison to the Pacific; *Deep Sea Res.* (Submitted)

Kusakabe M, Ku T L, Vogel J, Southon J R, Nelson D E and Richards G 1982 ^{10}Be profiles in seawater; *Nature (London)* **299** 713–714

Kusakabe M, Ku T L, Southon J R, Vogel J S, Nelson D E, Measures C I and Nozaki Y 1987a Distribution of ^{10}Be and ^{9}Be in ocean water; *Nucl. Inst. Meth. Phys. Res.* **B24** 306–310

Kusakabe M, Ku T L, Southon J R, Vogel J S, Nelson D E, Measures C I and Nozaki Y 1987b Distribution of ^{10}Be and ^{9}Be in the Pacific Ocean; *Earth Planet. Sci. Lett.* **82** 231–240

Measures C I and Edmond J M 1982 Beryllium in the water column of the central north Pacific at MANOP site R; *Nature (London)* **297** 51–53

Measures C I and Edmond J M 1983 The geochemical cycle of beryllium; a reconnaissance; *Earth Planet. Sci. Lett.* **66** 101–110

Measures C I, Grant B, Khadem M, Lee D S and Edmond J M 1984 Distribution of Be, Al, Se and Bi in the surface waters of the western North Atlantic and Caribbean; *Earth Planet. Sci. Lett.* **71** 1–12

Prospero J M 1981 Eolian transport to the world ocean; in *The sea*, Volume 7 (ed.) Emiliani (New York: Wiley) pp. 801–874

Asymptotes of coastal upwelling

P K DAS* and A P DUBE

Centre for Atmospheric Sciences, Indian Institute of Technology, New Delhi 110016, India
* Present address: Department of Ocean Development, Lodi Road, New Delhi 110003, India

Abstract. We present asymptotic solutions for small and large times of the oceanic response to forcing by atmospheric winds. They are solutions of Lighthill's equation for the propagation of a disturbance from the coast. Laplace transforms have been used, and the transform has been inverted numerically for long times. We find that the response need not synchronize with a change in wind stress.

Keywords. Asymptotes; coastal upwelling; Lighthill's equation; Laplace transforms.

1. Introduction

The response of the ocean, especially the upper mixed layer, was the topic of a remarkable paper by the late Professor Jule Charney in 1955. In this paper, he showed how coastal currents could be simulated if the atmospheric wind stress was treated as a body force acting over the depth of the mixed layer. This is relevant for the Indian summer monsoon because of a narrow inertial current off the east coast of Africa. It is often referred to as the Somali current. Recent observations suggest that the Somali current has a two gyre configuration. Initially, a single gyre, called the southern gyre, develops south of 4°N. This is attributed to the background wind. At a later stage, a second gyre, called the Great Whirl, develops between 4° and 9°N. The Great Whirl is the oceanic response to a narrow band of strong winds at an altitude of 1·5 km. This is the Findlater jet. Strong upwelling and cold surface waters in the form of wedges are observed on the northern sides of both gyres. With the progress of the monsoon the southern gyre moves northward and coalesces with the Great Whirl in the early part of September.

McCreary and Kundu (1988) modelled the two-gyre configuration of the Somali current using a $2\frac{1}{2}$ layer numerical model. This includes entrainment of colder water into the upper layer from the deeper parts of the ocean. A number of interesting problems were raised by their model. They find, for example, that if the western boundary was slanted, and there was a patch of strong winds north of the equator, then the Somali current slowly changed from a single gyre to a double gyre.

The purpose of the present paper is to examine the stability of the two-gyre system by examining the mixed layer response to (i) small and (ii) a long time after the wind stress is switched on. Asymptotic values of the Laplace transform have been used for this purpose. The propagation of a disturbance emanating from the coast has been used. This was derived in a celebrated paper by Lighthill (1969). Similar solutions have

---

* To whom correspondence should be addressed.

been found by Crepon and Richez (1982) and by Delecluse (1983). The former found the asymptotic response for long times, while the latter was concerned with different coastal boundaries and transient forcing in the form of a step function. Crepon and Richez (1982) were also concerned with transient wind stresses that were not necessarily realistic for the Somali coast. In agreement with our earlier papers on storm surges and a steady-state response of the mixed layer (Das 1981; Das et al 1983) we will consider both the divergent and the rotational parts of the wind stress. The two are of the same order of magnitude in regions near the equator. But, for the present we will not consider entrainment of cold water into the upper mixed layer, because our aim is to study the stability of a two gyre system that is generated by the wind forcing.

2. Governing equations

We consider a two-layer ocean with the lower layer at rest. The linearised shallow water equations for the upper layer are

$$u_t - fv = -g'h_x + (\Delta F/\rho), \tag{1}$$

$$v_t + fu = -g'h_y + (\Delta G/\rho), \tag{2}$$

$$h_t + H(u_x + v_y) = 0. \tag{3}$$

The zonal and meridional currents are u, v and the subscripts denote partial derivatives. The elevation of the upper free surface from its initial position is represented by h, while H is the thickness of the upper mixed layer. For simplicity we have neglected variations in the thickness of the deeper layer, but it is assumed that H is much smaller than the thickness of the deep ocean. The reduced gravity is

$$g' = g(\Delta\rho/\rho), \tag{4}$$

where $\Delta\rho$ is the difference in density between the upper and lower layer. ρ stands for the density of the upper layer, and f is the Coriolis parameter. We thus have three equations for the three unknowns u, v and h.

F and G represent the zonal and meridional components of the frictional stress. The net forcing is the difference between the wind stress at the ocean surface and the interfacial stress at the base of the mixed layer. This is represented by ΔF and ΔG. It will be assumed that the interfacial stress is small compared to the surface stress. This assumption is a little unrealistic, because the circulation is determined to some extent by the interfacial stress. But, if we assume, as observations suggest, that this is small, then our conclusions need not be invalid.

By eliminating u, v we have the following equation (Lighthill 1969) for h

$$\left[c^2\nabla^2 - \left(\frac{\partial^2}{\partial t^2} + f^2\right)\right]h_t = \frac{1}{\rho}(\nabla\cdot\tau)_t + \frac{f}{\rho}\hat{k}\cdot\nabla\times\tau, \tag{5}$$

where

$$\tau = \hat{i}\Delta F + \hat{j}\Delta G \tag{6}$$

is the frictional stress.

The speed with which an arbitrary disturbance moves away from the coast is measured by

$$c^2 = g'H, \tag{7}$$

and $\hat{k}$ is a unit vector along the vertical axis. The radius of deformation (L_R) is thus expressed by

$$L_R = c/f. \tag{8}$$

Equation (5) may be expressed in a non-dimensional form by using a rotational Froude Number (F)

$$F = f^2 L^2/c^2 = (L/L_R)^2, \tag{9}$$

where L represents a characteristic length. We put

$$X_* = X/L, \quad Y_* = Y/L, \quad t_* = ft, \quad h_* = h/H, \tag{10}$$

and

$$U_*^2 (\tau/\rho)_* = (\tau/\rho), \tag{11}$$

where U_* stands for a friction velocity.

On dropping the asterisks for convenience the non-dimensional form of (5) is

$$\left[\nabla^2 - F\left(\frac{\partial^2}{\partial t^2} + 1\right)\right] h_t = \lambda[(\nabla \cdot \tau)_t + \hat{k} \cdot \nabla \times \tau]. \tag{12}$$

The frictional coefficient λ is expressed by

$$\lambda = U_*^2/f^2 LH. \tag{13}$$

Let us next consider the Laplace transforms of h and τ. Putting

$$\hat{h}(s) = \int_0^\infty h \exp(-st) dt, \tag{14a}$$

$$\hat{\tau}(s) = \int_0^\infty \tau \exp(-st) dt, \tag{14b}$$

in (12) we find

$$[\nabla^2 - F(s^2 + 1)]\hat{h} = \lambda F[\nabla \cdot \hat{\tau} - \nabla \cdot \tau(0)/s + \hat{k} \cdot \nabla \times \hat{\tau}/s]$$

$$+ \left[\frac{1}{s}(\nabla^2 - (s^2 + 1))h(0) - h'(0) - h''(0)/s\right]. \tag{15}$$

We have denoted the initial values of h and τ by $h(0)$ and $\tau(0)$. The forcing terms in (15) are made up of the divergent and rotational parts of the wind stress and the initial values of h and τ. The primes on the right of (15) stand for time derivatives.

3. Solution for small time (t)

To simplify matters we will assume there is no background wind. Thus, $h(0)$ and $\tau(0)$ vanish and the mixed layer circulation is driven by only the divergence and the curl of the wind stress. For small time we make $s \to \infty$ so that only the divergent part of the wind stress drives the circulation. With this constraint (15) is reduced to

$$[\nabla^2 - Fs^2]\hat{h} = \lambda F(\nabla \cdot \hat{\tau}), \tag{16}$$

because all terms which have s in the denominator vanish. The solution of (16) is expressed by

$$\hat{h}(s) = \lambda F \iint_{-\infty}^{\infty} \nabla \cdot \hat{\varepsilon}(x_0, y_0) K_0(\sqrt{F} sR) dx_0 \, dy_0, \tag{17}$$

where X_0, Y_0 is the location of the disturbance and

$$R = [(X - X_0)^2 + (Y - Y_0)^2]^{1/2}. \tag{18}$$

$K_0(\sqrt{F} R s)$ is a modified Bessel function. This is the Green's function for a scalar wave equation in two dimensions, and it is the Laplace transform of a function $f(t)$ where

$$\begin{aligned} f(t) &= 0; \quad t \leqslant \sqrt{F} R \\ &= (1 \div (t^2 - (FR^2))^{1/2}); \quad t > \sqrt{F} R \end{aligned} \tag{19}$$

(Morse and Feshbach 1953). The Green's function is for an infinite ocean.

The spin-up time for specified values of F is thus $F^{1/2}R$. Considering the following representative values:

$$C = 1 \text{ m/s}, \qquad L = 100 \text{ km}$$
$$f = 10^{-5} \text{ s}^{-1}, \qquad F = 1 \cdot 0$$

and a disturbance on the coast ($x_0 = y_0 = 0$), the spin-up time at different distances from the coast are shown in table 1.

One may thus infer that the basin would begin to respond, at distances within 250 km from the coast, two to three days after the wind was switched on. Schott (1983) reported that the coastal currents are set up within days after the onset of monsoon winds. The spin-up times shown in table 1 support this observation.

4. Solution for long time (t)

We will next evaluate a solution of (15) for long times by making $s \to 0$. This will be done numerically. Assuming no background wind as before, $h(0)$ and $\tau(0)$ vanish and

Table 1. Spin-up times (T_c).

R (km)	T_c (days)
50	0·5
100	1·2
150	1·7
200	2·3
250	2·9

(15) is now

$$[\nabla^2 - F]\hat{h} = \lambda F\left[\nabla\cdot\hat{\tau} + \frac{\hat{k}\cdot\nabla\times\hat{\tau}}{s}\right]$$
$$= \phi(x,y,s). \tag{20}$$

We have retained the divergent part of the wind stress on the right of (20), because its Laplace transform which depends on the form of τ is not known. The numerical procedure for inverting $\hat{h}$ should be by a numerical solution of the scalar wave equation (20). But, a partial answer may be obtained by considering an idealized stress. The solution may be expressed in terms of a Green's function by

$$\hat{h} = \iint_{-\infty}^{\infty} \phi(x_0,y_0,s) K_0(F^{1/2}R) dx_0\, dy_0. \tag{21}$$

Comparing (21) with (17) we see that the Green's function in (21) does not involve s, which now appears only in the transform ($\hat{\tau}$). $K_0(F^{1/2}R)$ decreases with R, which means that it damps a disturbance with increasing distance (R) from its centre.

The mixed layer depth (h) can be computed by inverting $\hat{h}$. This can be done numerically if we know the transform of the stress.

Some insight may be gained if we separate the space (x,y) and time variations. Let

$$\tau = \tau(x,y)T(t)$$
$$h = h(x,y)\psi(t) \tag{22}$$

whence

$$\hat{\tau} = \hat{T}(t)$$
$$\hat{h} = \hat{\psi}(t) \tag{23}$$

provided the Laplace transforms exist.

At any fixed point in space, the increase of h with time will be obtained by inverting

$$\hat{\psi} \sim \hat{T} + \hat{T}/s$$
$$\sim \left(\frac{1+s}{s}\right)\hat{T}. \tag{24}$$

If we put

$$\hat{T} = [1 - \exp(-\alpha s)]/s^2, \tag{25}$$

then this will be the transform of a stress that increases linearly with time up to α, and then becomes steady as shown in figure 1. Combining (25) with (24) we now need to invert the integral

$$f(s) = \int_0^\infty h(t)\exp(-st)dt = \frac{(1+s)[1-\exp(-\alpha s)]}{s^3}. \tag{26}$$

It is possible to do this numerically (Doetsch 1961) by using the values of $f(s)$ at equidistant points. Putting

$$s = (2n+1)\sigma, \tag{27}$$

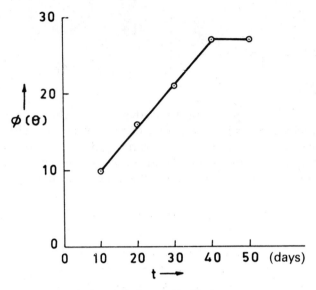

Figure 1. Variation of wind stress (τ) with time.

where σ is an arbitrary positive number and $n = 0, 1, 2, \ldots$, the integral (26) is transformed to a convenient form by the substitutions

$$h(t) = h\left(-\frac{1}{\sigma}\log\cos\theta\right) = \phi(\theta) \tag{28}$$

and

$$\exp(-\sigma t) = \cos\theta. \tag{29}$$

The transformed integral (26) is:

$$\sigma f(s) = \int_0^{\pi/2} (\cos\theta)^{s/\sigma - 1} \sin\theta\, \phi(\theta)\, d\theta. \tag{30}$$

Using (27) we have

$$\sigma f[(2n+1)\sigma] = \int_0^{\pi/2} \cos^{2n}\theta \sin\theta\, \phi(\theta)\, d\theta, \tag{31}$$

$\phi(\theta)$ may now be expanded by a Fourier series. We put

$$\phi(\theta) = \sum_{n=0}^{\infty} C_n \sin(2n+1)\theta, \tag{32}$$

where the coefficients are given by the following recurrence relations for $n = 0, 1, \ldots$

$$C_0 = \frac{4}{\pi}\sigma f(\sigma),$$

$$C_0 + C_1 = \frac{4^2}{\pi}\sigma f(3\sigma),$$

$$2C_0 + 3C_1 + C_2 = \frac{4^3}{\pi}\sigma f(5\sigma). \tag{33}$$

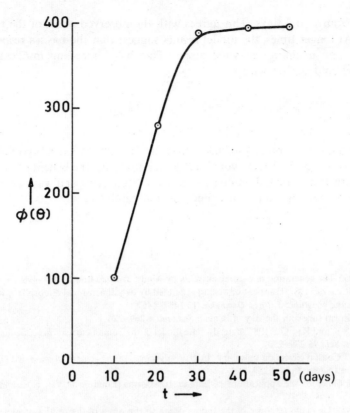

Figure 2. The basin response and its variation with time.

Further details are available in Doetsch (1961) and will not be repeated here. By using a finite number of coefficients (C_n) a partial sum, or a first approximation to $\phi(\theta)$ is obtained by (32). The appropriate value of θ for a specified time (t) is obtained from (29). For long periods ($t \to \infty$), the selected value of σ should be small because $s \to 0$.

We have inverted $\hat{h}(t)$ by computing $\phi(\theta)$ for $\alpha = 30 \cdot 0$ and $\sigma = 0 \cdot 1$. The result is shown in figure 2.

It is interesting to note that the wind stress and its growth represented by (25) is different from the response (26). This difference occurs because of the factor $(1 + s)/s$ in (24). It occurs because of the inclusion of wind stress curl.

Figure 1 shows that while the peak stress occurs after 40 days, the basin's response is reached after 30 days, i.e. about 10 days earlier. The reason for this difference is not clear, but it could be because the forcing is made up of (i) the time derivative of the stress divergence and (ii) the curl of the stress. Thus, (i) has time dependence but (ii) has no time derivative. This is being investigated.

5. Conclusion

The main results of this investigation are: (i) We have found solutions of Lighthill's equation for the propagation of a disturbance from the coast for (a) small and (b) large time after the stress was switched on. (ii) For small times the spin-up time for the basin varies from 1 to 3 days. This agrees with the observed onset of the monsoon current. (iii) At longer times, the model results suggest that the basin's response is a little earlier than a change in wind stress. This has interesting implications for upwelling and cold surface waters.

Acknowledgements

We are very happy to contribute this article on the sixtieth birthday of Professor Devendra Lal. One of us (PKD) would like to acknowledge the benefit of discussions he has had with Professor Lal on many occasions. Professor Lal always helped those who came in touch with him by his generous and candid views.

References

Charney J G 1955 The generation of oceanic currents by winds; *J. Mar. Res.* **14** 477–498
Crepon M and Richez C 1982 Transient upwelling generated by two dimensional atmospheric forcing and variability in the coastline; *J. Phys. Oceanogr.* **12** 1437–1457
Das P K 1981 Storm surges in the Bay of Bengal; *Sadhana* **4** 269–276
Das P K, Dube S K and Rao G S 1987 A steady state model of the Somali current; *Proc. Indian Acad. Sci. (Earth Planet. Sci.)* **96** 279–290
Delecluse P 1983 Coastal effects on upwelling; In *Hydrodynamics of the equatorial ocean* (ed.) C J Nihoul (Amsterdam: El Sevier Publn) 259–279
Doetsch G 1961 Guide to the applications of Laplace transforms (London: D Van Nostrand Co. Ltd.) 255 pp
Lighthill M J 1969 Dynamic response of the Indian ocean to the onset of the southwest monsoon; *Phil. Trans. R. Soc. London* **A265** 49–93
Morse P M and Feshbach H 1953 Methods of theoretical physics (New York: McGraw Hill) 1360 pp
McCreary J P and Kundu P K 1988 A numerical investigation of the Somali current during the southwest monsoon; *J. Mar. Res.* **46** 25–58
Schott F 1983 Monsoon response of the Somali current and associated upwelling; *Prog. Oceanogr.* **12** 357–381

Hydrogeology: A new focus

T N NARASIMHAN
Earth Science Division, Lawrence Berkeley Laboratory, 1 Cyclotron Road, Berkeley, California 94720, USA

Abstract. Hydrogeology is concerned with those geological processes that are influenced by water in the earth's crust. Although earth scientists have for long given consideration to the presence of water and other fluids in understanding earth processes, a perception has been gradually emerging that new insights could perhaps be gained by focussing attention on the fluids themselves and the manner in which they interact with the host rocks. In this essay, an attempt is made to briefly review current status of hydrogeologic knowledge as evidenced by key developments in several earth science disciplines. Field observations on systems with such widely varying characteristics as near-surface soils on the one hand and the oceanic crust beneath great depths of water on the other, together attest to the important role played by water and other subsurface fluids in influencing the physico-chemical evolution of the earth's crust. Current trends suggest that hydrogeologic processes will, in the future, receive a more integrated attention from earth scientists than has hitherto been the case. Remarkable developments in field instrumentation promise to provide new data from hitherto inaccessible geologic environments shedding light on the evolution of the crust. Developments in computational technology will enable the analysis of coupled nonlinear processes of transport and deformation that have so far defied quantification. If present trends continue, a broad-based, process-oriented hydrogeologic curriculum should evolve in the teaching of the earth sciences, facilitating strong interdisciplinary collaboration among earth scientists with subsurface fluids providing the motivation for synthesis and unification.

Keywords. Hydrogeology; oceanic crust; groundwater hydrology; groundwater contamination; waste disposal; geomorphology; watersheds; weathering; pedology; regional groundwater systems; deformational phenomena; plate margins; instrumentation; migration of fluids; mass and energy transport; fluid-rock interactions.

1. Introduction

Earth scientists have for long understood the important role of water in influencing geological processes in the earth's crust. Indeed, workers in the traditional branches of geology such as geomorphology, petrology, mineralogy, geochemistry, structural geology and economic geology have always had to account for the role of water in interpreting various observed phenomena. Nonetheless, over the past 25 years or so a perception has been growing among some earth scientists that perhaps new insights into earth processes could be gained by focussing attention on the role of the fluid. We will here use the term *hydrogeology* to denote the study of those geological processes that are influenced by water in particular and by other subsurface fluids in general. This definition is very broad in its scope. Yet, it truly reflects the enormous influence exerted by these ubiquitous fluids on the evolution of the earth's crust.

The acquisition of field data and the development of new ideas regarding hydrogeological processes are now so rapid that even an overview of the field will require considerable effort far beyond the scope of this work. Nonetheless, it is of value to pause and survey the field, even though briefly, to obtain a feel for the overall picture in order to synthesize seemingly unconnected knowledge. Towards this modest purpose, this paper is written as an essay in which emerging trends in hydrogeology, as perceived by the author, are outlined. As will be obvious from the following pages, the thoughts and ideas discussed are so freely drawn from the literature that a deliberate decision is made to avoid any formal citation of references.

An important goal of this paper is to argue that water and other subsurface fluids constitute a unifying theme for understanding diverse processes in the earth's crust. In support of this argument, the paper starts with an outline of key observational evidences drawn from several disciplines. This is followed by a brief description of the physico-chemical processes which provide the theoretical basis for integration. The practical implementation of the integration has received an enormous impetus within the past two decades from spectacular developments in computational technology, instrumentation, data acquisition systems and graphical portrayal of information. Aspects of these developments are discussed next in the context of quantifying hydrogeologic processes. The paper concludes with some thoughts on expected future developments in hydrogeology.

The motivation for venturing into this reflection arises from a desire to honour a distinguished scientist, Professor Devendra Lal, for his contributions to isotope geochemistry.

2. Observational knowledge

Our present knowledge of hydrogeologic systems is a result of collective observational wisdom from a large number of geologic disciplines. It is therefore appropriate to survey the current status of the field by examining recent developments in some of the disciplines of interest.

2.1 *Groundwater hydrology*

The nucleus for the thought that hydrogeology should be a discipline in itself perhaps had its origin in the work of groundwater hydrologists early in the twentieth century. Given the task of developing water supplies for communities, for agriculture and for industrial needs, these workers identified patterns of regional groundwater motion, recognized the analogy between motion of fluids in geologic systems and the conduction of heat in solids, and systematized the physics of transient fluid flow in porous media in terms of potential theory. Until about the 1970's when their attention began to be diverted towards chemical aspects, the main focus of groundwater hydrologists rested on the physical aspects of the resource, its occurrence in situ, the economics of its extraction and the physical consequences of its removal. An impressive array of solutions to the parabolic partial differential equation describing transient fluid flow in geologic media were developed and applied with success to quantify the flow of water to wells.

Even during the early stages of large scale groundwater development around the world, groundwater hydrologists recognized that deeply buried artesian aquifers released water from storage through an elastic decrease in pore volume, despite the fact that the actual strains involved were very small and seemingly negligible. That these small strains were not in fact negligible became evident during the 1940's when noticeable land subsidence was observed in many sedimentary basins around the world when subjected to large scale withdrawal of groundwater. In some arid areas, differential land subsidence, controlled by bed-rock topography, led to spectacular opening of earth fissures at the land surface.

With systematic observations of water levels in wells, evidence continued to accumulate about the elastic deformation behaviour of water-bearing formations. Unequivocal knowledge became available that even aquifers lying at depths of a few kilometers were systematically responding to the very small strains induced by earth tides or that water levels in wells may be affected by earthquakes with foci located thousands of miles away. Thus, water wells came to be recognized as sensitive strain gauges. Groundwater hydrologists are now in a position to interpret the responses of wells to earth tides and barometric tides and estimate the in situ hydraulic properties and elastic properties of aquifers with considerable confidence. However, quantitative interpretation of response of wells to earthquake radiation is still not within the realms of quantitative interpretation.

The region between the land surface and the water table is known as the *vadose zone* or *the zone of aeration*. This region, in which water and air coexist, is of enormous importance because meteoric water must pass through this zone before reaching the water table. Moreover, this is the region from which plants derive water for their sustenance. The physics of fluid flow in the vadose zone is quite complex, being governed by the presence of multiple fluid phases such as water, vapour and air, by capillary forces at fluid interfaces, by energy transport, and by evapotranspiration effects. Starting from the first decade of this century, a considerable theoretical basis has developed to devise novel methods of observation and analysis of the hydrology of the vadose zone. The vadose zone impacts the human environment in many ways. Agricultural scientists, among others, have devoted the greatest attention to the vadose zone, focusing on the interactions between the vadose water and plants. Their goal has been to improve irrigation methods, optimize plant growth and improve water usage.

Historically, groundwater hydrologists and petroleum engineers have shared many parallel interests in the fundamental processes and mechanisms controlling the dynamics of fluids in porous media. Together, they have made considerable progress in what is popularly known as *pressure transient analysis*, which is concerned with the analysis of the history of fluid pressure variations around water wells, oil wells and gas wells under varying conditions of fluid extraction. Whereas groundwater hydrologists are generally concerned with open, predominantly single-phase fluid flow systems, petroleum engineers are preoccupied with the flow of multiple fluid phases in closed systems.

2.2 *Groundwater contamination and waste disposal*

During the first half of the twentieth century the attention of groundwater hydrologists was largely devoted to exploring new sources of groundwater and to devise efficient methods of exploiting those resources. However, over the past quarter of a century, their attention has gradually but surely been redirected to a more ominous environmental problem relating to groundwater. Throughout the industrial world, groundwater reservoirs, shallow and deep, are being contaminated by a variety of human activities: agricultural practices including irrigation, fertilization and application pesticides and soil amendments; septic tanks, sanitary wastes and landfills; abandoned mines and mine wastes; industrial waste disposal through surface facilities; and deep-well disposal of toxic wastes. To this list we may also add the degradation of groundwaters by the intrusion of salt water in coastal areas caused by over exploitation of freshwater and the consequent decline in freshwater heads.

The health hazards from these activities arise from a variety of chemical species occurring at unacceptably high levels in the groundwater both as dissolved species or as coexisting immiscible constituents. They include heavy metals, a large array of organic constituents and radioactive elements. Remediation or clean-up in emerging as a venture that must go hand in hand with the task of identifying and quantifying contamination. Accumulating field evidence indicates that the effects of contamination are dictated by very complex and sometimes very longlived physicochemical interactions between the aqueous phase and the solid phases on the one hand and between the aqueous phase and other fluid phases on the other. The interactions are often mediated by biological activities. As a consequence, remediation is coming to be recognized as an extremely difficult, time-consuming and expensive undertaking.

Closely related to the problem of groundwater contamination is that of waste disposal, especially of high level radioactive wastes and certain other highly toxic wastes. One of the preferred methods of radioactive waste disposal is that of burial in mined caverns within favourable geological formations. A candidate geological formation must necessarily act as a barrier to the migration of groundwater. Therefore, interest has developed in recent years in identifying geological formations with extremely low permeability. Considering the fact that the early focus of the field of groundwater hydrology was towards identifying and quantifying aquifers of high productivity, this new desire for identifying non-productive, very low permeability formations has led to significant changes in problem perceptions and methodologies of analysis. Typical examples of such formations of very low permeability include massive, poorly fractured igneous rock masses such as granites; dense formations of rock salt; beds of dense clay; and thick vadose zones in arid regions.

Because of the health threats posed by the highly toxic wastes, a candidate geological formation should be capable of preventing the migration of the contaminant to the accessible biological environment on a time scale of at least a thousand years or more. This time scale is far longer than is commonly encountered in engineering practice. The characterization of the geometry of the system, the physical properties of the rock and the nature of the forcing functions to enable predictions on this time scale, are unprecedented scientific challenges.

The notion of a natural analog may provide a perspective on the long term behaviour of the system response by looking at the historical behaviour of an analogous natural system. Thus, for example, one could examine the ages of groundwater in a very low permeability candidate formation to evaluate its ability to act as a natural barrier to groundwater migration. Age-dating of groundwaters with the help of stable and unstable isotopes is extremely useful in this regard. Research is active in this field. Perhaps the chief challenge confronting the isotopic dating techniques of groundwaters is not so much the quantitative estimate of isotopic concentrations at very minute levels; highly sensitive mass spectrometers provide an ability to measure extremely low isotopic concentrations. Rather it is the interpretation of the isotopic analyses. The constituents of the aqueous phase in low permeability systems interact intimately with the solid phases. Hence the isotopic age of groundwaters could be much influenced by such processes as sorption, solid diffusion, ion exchange, precipitation and dissolution of participating isotopes.

2.3 Geomorphology and watersheds

The interactions between the atmosphere, the streams on the land surface and the shallow groundwater have long been of interest to hydrologic engineers and geomorphologists. Proper management of watersheds requires that the manner in which precipitation is apportioned between groundwater, stream flow and evapotranspirative losses be quantified in space and in time. Hydrologists have continuously attempted to improve their ability to empirically relate evaporation to precipitation and meteorological factors on the one hand and to relate overland flow and base flow in streams to storm intensities on the other. An essential component of these tasks in the dynamics of water flow within the geological materials immediately below the land surface. Geomorphologists are also interested in the hydrology of water sheds but from a somewhat different, long-term perspective. Watershed hydrology is a reflection of the processes that govern the evolution of streams, the associated processes of erosion and, ultimately, the long-term evolution of the landform. Erosion being an important phenomenon of interest, geomorphologists devote special attention to the transient response of fluid pressures in soil masses at the head of incipient streams immediately following strong storm events and the effect of these processes in causing failure of hill slopes. The subsurface fluid flow regime and the erosional process in the hill slopes are known to be significantly influenced by root growth on the one hand and the activities of burrowing animals on the other.

2.4 Weathering and pedology

Weathering is the process by which rocks near the land surface are altered in situ through interactions with the atmosphere. Closely related to weathering is the process of soil formation from host rocks. Apart from the purely mechanical role that water might play in these processes by way of disintegration and transport, it also plays a profound role in controlling the chemistry of weathering and soil formation. The zone of weathering, which coincides with the vadose zone, derives one of its most important chemical attributes from its free access to oxygen in the atmosphere. Oxygen dissolved in water renders water the ability to corrode and leach multivalent metallic ions from the solid phase and mobilize them into the aqueous phase. The ability to leach is almost always enhanced by the presence of accessory sulphide minerals such as pyrite in the host rock, which, upon oxidation, produce highly corrosive sulphuric acid. In addition to oxygen, carbonic acid generated by the dissolution of carbon dioxide in water and organic acids such as humic acid generated by decaying vegetation, also contribute to the mobilization of metallic ions in the weathered zone. Once leached, the transported species may either move downward with the infiltrating water or may even be drawn up to the land surface by capillary action in conjunction with evaporative processes. The zone of weathering is thus an important link in the remobilization history of chemical elements that had previously been fixed in rocks. The chemical changes in the zone of weathering influence, in turn, the nature of water movement by altering the physical parameters affecting fluid transport such as porosity, permeability and saturation. Thus, the weathered zone is of basic importance in understanding the overall evolution of the earth's crust.

2.5 Regional groundwater systems

After crossing the vadose zone, meteoric water enters the phreatic zone at the water table. From within the phreatic zone water may return to the land surface after a relatively short time by way of evaporation or as base flow in streams. Or it may follow a tortuous path deep into the earth's crust, and may never reach the land surface for tens of millions of years. That water circulates to very great depths in the earth's crust, perhaps down to tens of kilometers, has been well documented from the study of groundwater basins, petroleum deposits, geothermal reservoirs and hydrothermal mineral deposits around the world. The notion of *regional groundwater systems* provides a framework that enables a unified appreciation of a variety of processes that are otherwise studied with limited objectives.

Meteoric water derives much of its drive for movement from potential energy in the gravitational field. Impelled by gravity, it moves down into the crust towards regions of lesser energy potential, its path being modified and channelled by the presence of formations of high resistance to fluid flow. Recent field work suggests that not only does gravity driven flow persist to great depths (several kilometers) but also that fluid flow at great depths continuously adjusts itself to the changes of loads at the land surface due either to deposition of sediments or to erosion. Whereas deposition and erosion give rise to stress changes predominantly in the vertical direction, fluid potentials in sedimentary basins also dynamically respond to changes in horizontal stresses arising from tectonic causes.

The analysis of fluid flow patterns in regional groundwater systems, especially those comprising large sedimentary basins, is a challenging task. Even as gravity drives fluids down into the earth's crust, forces derived from other processes oppose this drive and impel fluids upward, back towards the land surface. Such processes include: influx of thermal energy from depths; expulsion of water caused by sediment compaction; and the release of water from the breakdown of clay minerals during low-grade metamorphism. Zones of abnormally high fluid pressures observed in sedimentary basins around the world attest to the existence of these driving forces. The response of the overall system to these forces is dictated by the heterogeneities in the distribution of physical properties of the host rocks. Complementing these zones of anomalously high pore fluid pressures within the crust are regions of anomalously low pressures, which may act as sinks for the accumulation of fluids. These zones may owe their existence to a variety of causes.

Careful study of many mineral deposits and petroleum reservoirs provides persuasive evidence that localization of these deposits can be attributed to favourable combination of physico-chemical events involving oxidation-reduction, local minima of potential energy and host rock heterogeneity. Recent field evidence also suggests that in deep systems, ore-forming fluids may be released in episodic pulses, due to the influence of precipitation/dissolution on rock porosity, permeability and fluid pressure on the one hand, and the complementary effect of fluid pressure on rock strength and fracturing on the other.

Geothermal reservoirs around the world constitute a very interesting class of regional groundwater systems. These systems simultaneously combine features of groundwater reservoirs, incipient hydrothermal ore deposits and economic sources of thermal energy. Most of the currently exploited geothermal systems represent an active interaction between circulating meteoric water and influx of heat from depths in regions of anomalously high geothermal gradients. Due to strong rock-water interactions that prevail at high temperatures, geothermal waters are often rich in dissolved mineral constituents. A majority of the geothermal systems in the world are dominated by the presence of the liquid phase, at least before exploitation. A few geothermal systems in the world are characterized by the presence of almost pure vapour phase, in conjunction with anomalously low fluid pressures. Although it has been suggested that these peculiar deposits may be related to the existence of special conditions of fractured porous rock, their genesis has yet to be finally established.

Whereas regional groundwater systems associated with sedimentary basins are characterized by deep downward circulation, the thick vadose zone in many arid regions of the world constitute a distinct type of regional systems characterized a combination of upward movement of water (due to evaporation cycles) and long-term downward infiltration of water. The depths of these vadose zones may sometimes exceed several hundred meters. The physico-chemical processes which dictate the long-term evolution of these vadose zones are in themselves very interesting. In addition, the genesis of certain evaporite mineral deposits and of certain enriched deposits of copper is associated with these vadose zones.

The presence of very strong brines is common in many regional systems. Their occurrence may be related to long residence times and elevated temperatures, both of which are conducive to mineral dissolution. In shallow near-surface environments of playas and dry depressions in arid regions, strong brines may result due to pronounced evaporative losses of water. The complex thermodynamic characteristics of these brines and the strong dependence of brine density on temperature introduce special challenges in the analysis of transient flow patterns within these.

2.6 Deformational phenomena

Rock deformation is an essential and intrinsic process governing the transient migration of water in geologic media. Change in pore volume due to deformation of rock fabric caused by change in fluid pressure gives rocks a capacitance property to rocks in relation to fluid conduction. Indeed, recognition of this capacitance property as the cause of groundwater storage in aquifers and of the analogy between this capacitance and heat capacitance of thermal conduction were milestones in the field of groundwater hydrology during the 1930's. This non-catastrophic deformation has since been used to interpret such observed phenomena as land subsidence and earth tide response of aquifers. Apart from these, several interesting geologic processes related to catastrophic deformation (failure) are strongly influenced by the role of water and other fluids in the crust.

Dramatic correlations observed during the 1960's between deep well fluid injection patterns and earthquake swarms established the reality of *induced seismicity or triggered earthquakes*. This phenomenon has since been confirmed by the strong correlation of earthquake swarms and water level changes in man-made reservoirs in many areas around the world. These observations have lent credence to the applicability of the classical theory of failure of materials to the failure of geologic media. This theory states that sliding failure occurs along planes of weakness when shear strength is exceeded by shear stress acting on the plane.

That pore fluid pressures may in fact influence the occurrence of natural earthquakes in particular and large displacements along major thrust planes in general, had been suggested prior to the 1960's based on the premise that excessive pore pressures may weaken a fault plane. Based on the same reasoning pore pressures are also interpreted to be the cause of land slides on hill slopes, when triggered by heavy rainfall.

The inspiration to the above paradigm of weakening associated with pore pressure is the principle of *effective stress* postulated in the early 1920's. Based on careful laboratory experiments, this principle established that fluid pressures, acting outwards on the inner walls of pores cause pores to dilate and shear strength to decrease. Although the role of pressure in causing induced seismicity and land slides is reasonably well established, the magnitude of its influence in causing major earthquakes and large displacements along major thrust faults is still an unresolved issue.

2.7 *The oceanic crust and plate margins*

The spectacular developments in the earth sciences during the post second world war era, culminating in the establishment of the plate tectonic theory owe much to the early recognition by geoscientists that the oceans held a key to understanding the structural evolution of the earth. Thanks to this recognition, there has been an explosive accumulation of exciting geophysical and geological data from the oceanic crust over the past three decades. These data include information on fluid pressures, temperature distributions, heat flow patterns, water chemistry, porosity and permeability variations, and, seismic profiles of refraction and reflection within the oceanic crust. Much is now known about the structure and dynamics of various types of sedimentary prisms and detachment planes associated with plate margins as well as regions of rift and extension. Rudimentary hypotheses have been formulated about the evolution of convective hydrothermal systems within the oceanic plates, as these plates move away from the rift and are buried under later sediments. Correlations between some of these convective cells and the occurrence of ore bodies of copper and manganese have been noticed. The observation of very high pore pressures in accretionary prisms as well as the observation of zones of shear wave attenuation in seismic profiles have led to speculation that perhaps high pore pressures are important in the mechanics of plate subduction and that they influence the zones of seismicity and aseismicity beneath accretionary prisms. Attempts are already being made to interpret field data in terms of regional groundwater flow patterns within the oceanic crust, integrating fluid pressure data, heat flow measurements and aqueous geochemistry. Speculations are being made to relate these flow patterns to the known occurrence of mineral deposits within the oceanic crust.

2.8 *Instrumentation*

It is appropriate to conclude the discussion on field lore with a brief look at recent developments in instrumentation and methodologies of observation. Essentially, quantified observations available to hydrogeologists include information on fluid pressure, fluid temperature, and fluid samples for geochemical analysis. Occasionally, these may be augmented by in situ strain measurements in fluid-filled rocks. Thanks to new developments in piezoelectric technology and integrated circuitry, it is now possible to make fluid pressure measurements in hostile geologic environments. Currently available pressure transducers enable fluid pressure measurements accurate to a hundredth of a psi at a total of 10,000 psi and at downhole temperatures in excess of 300°F. Other pressure gauges, capable of lesser resolution may tolerate even higher temperatures. Time resolution of the order of a microsecond in the acquisition of fluid pressure data is currently possible. In addition, it is possible to collect geochemical samples from deep wells where fluid pressures approach lithostatic pressures and temperatures exceed several hundred degrees. Spinner-type flow meters are commonly used to measure vertical flow velocities within water wells and to draw inferences on interflow of water between formations. New developments in borehole geophysical logging enable estimation of formation porosity, fluid saturation and permeability at depths of several kilometers. Borehole televiewers and acoustic imaging instruments aid in the detection of fluid conducting fractures and their orientation.

3. The processes

3.1 *General*

Hydrogeologic phenomena are governed by a combination of several processes that occur simultaneously. These include the transient migration of fluids, the transport of energy, dissolved chemical species and suspended solids by the flowing fluid, the deformation of the host rock due to interactions between rock stresses and fluid pressure and the chemical interactions between the fluids and the rock matrix. These process interactions are constrained by spatial heterogeneities of physical properties within the system. The challenges in quantifying hydrogeologic phenomena arise not only because of the coupling that exist between several dynamic processes but also because of the profound manner in which spatial heterogeneities can give rise to a wide variety of system response.

3.2 *Migration of fluids*

Basic to analysing hydrogeologic systems is the transient flow of fluids in deformable media in which saturation of fluid phases changes with time. The fluids may be slightly compressible (water or oil) or highly compressible (e.g. air, carbon dioxide, methane). In the presence of the solid phase, interfacial forces come into play, which, coupled with forces of preferential adhesion between fluids and solids, give rise to pronounced capillary effects. The equation of state for the general multiphase system, expressed in terms of phase saturation, phase pressure, and external stresses is thus extremely complex. The motion of fluids is dictated in general by balance between forces that impel the fluid to move and resistive forces that oppose fluid motion. In gravity-driven systems, impelling forces arise primarily due to location in the gravitational field and due to compressional elastic energy stored in the fluids consequent to fluid pressure. In systems where energy transport is relevant or where fluid density changes appreciably with chemical concentration, buoyancy forces may give rise to circulation and convective motion. In some special situations when highly saline solutions are in juxtaposition with dilute solutions in media with very fine pores (dense clays or shales) osmotic potential could be important in modifying fluid motion.

The forces that resist fluid motion arise primarily from fluid viscosity and the associated frictional stress imposed on the solid surface of the porous material by viscous drag. There is reason to believe that in almost all hydrogeologic situations fluid velocities are sufficiently small that fluid flow may be assumed in general to be laminar. The equation of motion, generally known as Darcy's Law, assumes that the steady state fluid flux through a flow tube is directly proportional to the drop in potential over the flow tube, the constant or proportionality being the reciprocal resistance to flow. The resistance of the flow tube is a function of the microscopic pore geometry of the medium, the viscosity of the fluid and the macroscopic geometry of the flow tube.

The equation governing fluid flow in a subsurface geologic system is usually expressed as an equation of mass conservation, represented by a partial differential equation. Alternatively, it is convenient to express the governing equation for fluid flow in a general heterogeneous medium in terms of a variational statement. Under steady laminar flow conditions, this variational statement postulates that the flow geometry within the system will adjust itself to boundary conditions in such a fashion that the rate of energy dissipation by frictional heat loss integrated over the flow domain is a minimum. This statement could be extended to transient flow systems by adding to the rate of frictional heat loss, the rate at which energy is stored in the flow domain consequent to the changing storage of fluid in the system.

3.3 *Mass and energy transport*

By the phrase mass transport we here imply the transport of dissolved or suspended species by the flowing fluid. The importance of the transport of suspended colloidal materials and biological organisms in groundwater systems has come to be recognized only recently, with the study of chemical contamination problems. As such, we know very little about the transport processes involved. It is known from available field data that suspended particles are primarily transported by advection in the flowing water. Since colloidal particles have electrically active surfaces, they may sometimes be repelled from the pore walls and be concentrated within the central, axial part of the aqueous phase as they migrate by advection. When this occurs, the colloids may actually migrate faster than what the average bulk porous medium flow might suggest. However, they may also flocculate and be strained out of the aqueous phase.

Chemical constituents dissolved in liquid water are transported by three different mechanisms: advection, hydrodynamic dispersion and molecular diffusion. *Advection* is the process by which a dissolved constituent is transported at the average flow velocity of the bulk fluid motion. *Hydrodynamic dispersion* is an equivalent diffusion-type process that is used to represent the spreading of dissolved species on a macroscopic scale caused by the variation of pore fluid velocities on a microscopic scale. Although the validity of this representation is often questioned, it is presently the only viable model to quantify dispersive spreading of solutes in groundwater systems. *Molecular diffusion* of solutes in the aqueous phase is the third transport process. Diffusion arises from a purely random motion of ions in the aqueous phase and results in a macroscopic flow driven by concentration gradients. As a dissolved species is being transported, it may be removed from the aqueous phase either due to interactions with the solid phase or due to spontaneous decay. We will consider this process in the next section.

The transport of energy (heat) in hydrogeologic systems is quite analogous to that of chemical transport. Energy is also transported by advection, hydrodynamic dispersion and heat conduction. Because of the fast temperature equilibration between the aqueous phase and the solid phase, thermal fronts are much more strongly retarded in general than chemical fronts. Thus, hydrodynamic dispersion is not as important a process in thermal systems. On the other hand, density variations induced by temperature changes as well as modification of fluid pressures caused by thermal expansion associated with changes in fluid temperature both cause a feedback of energy transport on fluid flow. In contrast, except in situations when strong spatial variations in chemical concentrations exist (such as when dense brines are involved, or in strong evaporative regimes) the chemical transport seldom influences the fluid flow process directly. In other words, the fluid flow process and the energy transport process are more strongly coupled than the fluid flow process and the chemical transport process. In systems involving dense brines, chemical transport may influence fluid flow in two ways: either spatial variations in fluid density may cause buoyancy effects or chemical potentials may provide an osmotic drive.

3.4 Fluid-rock interactions

The interactions between the aqueous phase and the solid phase may be physical, relating to stress, strain and deformation or may be chemical. The transient motion of fluids, especially slightly compressible fluids, is invariably accompanied by changes in the pore volume occupied by the fluid. In so far as the scalar fluid pressure is concerned, the fluid merely responds to volumetric changes in the voids. But, the solid skeleton deforms only due to changes in the three-dimensional skeletal stress field. Clearly, a given pore volume change may arise due to a variety of combinations of stress changes in three dimensions. Therefore, to ensure complete consistency between fluid pressure changes and skeletal stress changes, the stress-strain equation and the fluid flow equation must be coupled. The theoretical basis for achieving such coupling was proposed almost half a century ago and is now widely in use.

A variety of processes need to be considered in regard to chemical interactions between the rock and the fluid. These include precipitation or dissolution due to oxidation-reduction and acid-base reactions; sorption by adsorption or ion-exchange and complexation; and diffusion into the solid phase. Theoretical geochemical models, developed over the past twenty years, enable analysis of the dynamic paths of reaction of several chemical species that react in the presence of known solid phases. Application of these models to field conditions has shown that mineral assemblages observed in the field can be reasonably reproduced by these models, thereby rendering credibility to the models. Perhaps the greatest challenge in the realistic implementation of these models pertains to developing a reliable thermodynamic data base for the multitude of participating aqueous species and solid phases. It is becoming increasingly clear from field evidence that many rock-water interactions in shallow systems are influenced by biological processes and by organic chemical reactions.

The concept of *kinetics* or *non-equilibrium* interactions between the fluid and the rock matrix is of considerable importance. In analysing physical interactions such as that between fluid pressure and rock deformation or heterogeneous chemical interactions, it is customary to assume, as an approximation, instantaneous equilibrium. However, this assumption of equilibrium is unacceptable in many field situations. For example, laboratory experiments have shown that rocks subjected to constant loads over prolonged periods of times (several months) gradually creep. Analogously, as chemically reacting fluid-rock systems evolve from one equilibrium state to another through paths involving phenomena such as precipitation/dissolution and hydrolysis, a finite amount of time is invariably needed to establish the equilibrium. Depending on the conditions of a particular system, this time interval may be very short, involving but a few seconds or it may be very long, involving hundreds of years or more. Accumulating evidence from the field suggests that there are many natural systems of interest in which physico-chemical equilibration is sufficiently slow to warrant the need for incorporating kinetic effects in coupling fluid transport processes with fluid rock interactions.

3.5 *The coupling*

Under the most general conditions, hydrogeologic systems are characterized by coupling between four processes, namely, transient fluid flow, stress-strain, chemical transport and energy transport. Not all couplings are equally strong, however. Perhaps, the strongest coupling relates to that between fluid flow and deformation. In this case the coupling operates both ways, fluid pressure influencing deformation and deformation in turn controlling fluid pressure changes. The coupling between fluid flow and chemical transport is often one way in that fluid flow influences chemical transport through the advective process. The reverse influence of chemical transport on fluid flow is only important in those situations where chemical concentrations appreciably change fluid density or when membrane processes may be operative. The coupling between energy transport and chemical transport is primarily one way in that temperature changes, in conjunction with fluid pressure changes significantly influence the magnitude of reaction coefficients. In the case of energy transport by fluid migration, the advective transfer of heat is strongly influenced by the nature of the fluid flow field. The fluid flow field in turn is influenced to a lesser extent by the fluid pressure changes accompanying changes in fluid temperature.

4. Quantification

Modern hydrogeology, like all other sciences, is in an era of quantification. Realistic field problems, which were too complex to be represented by closed-form solutions are now within the realms of quantified analysis using computer-based numerical methods. With the application of this powerful tool, new challenges have arisen by way of increased data needs, criteria for model verification and model validation and, improving computational efficiency.

The classical framework of the partial differential equation still forms the foundation for hydrogeological analysis. However, the four governing equations with various degrees of mutual coupling must still be solved. Considerable progress has already been made in developing appropriate numerical algorithms to solve these coupled problems. Founded on the principles of Newtonian mechanics, the classical differential equation is a deterministic statement of the physical problem. However, as the hydrogeologist seeks quantification amidst uncertainties associated with natural earth systems, new perspectives have recently evolved with a need to superimpose stochastic processes on the deterministic equations in order to quantify uncertainty.

In solving the classical differential equation using analytical methods (as opposed to numerical methods), one invariably considers either homogeneous media or very simple heterogeneities, in which the flow geometry is essentially prescribed *a priori*. However, by resorting to numerical techniques this constraint is removed and, fairly general patterns of heterogeneity can be treated. This new ability is in fact essential to quantification because the attributes of unusual hydrogeologic systems can often be explained by the particular heterogeneity associated with the system. There is therefore an enormous incentive now to devise field methods that help to characterize heterogeneity. For example, in the field of petroleum reservoir engineering, it is known that the efficiency of enhanced recovery operations is intimately linked with the spatial distribution of sand and shale lenses or that of fractures and fracture zones. Three-dimensional imaging of reservoir rocks using seismic-, acoustic- and radar tomographic techniques is an emerging technology in this regard. On a smaller scale, imaging techniques using nuclear magnetic resonance techniques are being used to visually study the dynamic disposition of multiple fluid phases in soils and rocks.

Granting that the geometry of the heterogeneities can now be described in detail, the next task is to specify the physical properties of individual entities. Almost all the physical properties used in hydrogeology are macroscopic averages and hence their magnitude are subject to the chosen scale of observation. Physical properties estimated from laboratory measurements may often be orders of magnitude different from corresponding estimates from field observations. Furthermore, some of these estimates may be significantly influenced by the disposition of the instruments used for observation or by the specific models used for inverting field data. The role of scale hierarchy on the magnitude of physical properties is an area of active research interest.

The mathematical models of hydrogeology are being extended and embellished by developments in the fields of probability and statistics. Although the Newtonian basis of cause and effect still constitutes the foundation for analysis, probability and statistics help quantify uncertainties and perturbations about a mean deterministic behaviour.

Statistical concepts are used in a variety of ways in modern hydrogeology. Hydrologists involved with long-term flood forecasting or environmental engineers concerned with the prediction of the rates of migration of contaminant plumes use statistics to quantify reliability of prediction from uncertain input data. Physical characterization of heterogeneities in hydrogeologic systems for purposes of numerical modelling is often advantageously handled through generating hypothetical heterogeneous flow domains generated as statistical realizations. Statistical models are also sometimes used as a basis for defining physical parameters. A good example of this application is the use of statistically described water velocity distributions in heterogeneous media to define the coefficient of *hydrodynamic dispersion*. This coefficient represents, in a statistical sense, an equivalent diffusion-type parameter for quantifying what is actually an advection-dominated process on a microscopic scale.

With the application of hydrogeologic models to practical problems of engineering interest, the credibility of numerical models is a matter of practical interest. The task of *verification* of a mathematical model consists in asserting that the solution it generates is in fact mathematically consistent with the governing equations it seeks to solve. The task of *validation*, on the other hand, consists in showing that the mathematical model indeed is a faithful representation of the physical system it seeks to mimic. Closed form solutions to differential equations are easily verified by simply differentiating the functions in space and in time and checking the equivalence of the derivatives. No such simple approach exists for verifying numerical solutions. Because numerical methods are traditionally treated as approximate solvers of the differential equation(s), the current philosophy is to verify numerical models against known analytic solutions, whenever such solutions are available. The fact that natural hydrogeologic systems are intrinsically heterogeneous and discrete, that we observe natural systems with discrete instruments which affect the system by the very act of observation and that modern computers provide us unprecedented abilities to manipulate numbers suggest that perhaps we need to explore alternatives to the differential equation as a means of representing hydrogeologic systems. The task of validating mathematical models is even more difficult than that of verification. Clearly, our present abilities of computational simulation far exceed our abilities to carry out controlled experiments. As a result, a given experiment may help validate only limited aspects of a mathematical model, as constrained by the number of parameters that can be reliably observed during the experiment.

Hydrogeologic models can be of use in a variety of ways. Where refined data are available, they can be used for accurate model calibration and prediction. There are situations, however, in which even the most sophisticated model currently available can be of help only in gaining semi-quantitative insights. Such indeed is the case in analysing regional hydrologic systems on a global scale, for which data are sparse and the region to model is very large, within which physical properties are averaged over very large volumes. Yet, in these cases mathematical models have been applied with success in semi-quantitatively identifying the hydrogeologic constraints of system evolution.

5. The future

As man probes the earth deeper and in ever greater detail, he finds that fluid pressures, fluid temperatures and fluid chemistry are the quantities most amenable for measurement. Thus, subsurface fluids indeed constitute an important window through which man is able to observe the evolving earth. This recognition is now quite pervasive among earth scientists. As a result, serious efforts are already being made to cultivate interdisciplinary collaborations in understanding earth processes mediated by subsurface fluids. Many academic institutions have already begun to provide a hydrogeology focus in their curriculum. It is reasonable to expect that as these curricula evolve, students in the earth sciences will have an opportunity to gain a broad-based, process-oriented training on the nature of subsurface fluids and the manner in which they may influence various processes in the crust. Such a training should prove to be very conducive for handling the interdisciplinary tasks of the future.

If the present is any indication, degradation of the quality of groundwaters by chemical contamination will constitute the single most important hydrogeologic issue of the next decade. The practical need for detecting the existence of contamination will lead to ingenious methods of sampling subsurface fluids and to methods of chemical analysis that enable unprecedented sensitivity of measuring chemical concentrations. Accumulation of voluminous data of remarkable precision will pose new challenges in regard to data interpretation. The physical characterization of the site (the geologic domain of transport) will continue to be a major problem that defies a satisfactory solution. Even if one manages to describe the geometry and the heterogeneity of the system in great detail, the existing conceptual models for chemical transport will still prove to be inadequate to quantify transport of contaminants in a rigorous way. There is a growing concern in the literature that the notion of a continuum, which is critical to the framework of the differential equation, may not be appropriate for quantifying transport in heterogeneous media in which observations are made using instruments of a finite size and a finite reaction time. As an alternative to the differential equation, approaches based on integral equations and on stochastic concepts are already being tried. It is too early to conclude if these alternate approaches will prove to be successful or if radically new approaches will have to be developed to quantify transport of contaminants in heterogeneous media.

Purely from a process point of view, the reality of heterogeneity will very much underlie future developments in hydrogeology. Fractures of all scales constitute an important class of heterogeneities that exert a profound influence on the evolution of the crust. Within the past decade there has emerged a spurt of basic research activity in understanding the transport properties of fractures relating to fluid flow, chemicals and energy. Some researchers would consider that research is still in the preliminary stages in regard to the understanding of the transport properties of fractures. If so, the next decade may witness very sophisticated research efforts in understanding fractures on all scales.

A global awareness has already emerged about the need to study atmosphere-biosphere interactions, towards a proper management of the human habitat. A key component in this linkage is the vadose zone. The physical processes that take place in the vadose zone are very complex due to multiphase flow and energy transport. The problem is further complicated by the highly transient coupling between the soil and the atmosphere. The global awareness of the environment may provide a very strong impetus towards understanding the vadose zone. In many arid regions of the world learning to manage the vadose zone may prove to be critical to judicious water management.

From the point of view of pure curiosity of understanding how the earth works and evolves, spectacular developments are likely to occur in the future. Deep drilling in the continental crust as well as in the oceanic crust will continue to provide a wealth of new data collected under extremely hostile conditions of pressure and temperature. A concerted effort is being made to make these data freely available for analysis and interpretation. Latent in these data is a wealth of new knowledge awaiting interpretation and discovery.

There is little doubt that hydrogeology, as an emerging focus, has an exciting decade to look ahead to.

Acknowledgements

I would like to thank J A Apps for many valuable criticisms of the manuscript. This work was supported by the U.S. Department of Energy through Contract No. DE-AC03-76SF00098, by the Assistant Secretary for Energy Research, Office of Basic Energy Sciences, Division of Engineering and Geosciences.

Geomagnetic paleointensity from quaternary sediments: Methodology, results and future prospects

SUBIR K BANERJEE

Department of Geology and Geophysics, University of Minnesota, 310 Pillsbury Drive, SE, Minneapolis, MN 55455, USA

Abstract. Sediments yield only a relative paleointensity record but even so, no single method has yet proved to be highly reliable for all sediments from oceans or lakes, dry or wet. Comparisons of different normalization techniques, used to counter fluctuations in the content and/or grainsize of magnetite, the carrier, show that techniques based on ARM (Anyhsteretic Remanent Magnetization) are best suited to mineral-rich sediments and provide highly reliable data. Normalizations by the method of StRM (Stirring Remanent Magnetization) are better than ARM if there is appreciable organic or clay content. A large clay content may introduce appreciable errors unless StRM experiments are carried out in solutions that simulate accurately the electrical conductivity of the original seawater.

The best of the available timeseries show that during the present normal polarity chron (Brunhès) the field has been periodic with multiple components. Maximum fluctuation has been ±50% of the average field. Although it is unclear yet if there has been a strong 10,000 year periodicity, there is considerable power in the spectrum between 10,000 and 50,000 years. Paleointensity appears to show another periodicity between 2,000 and 3,000 years which has been postulated by Olson and Hagee on the basis of directional data alone. This may be a thermally driven dynamo wave at the coremantle boundary. Future work on ^{14}C and ^{10}Be productivity may help to separate dipole components of the field from non-dipole components in a novel and unique manner.

Keywords. Geomagnetic field; paleointensity; periodicity; secular variation; quaternary sediments; sediment matrix control.

1. Introduction

For a complete description of the paleomagnetic field (i.e. the geomagnetic field from pre-historic and geologic past) it is imperative that both intensity *and* directions of the paleomagnetic vector be known. Yet it is true to say that perhaps only a few percent of the current paleomagnetic database includes paleointensity values. Their determination, qualitatively and quantitatively, is a more complex and more time-consuming problem than the determination of paleodirections of the paleomagnetic vector. In this article, however, I propose to review for the first time the presently available worldwide database of the more reliable records of geomagnetic paleointensity from the quaternary period (approximately the last 2 million years) and indicate the future directions that such research may take in the next decade or so.

There are at least 3 major benefits to global geophysics and geochemistry studies that can be accrued from a more robust quaternary paleointensity database than what is now available. They are: (i) Complete vector modelling of geomagnetic time variations which fall under the rubric of paleo-secular variation (PSV) and dipolar reversal; (ii) long duration and high resolution timeseries descriptions and frequency content determination of PSV and dipolar reversal leading to important constraints on theoretical models of the geodynamo and (iii) tests of models claiming geomagnetic field control of some parts of the geologic record of ^{14}C and ^{10}Be isotopes on Earth.

A reference to models of geomagnetic field (Lund and Olson 1987) and its source, the geodynamo, will reveal quickly that nearly 90% of the field at the Earth's surface is dipolar, appearing to be due to an imaginary magnetic dipole of intensity 8.5×10^{22} Am^2 located at the centre of the Earth and oriented (at the moment) at an angle of roughly 11° to the spin axis. The 11° departure is, however, ephemeral because when averaged over a suitably long time-window, the spin and rotation axes coincide. This fact leads to the well-known paleomagnetic formula for the determination of the paleolatitude (λ) of a plate from its geomagnetic paleoinclination (I), according to the formula:

$$\tan I = 2 \tan \lambda.$$

It is this dipole field whose intensity partially controls the fluctuations of the solar cosmic ray (SCR) in the earth's magnetosphere and in so doing provides a potential source for the *in situ* variations of the isotopes ^{10}Be and ^{14}C (Damon 1970; Lal 1988; Sharma *et al* 1983). A correct isolation of the dipole-controlled part of the production record of ^{10}Be and ^{14}C provides a very important datum for geomagnetists: long-term record of the fluctuations of the purely dipolar component of the paleomagnetic field which *cannot* be obtained from local, spot readings of paleodirection and paleointensity on the earth's surface. In the last section of this article, I will indicate how this fact can lead to new information about the behaviour of the geodynamo, an approach that has remained fallow up to now.

The rest of the article is divided into three sections. The first deals with the methodology of extracting relative paleointensity fluctuations from continuously deposited sediments. The second deals with the available relative paleointensity timeseries for the quaternary period (2×10^6 yrs BP), including its Holocene (1×10^4 yrs BP) part, and where available, comparisons between such continuous but *relative* paleointensity data and *absolute* paleointensity data points obtained from igneous rocks or archeological material. In the third and final section I will summarize the implications of the paleointensity data for both the geodynamo and for potential contributions from isotope geochemistry.

2. Methodology

2.1 *General remarks*

In this section I will look critically at the approaches that have been employed to extract relative paleointensity from sediments, problems associated with them and the prospects for improved techniques for the future.

A variety of approaches have been used to normalize the observed variation in natural remanent magnetization (NRM) so as to take into account the fluctuations in magnetic mineral content from sample to sample and depth to depth. The normalized NRM, i.e. NRM divided by a concentration-dependent magnetic parameter is then used as the relative paleointensity timeseries versus depth (time). There is a considerable difference of opinion among workers as to which magnetic parameter yields the most reliable value. However, many have not clearly understood or expressed the complexity inherent in extracting relative paleointensity from sediments.

The process of sediments acquiring NRM is best understood as a linear system with an intrinsic filter and a scrambler (or a noisemaker) delivering an output (the NRM) which may have a vastly different amplitude and phase relationship from the input (the fluctuating geomagnetic field) (Hyodo 1984; Banerjee et al 1987). For a high fidelity reproduction of true field fluctuations, it is necessary to correct for at least three non-field related phenomena which may not remain constant down a sediment core: (a) concentration and size of the magnetic grains, (b) composition and texture of the non-magnetic matrix and (c) compaction, authigenesis and diagenesis after the orientations of the magnetic grains have been "locked-in". Of these only the first question has been considered in some detail up to now, the second and the third are only beginning to be addressed. Let us consider concentration and grainsize variation first.

2.2 Approach to the problem of varying magnetite content

The importance of subtracting the influence of magnetite of varying concentration and grainsize from the observed NRM variations was recognized early and various parameters, e.g. low field susceptibility (χ), isothermal remanent magnetization (IRM) or saturation isothermal remanent magnetization (SIRM) and anhysteretic remanent magnetization (ARM) have all been tried at some time or other. It was, however, Levi and Banerjee (1976) who first discussed the question of the "best" normalizer among these and made two recommendations: (i) Since NRM is a remanent and not a field-induced property, the normalizer must also be a remanent magnetization and not, for example, low field susceptibility (χ) which is measured in the presence of a field. (ii) Since neither IRM, SIRM or ARM is an exact analogue of PDRM (post-depositional remanent magnetization), one should choose from among these three laboratory-imparted remanences the one that on AF demagnetization displays the closest behaviour to the AF demagnetization of the NRM of a sediment. Another way of saying this is that the relative fraction of remanences carried by the different coercivity (H_{cr}) fractions should be the same for the NRM and for the normalizer remanence.

These provisos have not been explicitly adhered to by many later workers, some of whom have used ARM normalization and considered it to be the preferred *universal* method recommended by Levi and Banerjee (1976) because for the particular sediments that these latter workers had used, ARM was indeed shown to have the closest resemblance to NRM so far as the distribution function of H_{cr}, i.e. $f(H_{cr})$ went. A marked improvement in the determination of $f(H_{cr})$ of ARM has now been made by Jackson et al (1988) who have developed a direct, and less error-prone method, that of measuring directly the partial ARM for distinct windows of H_{cr} or ΔH_{cr}.

The next major improvement in ARM normalization was made by King et al (1983) who established the criterion for allowable extremal fluctuations in concentration and grainsize of magnetite in a given suite of samples. This is so because both ARM and PDRM have separate and different dependences on grainsize. From the existing literature King et al (1983) found, however, that the *ratio*, PDRM/ARM, was invariant for magnetite grains of 1 μm to 20 μm, roughly the pseudo-single domain (PSD) grainsize range. Unfortunately, in the enthusiasm for extraction of relative paleointensities, this conclusion has also been mistakenly interpreted to mean that

ARM normalization will yield true relative paleointensity if it can be shown, directly or indirectly, that the magnetite content is between 1 and 20 μm. Rarely do these other authors spend enough effort to test the second criterion of Levi and Banerjee (1976) that the $f(H_{cr})$ of the NRM and ARM are indeed similar.

To summarize, the concentration and size variation of magnetite can be "normalized out" by first discovering a laboratory remanence (IRM, ARM) that has the same or similar $f(H_{cr})$ as that of the NRM. Secondly, the extremal range of grainsizes should be determined for the suite of samples and either new or archival data should be cited for a relatively slow (or zero) variation of the ratios, e.g. PDRM/ARM or PDRM/IRM, with change in grain size within the observed grainsized window.

2.3 Sediment matrix control of the derived paleointensity

In the previous discussion it has been implicitly assumed that the recorded NRM intensity is only a function of the geomagnetic field and the magnetite concentration and grainsize, and not controlled by the non-magnetic sediment matrix composition and texture. Of course, this is patently untrue when one considers the *absolute* paleointensity. For *relative* paleointensity down the length of a sediment core, it may be assumed as a first approximation that whatever the influence of sediment matrix, it remains constant throughout the lifetime of a sediment basin, a lake or deep ocean. Recent studies show, however, that clay content and electrical conductivity of sediment-water slurry can be a very strong influence on the intensity of PDRM, as much as a factor of 3 (Lu *et al* 1988, 1989). This observation has a strong repercussion on relative paleointensity methods based on laboratory redeposition of natural sediments in a known magnetic field (Tucker 1981; Thouveny 1987). Conventional redeposition followed by evaporation was used by Johnson *et al* (1948) to obtain approximate relative paleointensities but the process was made more efficient by Tucker (1981) who showed that a small sample of dense sediment-water slurry could first be stirred at ~ 1 Hz and then if allowed to dry quickly, it would acquire a stirred remanent magnetization (StRM) of the same direction and order of magnitude of intensity as the NRM. This method has been used by Thouveny (1987) whose results will be discussed in the next section.

Workers will be attracted to StRM normalization for two reasons: (i) StRM appears to mimic the PDRM process of nature better than the laboratory remanences (IRM, ARM). (ii) Because the process of StRM is less time-consuming and less wasteful of material than ordinary evaporation, it is more reasonable to adopt it for a large suite of samples.

There are two problems with StRM that should be addressed in future. Firstly, it is necessary to test it against known laboratory or ancient geomagnetic intensity values. Secondly, as Lu *et al* (1988, 1989) point out, fine grains of magnetite, clay and possibly sand carry surface electric charges causing fluctuating (attractive or repulsive) force between charged magnetic and non-magnetic grains. Add to this the ever-present Van der Waals' force among submicroscopic grains. The net coupling can be stronger than the magnetic torque and, therefore, it is necessary to perform control StRM experiments whenever clay/sand or clay/calcite ratio varies in a suite of sediments. Such control experiments in known electrical conductivity solutions and simulated ocean water conditions would yield the correct factor for extracting relative paleointensities.

2.4 Acceptability criterion for relative paleointensity methods

Whether a method "works" or not can only be determined by comparing the relative paleointensity to the known laboratory fields or the absolute paleointensities determined from lava flows, baked clays or archaeological samples. In no case should the decision be made on the basis of subjective criteria such as the expected frequency content in a timeseries. But this is where a big hindrance lies. Only for the Holocene period (10,000 yrs BP) does such absolute paleointensity data exist and when feasible, excellent correlations can be seen (King *et al* 1983). Beyond 10,000 yrs BP we have to depend on a few calibrated points and/or redeposition in the laboratory. The latter technique will always have the onus of proving that the laboratory conditions (water content, electrical conductivity, approximate field magnitude etc.) do, in fact, come close to the natural conditions.

3. Paleointensity data

3.1 *General remarks*

By absolute paleointensity we usually mean paleointensity values derived from heating and cooling a sample in a known laboratory field and then comparing such a thermoremanent magnetization (TRM) with a similar thermally imparted NRM (Thellier and Thellier 1959). Examples of such samples are archaeological bricks, potsherds and lava flows. Under the proviso that the laboratory reheating and cooling does not cause any chemical or microstructural alteration, and that the TRM acquisition is not highly anisotropic, the Thellier and Thellier method can yield absolute, that is, numerically exact paleointensities. Recently published strong exchanges among practitioners of this method, however, lead me to urge caution to would-be experimentalists (Aitken *et al* 1988a, b; Walton 1988a, b). It seems that the question of how to correct for small but ubiquitous thermochemical alteration in natural samples is still far from satisfactorily answered.

Be that as it may, radiocarbon (^{14}C) dated samples have been used to obtain absolute paleointensities which have been reviewed by McElhinny and Senanayake (1982), who have constructed interpolated curves of intensity fluctuation of the earth's virtual dipole moment (VDM). As it must be clear from the discussion in §1, paleointensity data obtained from a single site can be used to calculate the magnetic moment of a virtual dipole at the earth's centre if the original orientation of the sample can be determined. From the measured value of I, using the dipole formula in section 1 the paleolatitude λ is first calculated. Then this value of λ and the observed paleointensity (F) lead to the magnitude (M) of the dipole moment according to the formula:

$$M = (4\pi r^3 F/\mu_0)(1 + 3\sin^2\lambda)$$

where r is the average radius of the earth and μ_0 the permeability of free space.

This is one way of reducing observations from different paleolatitudes to a common datum. An alternative is to use another version of the dipole formula to calculate paleointensity at the paleoequator (F_{eq}) from a knowledge of paleolatitude. When, however, the sample's paleo-orientation is unknown, the derived paleointensity cannot be directly compared with that determined for another site. One solution that is often chosen is to plot the timeseries corresponding to the accessible ages and then superpose such different timeseries for *different* sites but the *same* time-window to obtain the mean intensity fluctuation which is that of the dipole component alone. One thing is clear, whether it is absolute or relative paleointensity that has been determined from a particular site on earth, it is an undetermined combination of dipole and non-dipole components. If, on the other hand, it is correct that the field-controlled fluctuations in ^{14}C and ^{10}Be production can be separated from other causes, such data sets can be inverted to yield true dipole field fluctuations with time (Damon 1970).

3.2 *Holocene data, comparison of absolute with relative paleointensity*

Although on one or two occasions there have been published comparisons between absolute, for example, archeointensity, and relative paleointensity series determined by SIRM normalization technique (Hyodo 1984), the more robust comparisons have usually been made with ARM normalization. King *et al* (1983) did this first for the last 4,000 yrs for North America and Constable (1985) for eastern Australia. Tauxe and Valet (1989) have recently compared archeointensity of Czechoslovakia with the relative paleointensity (ARM-normalized) for the same time-window for North Atlantic sediments. The agreements are excellent but it is also clear that there was little or no variation in the magnetite grainsize in the sediments, and in two cases out of three, the sediments were relatively free of organics and rich in minerals. Under these two provisos, ARM normalization seems to work exceedingly well.

It is also quite clear that each globally separated site produces distinctly different patterns of paleointensity variation. Therefore, we need many more critically located sites before we can construct continuous dipole intensity variations for the whole earth. In north America, there are two maxima in paleointensity at 1,000 and 2,000 yrs BP. In Europe-North Atlantic there is an indistinct peak at 2,000 yrs BP and a clear maximum at 8,000 yrs BP. In eastern Australia, on the other hand, there is no discernible peak at 2,000 yrs BP but there is one between 3,000 and 4,000 yrs BP. In all the cases the peak-to-trough fluctuation is of the order of 50% of the mean value. It must also be mentioned that if we accept the McElhinny-Senanayake (1982) tabulation and averaging of global archeointensities, none of the continuous relative paleointensity timeseries above appear to mimic it. This merely confirms the importance of long-standing non-dipole contributions to continent-sized sources of geomagnetic intensity.

3.3 *Rest of the quaternary*

It seems natural to continue the discussions of §3·2 into the time period beyond the last 10,000 yrs. The lake sediment data from eastern Australia is a composite of data from two lakes about 5 km apart, whose sediments were ^{14}C-dated. The data range from 1,000 yrs BP to 14,500 yrs BP. Except for the period 12,000 to 14,000 yrs BP when the field fell to half of the average value of the more recent period, the field shows gentle oscillations with periodicities lying within a few hundred years. There are discernible but weaker thousand year periods, but without proper numerical analysis they cannot be identified further. The North Atlantic data are from box cores which span the time period 2,000 to 22,000 yrs BP (Tauxe and Valet 1989). Except for the previously mentioned maxima at 2,000 (I find this suspect) and 8,000 yrs BP there are no regular periodicities. The field decreases smoothly from 8,000 yrs BP to 22,000 yrs

BP reaching a value about 33% of the maximum seen at 8,000 yrs BP. In some ways this can be construed to be similar to the (poorly determined) global absolute paleointensity data of McElhinny and Senenayake but caution is urged because the time resolution of both data sets is far from ideal. An observed lack of periodicity can also be laid at the door of inadequate time resolution. Very recently Thouveny and Williamson (1988) have obtained an ARM-normalized relative paleointensity timeseries for the window 12,000 to 24,000 yrs BP from Cameroon in west Africa. This series also shows a near-stationary periodic behaviour with a 50% variation about the mean. Although no periodicity analysis was conducted, the dominant period seems to fall within 2,000 to 3,000 years.

Compared to the above, relatively robust, data sets from ARM normalization, there are not that many timeseries available from SIRM or StRM normalization. Hyodo (1984) used SIRM normalization for young ($<$ 6,000 yrs BP) marine sediments from Osaka bay and then used an inverse Fourier transform to deconvolve the data to correct for the effects of time-averaging or integration common to PDRM. The data show periodicities of a few hundred years but comparison with archeointensity gives poor results. Thouveny (1987) has done a painstaking and thorough comparison of susceptibility, IRM, ARM and StRM normalizations for western European lake sediments between 10,000 and 25,000 yrs BP. He prefers ARM and StRM techniques over the others and among these two, the StRM method because it appears to show to the author more high frequencies. I believe that in the absence of cross-comparison with absolute paleointensity or laboratory redeposition check, it is well-nigh impossible to make a clear choice. Be that as it may, Thouveny (1987) finds a stable series of superposed oscillations (about a steady average field) with 50% amplitude fluctuations. By inspection, the author discerns periodicities of 11,000 years, 3,000 to 3,500 years, and 1,500 to 2,000 years, respectively. It appears to me though that most of the power is in a periodicity between 2,000 and 3,000 years.

Finally, I intend to discuss the more coarse-resolution paleointensity timeseries that have been obtained for the present normal (Brunhès) chron or the last 730,000 years. Here we are forced to compare apples with oranges. Kent and Opdyke (1977) used ARM-normalization for a core from the North Pacific, Sato (1980) has used SIRM-normalization for four cores from the West Pacific while Wu and Tauxe (1990) have given an oral presentation of data from a core of the same age from the Ontong-Java plateau which was subjected to susceptibility-normalization. Because of the variable non-dipole content of each of the records, the timeseries *should* look non-identical and indeed they do, in detail. What is surprising, however, is that the average peak-to-trough amplitude is not too different from the 50% value, as mentioned before for the more recent period. An interesting fact is that, to the naked eye, it appears that there is a component with a nearly 100,000 year periodicity, as seen in Milankovitch cycle. Kent and Opdyke (1977) and Sato (1980) did carry out careful numerical analyses of their datasets. The one periodicity that Kent and Opdyke appear to believe most strongly in is 43,000 years. Sato found most of the power in the Blackman-Tukey spectra to lie between 10,000 and 50,000 year periodicities, but he does not feel convinced about their reality because they do not appear with the same relative power in an alternative auto-regressive method of timeseries analysis. Wu and Tauxe (1990) display very clear periodic timeseries without numerical analysis. But even on inspection by the eye, the data show considerable power in the region of 20,000 to 50,000 year periodicities.

In conclusion, I have to admit that none of the above Brunhès datasets can be utilized yet to make strong conclusions about geomagnetic periodicities. Although periods in the window between 10,000 and 50,000 years are present, we cannot provide an error estimate for any of them and one has to recall that none of the cores were shown to satisfy the second criterion of Levi and Banerjee (1976), overlapping $f(H_{cr})$ of NRM and normalizing remanence for a substantial number of samples.

4. Conclusion

Before making any firm conclusions it should be pointed out that a given timeseries is only as robust as its time axis. Radiocarbon dating, nanofossil dating, etc have been used mostly, varve chronology occasionally. The reader is advised to check thoroughly the age dating technique and its associated errors before trusting the numerical value of any observed periodicity.

It is probably true to say that throughout most of the present polarity chron the field appears to fluctuate by $\pm 50\%$ about a mean value. This behaviour was assumed by Cox (1968) to be true for all times based only on the data for the last 10,000 years. That this is indeed true for the last 730,000 years is reassuring but still a bit surprising to me. Cox's next assumption, that if we neglect short periods like a few hundred years due to secular variation, the dominant next higher period is 10,000 years, has not been confirmed dramatically in this study. McElhinny and Senanayake (1982) claim that the 10,000 year period is absent but their conclusion is based on a rather small data set with large errors. Further numerical work needs to be done to discover the dominant period(s) in the 10,000 to 50,000 year band which seems to have considerable power. Kent and Opdyke's (1977) observation of a 43,000 year period, supposedly precessional, needs clearly to be investigated further for its accuracy. As Sato's (1980) detailed timeseries analysis makes it clear, we do not have clean enough data now to distinguish a 43,000 year period from the other nearby values in this broad spectrum between 10,000 and 50,000 years.

Most interesting is the observation that when the datasets are clean (King *et al* 1983; Constable 1985; Thouveny 1987; Thouveny and Williamson 1988) it is possible to identify a periodicity around 2,000 to 3,000 years. It will be recalled that such a periodicity was first clearly identified from robust numerical analysis of *directional* data in North America (Lund and Banerjee 1985) and has now been discovered by Olson and Hagee (1987) in the data of Verosub *et al* (1986) and Constable and McElhinny (1985); this has been modelled by Olson and Hagee with a thermal wind-type dynamo wave in the 100 km-thick region below the core-mantle boundary. The presence of such a periodicity in paleointensity as well appears to be an important verification of the Olson and Hagee claim, and provides a new model for the origin of secular variation with 2,000 to 3,000 year periodicity.

Past studies involving production variation of ^{14}C and ^{10}Be on the one hand and paleointensity on the other have been somewhat of a bootstrap effort with neither dataset appearing to be the quantitative standard against which the other can be tested. Recently Lal (1988) has, very appropriately, encouraged workers to investigate this area of field control of isotopic production. At the present state of both datasets, neither type can be called robust and, instead of trying to correlate the two, it is essential for paleomagnetists and geochemists to spend efforts unilaterally to provide the best paleointensity and isotopic datasets, respectively.

For the isotope geochemist, it has always been clear that it is interesting to study the midpoint of a field transition (normal to reverse, or vice-versa) when intensity may be 10% (or lower) of the full value. But what has not been emphasized is that while local paleointensity data is always a mixture of dipole and non-dipole sources, the isotopic datasets yield purely dipolar field fluctuations. This means that we have an elegant method at our disposal to study the *relative* variation of dipole and non-dipole components with time, essential for the study of regeneration and reversal of the geomagnetic field.

Acknowledgements

This article expresses my thanks to Professor Devendra Lal who encouraged me to look at the interplay between the two fields of research mentioned in the preceding section. I will also remain grateful to the late Professor Allan Cox who taught more than one rock magnetist to take an interest in the geomagnetic field's origin and evolution. Conversations with my colleagues Steve Lund, Ron Merrill, Peter Olson and Lisa Tauxe continue to be highly educational. The US National Science Foundation has kindly supported my work for the last 20 years through awards of research grants.

References

Aitken M J, Allsop A L, Bussell G D and Winter M B 1988a Determination of the intensity of the earth's magnetic field during archaeological times: Reliability of the Thellier technique; *Rev. Geophys.* **26** 3–12

Aitken M J, Allsop A L, Bussell G D and Winter M B 1988b Comment on "The lack of reproducibility in experimentally determined intensities of the earth's magnetic field" by D Walton; *Rev. Geophys.* **26** 23–25

Banerjee S K, Butler R F and Schmidt V A 1987 *Problems and current trends in rock magnetism and Paleomagnetism: Report of the Asilomar Workshop of the Geomagnetism and Paleomagnetism Section of the Geophysical Union*, (Washington D.C.: American Geophysical Union) pp. 12

Constable C G 1985 Eastern Australia geomagnetic field intensity over the past 14000 yr; *Geophys. J. R. Astron. Soc.* **81** 121–130

Constable C G and McElhinny M W 1985 Holocene geomagnetic secular variation records from northeastern Australian lake sediments; *Geophys. J. R. Astron. Soc.* **18** 103–120

Cox A 1968 Lengths of geomagnetic polarity intervals; *J. Geophys. Res.* **73** 3247

Damon P E 1970 Climate vs magnetic perturbation of the carbon-14 reservoir. In *Radiocarbon variations and absolute chronology*, (ed.) I U Olsson (Stockholm: Alinqvist and Wiksell) pp. 571

Hyodo M 1934 Possibility of reconstruction of the past geomagnetic field from homogeneous sediments 1984; *J. Geomagn. Geoelectr.* **36** 45–62

Jackson M, Gruber W, Marvin J and Banerjee S K 1988 Partial anhysteretic remanence and its anisotropy: applications and grain size dependence; *Geophys. Res. Lett.* **15** 440–443

Johnson E A, Murphy T and Torrenson O W 1948 Pre-history of the earth's magnetic field; *Terr. Magn. Atmos. Electr.* **53** 349

Kent D V and Opdyke N D 1977 Paleomagnetic field intensity variation recorded in a Brunhes epoch deep-sea sediment core; *Nature (London)* **266** 156–159

King J, Banerjee S K and Marvin J 1983 A new rock magnetic approach to selecting sediments for geomagnetic paleointensity studies; Application to paleointensity for the last 4000 yr; *J. Geophys. Res.* **88** 5911–5921

Lal D 1988 *In-situ* produced cosmogenic isotopes in terrestrial rocks; *Annu. Rev. Earth Planet. Sci.* **16** 355–388

Levi S and Banerjee S K 1976 On the possibility of obtaining relative paleointensities from lake sediments; *Earth Planet. Sci. Lett.* **29** 219–226

Lu R, Banerjee S K and Marvin J 1988 The effects of clay mineralogy and solution conductivity on DRM acquisition in sediments; *Eos* **69** 1159

Lu R, Banerjee S K and Marvin J 1989 The effects of clay mineralogy and the electrical conductivity of water on the acquisition of DRM in sediments; *J. Geophys. Res.* (in press)

Lund S and Banerjee S K 1985 Late quaternary paleomagnetic field secular variation from two Minnesota lakes; *J. Geophys. Res.* **90** 803–825

Lund S P and Olson P 1987 Historic and paleomagnetic secular variation and the earth's core dynamo process; *Rev. Geophys.* **25** 917–928

McElhinny M W and Senanayake W E 1982 Variations in the geomagnetic dipole 1: the past 50000 yr; *J. Geomagn. Geoelectr.* **34** 39

Olson P and Hagee V 1987 Dynamo waves and paleomagnetic secular variation; *Geophys. J. R. Astron. Soc.* **88** 139–159

Sato T 1980 *Paleomagnetic field intensity obtained by use of deep-sea cores of quaternary*, Ph.D. thesis, Osaka University, Japan; see also *Rock Mag. and Geophys.* **7** 16–21

Sharma P, Bhattacharya S K and Somayajulu B L K 1983 Beryllium-10 in deep-sea sediments and cosmic ray intensity variation; *Proc. 18th ICRC, Bangalore* **2** 337–340

Tauxe L and Valet J-P 1989 Relative paleointensity of the earth's magnetic field from marine sedimentary records: a global perspective; *Phys. Earth Planet. Int.* (in press)

Thellier E and Thellier O 1959 The intensity of the geomagnetic field in the historical and geological past; *Akad. Nauk SSR. Izv. Geophys. Ser.* 1296

Thouveny N 1987 Variations of the relative paleointensity of the geomagnetic field in western Europe in the interval 25-10 kyr BP as deduced from analyses of lake sediments; *Geophys. J. R. Astron. Res.* **91** 123–142

Thouveny N and Williamson D 1988 Paleomagnetic study of the holocene and upper Pleistocene sediments from lake Barombi Mbo, Cameroun: first results; *Phys. Earth Planet. Sci. Int.* **52** 193–206

Tucker P 1981 Paleointensities from sediments: Normalization by laboratory redepositions; *Earth Planet. Sci. Lett.* **56** 398–404

Verosub K L, Mehringer P J and Waterstraat P 1986 Holocene secular variation in western North America: paleomagnetic record from Fish lake, Harney County, Oregon; *J. Geophys. Res.* **92** 3609–3623

Walton D 1988a Comments on "Determination of the intensity of the earth's magnetic field during archaeological times: Reliability of the Thellier technique"; *Rev. Geophys.* **26** 13–14

Walton D 1988b The lack of reproducibility in experimentally determined intensities of the earth's magnetic field; *Rev. Geophys.* **26** 15–22

Wu G and Tauxe L 1990 Paleointensity for the last 750 ka: marine sediment data from the Western equatorial Pacific; *J. Geophys. Res.* (in press)

Steps toward understanding the Earth's dynamic interior

PETER J WYLLIE

Division of Geological and Planetary Sciences, California Institute of Technology, Pasadena, CA 91125, USA

Abstract. During the two decades following the plate tectonic revolution, cartoons of mantle convection have become more refined and based on data from geochemistry and geophysics rather than on imagination. Successive Reports from the US National Academy of Sciences have given progressively greater prominence to the study of mantle convection. It was one of 8 topics selected as high priority studies in the 1983 report on "Opportunities for Research in Geological Sciences". It was a central topic in 1987 and 1988 Reports on "Earth Materials Research" and "Mission to Planet Earth". With recent advances in isotope and trace element geochemistry, seismic tomography, geoid studies, and computer models of convection and fluid mechanics, we may be on the verge of obtaining a three-dimensional picture of the flow pattern in the Earth's mantle. The internal dynamic history of the Earth exerts a major control on most aspects of geology. Interactions between the dynamic mantle, the lithosphere and the ocean-atmosphere-biota system are important for understanding even short-term global changes. Continuation of multiple approaches to mantle dynamics must continue to yield fundamental results.

Keywords. Basalt; geoid; isotope geochemistry; mantle convection; plate tectonics; seismic tomography.

1. Introduction

The Organizing Committee suggested that contributors to this volume should emphasize futuristic aspects and uncharted domains, pointing to entirely new lines of thinking and research, rather than presenting a traditional review of personal, specialized research. I have found this charge difficult to satisfy. My first response was that if I could dream up something like this I would already be working along those lines and would have submitted research proposals to follow them in more detail. When I began trying to formulate an article I found that most of the topics I considered trying to develop were marginal to my own area of research operations, and although they appeared enticing I was not familiar enough with details to proceed without spending too much time in study. Perhaps it was my ignorance of the details that made the topics appear enticing.

In fact, there have been quite a few formal reports prepared by committees and published under the aegis of the US National Research Council which do satisfy to some extent the charge given to contributors to this volume. Many of the topics that interest me have already been reviewed in detail by teams of experts. Although these reports are available, their distribution is commonly rather limited, which is unfortunate. I have concluded that the best contribution I could make in the spirit of this volume is to pull together a selection of the conclusions and recommendations presented in some of these reports. I have reproduced sentences and phrases as required to transmit the messages from some of these reports, trusting that this is recognized as respect for the prose produced by experts after many drafts, rather than as plagiarism.

One of the most recent, comprehensive, and forward-looking reports is the 7-volume set "Space Science in the Twenty-First Century: Imperatives for the Decades 1995–2015", prepared for NASA by the Space Science Board. Before citing items from this publication, however, I will reach somewhat further back in history to identify some robust recommendations and approaches that can be expected to continue as productive lines of research. These can be considered against the background of figure 1, a review of current interpretations of the internal structure of the Earth.

2. The dynamic Earth

The revolution in earth sciences which converted a fixist scientific society to the mobilist viewpoint occurred abruptly in the late 1960's. The revolution changed our approach from one involving many specialists gathering data which tended to foster yet more specialization into a global approach in which the Earth had to be treated as a single system. With the global approach, successive discoveries have been made in diverse fields, but their inter-relationships are more apparent. The kinematics of plate motions moving the continents around beneath the caresses of the fluid envelopes of hydrosphere and atmosphere require that convection occurs within the Earth's mantle. Study of the geochemistry of basalts and mantle nodules indicated the occurrence of separate rock reservoirs in the mantle which have been related to physically separate convective units. Seismic tomography is revealing the three-dimensional heterogeneous structure of the mantle, and directions of flow of mantle material have been determined from seismic anisotropy. Intriguing structures have been defined by seismology in the region of the core-mantle boundary. Laboratory studies of minerals at high pressures and temperatures exceeding conditions corresponding to the core-mantle boundary have provided a firmer basis for interpretation of variations in the Earth's internal properties revealed by seismic studies. Interpretation of the geoid and fluid dynamic models of mantle convection using modern computers have supplemented the seismic approach in a most satisfying fashion.

We have reached a stage where I feel a sense of anticipation that we are on the verge of far greater understanding of the Earth's behaviour. But we have not reached consensus about the dynamics of the Earth's interior.

Figure 1 shows four interpretations of mantle convection and its association with the major basaltic types. Most models include a mantle source region of peridotite described as "primitive", because it contains elements indicating that it has not experienced significant partial melting, and another source region termed "depleted" because its chemistry indicates that basalt has been extracted from it. Lithosphere peridotite, strongly depleted by extraction of MORB, is identified in black. In fact, the situation is far more complicated than this, with metasomatic events over-printing the mantle composition and "enriching" it in certain minor and trace elements (Allègre 1987). Basalt returned to the interior through subduction is converted to eclogite (white in figure 1). Mantle minerals experience a phase transition where they are converted essentially to perovskite at a conditions corresponding closely to the seismic discontinuity at 650 km depth (Takahashi 1986).

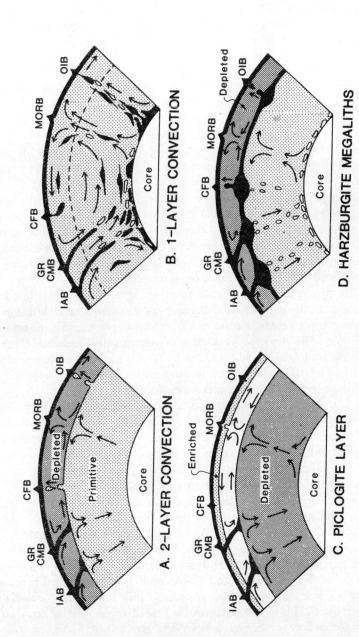

Figure 1. Four current interpretations of convection in the mantle, showing possible sources for the main basalt types. See text for advocates of the different interpretations. Primitive and depleted mantle is identified by light and dark stipple, respectively; lithosphere materials is black; eclogite/piclogite is white (perovskitite in lower mantle of figure D); IAB—island arc basalts; CMB—continental margin basalts; CFB—continental flood basalts; MORB—mid-ocean ridge basalts; OIB—ocean island basalts; GR—granitcid rocks (from Wyllie 1988a, b).

Figure 1A represents the two-layer model proposed by DePaolo and Wasserburg (1976) and developed by Jacobsen and Wasserburg (1981). The suggestion is that the upper reservoir, depleted by extraction of MORB, has grown downward through time. Enriched OIB and CFB are related to blobs of primitive mantle rising from the deeper layer. The two layers convect without significant mixing between the two.

An opposing interpretation of mantle-wide convection is presented in figure 1B. Hofmann and White (1982) proposed that subducted slab material sank to the mantle-core boundary where it was stored until a thermal plume lifted it to higher levels where chemical heterogeneities were transferred to the magmas generated. The physical interpretation involves increasing viscosity as a function of depth (Davies 1984; Hager and Clayton 1989), which causes reduced rates of flow and mixing in the deeper mantle. Despite the mantle-wide convection, the upper and lower mantle therefore tend to remain separate from each other through long periods. A combination of studies of the geoid, of mantle tomography and fluid mechanics of mantle convection by Hager and Clayton (1989) provide a best-fit with a non-layered mantle increasing in effective viscosity from the asthenosphere to the lower mantle by a factor of 300.

Anderson's (1982) more complex two-layer model for the recycling of oceanic crust is illustrated in figure 1C. He suggested that extensive partial melting of primitive mantle during accretion of the Earth caused formation of a cumulate layer of olivine eclogite (piclogite), overlying the residual, depleted peridotite. MORB is produced by melting of the piclogite, and this layer is continuously replenished by subduction of the oceanic crust derived directly from the layer some millions of years earlier.

Ringwood (1982) proposed a more elaborate two-layer model illustrated in figure 1D, with an additional reservoir occurring as megaliths of subducted lithosphere that flowed to form a layer separating the two convecting layers, and with eclogite playing a special role. The depleted peridotite of the megaliths reaches thermal equilibrium after 1–2 billion years, and then rises buoyantly in thermal plumes which yield OIB and CFB, with the enriched chemical signature having been supplied by partial melting of the entrained eclogite in the megaliths.

Allègre (1987) has interpreted mantle and crustal evolution on the basis of multi-isotope studies, treating the system as a transfer matrix made of transfer functions that describe the transfer of mass among reservoirs. His interpretation includes some aspects of figure 1A, but there are additional complexities involving a mesosphere boundary layer between upper and lower mantle, resembling figure 1D in geometry. This is the most comprehensive treatment that I am aware of.

The densities of the various masses depicted in figure 1, and the rates of heating of sunken cold material, are the key factors in controlling the occurrence of 2-layer or 1-layer mantle convection, and whether or not the subducted lithosphere slabs can penetrate the phase transition boundary near 650 km depth, while in downward or upward motion. Understanding the global dynamics of the Earth with consequent origin and evolution of the continents, and understanding the geological processes which occur at the boundary between (i) the convecting fluid envelopes of ocean and atmosphere, and (ii) the rigid lithosphere carapaces shifted around by the convecting mantle, require approaches combining geophysics, geochemistry, and petrology (chemistry and physics of minerals and rocks). This approach has been advocated strongly through many years, and promises great advances in the near future.

3. Research opportunities perceived in 1983

The internal processes interpreted in figure 1 were already topics of major concern and effort in 1983, as shown by the recommendations listed in "Opportunities for Research in the Geological Sciences", a 1983 report prepared by a Board of Earth Science committee chaired by William Dickinson, published by the National Research Council. The study was requested by Robin Brett, then with the National Science Foundation, to recommend policy and research priorities for the 1980's. The report presents a review of the major research challenges in the geological sciences, followed by a list of priorities each of which cuts across and encompasses a number of the specific research challenges. Although the list was numbered, no significance was attached to the ordering of topics after No. (i).

(i) A more detailed and accurate definition of the structure and composition of the continental lithosphere, including the continental margins.
(ii) Quantitative models for sedimentary basin evolution.
(iii) Improved understanding of magma generation and emplacement.
(iv) Knowledge of the physical and chemical properties of rocks.
(v) A better understanding of tectonic processes, the physical and chemical states that produce them, and the structures that result.
(vi) A model of convection in the Earth's interior.
(vii) Evolution of life.
(viii) Surficial processes.

The support needed to tackle these objectives included instruments, facilities, and major projects. The report noted the decadent state of instrumentation in the research laboratories of American universities which had been well documented for a decade or more, and recommended increased funding for instrumentation and a strategy for ensuring long-term stability in state-of-the-art instruments. The proposals included recommendation of expansion of the deep seismic reflection investigations exemplified by the Consortium for Continental Reflection Profiling (COCORP), and immediate planning for deep continental drilling to provide a means of measuring rock properties, establishing geological structure, and observing earth processes at exceptional depths. Improvement of the global seismographic observatory network was recommended by installation of digital instruments to replace the existing analog-recording systems.

Another 1983 report by a 12-person committee co-chaired by Charles Drake and Don Anderson was one of a series of research briefings developed by the Committee on Science, Engineering, and Public Policy (COSEPUP) for the White House Office of Science and Technology, and federal agencies. The idea was to identify research areas within each field that were likely to return the highest scientific dividends as a result of incremental federal investment. The committee based its recommendations on the "Opportunities" report discussed above, identifying five research areas for incremental support:

(i) Seismic investigations of the continental crust;
(ii) Continental scientific drilling;
(iii) Physics and chemistry of geological materials;
(iv) Global digital seismic array; and
(v) Satellite geodesy.

The instruments, techniques and approaches involved were reviewed, and then the scientific opportunities that could be exploited by the recommendations were focused on: (i) Structure and composition of the lithosphere; (ii) Dynamics of tectonic processes; and (iii) Evolution of the continental lithosphere.

4. COCORP, DOSECC, IRIS and GPS

Four of the areas recommended by COSEPUP were already existing as programmes, or were organized promptly. Topic (i) was covered by COCORP, mentioned above. Topic (ii) was furthered by DOSECC (Deep Observation and Sampling of the Earth's Continental Crust), a non-profit corporation organized by a university consortium to develop and operate a continental scientific drilling programme for the National Science Foundation in cooperation with other federal agencies. Another consortium, IRIS (Incorporated Research Institutions for Seismology), was formed to organize topic (iv) as well as supporting topic (i) and other seismological research. Many aspects of topic (v) were already under way in NASA, and the new Global Positioning Satellite (GPS) system promised to revolutionize both conventional and tectonic geodesy. Topic (iii) initially was not represented by any particular group, but the COSEPUP recommendations stimulated experimentalists to organize a workshop which led to the publication of a report in 1987 (see below).

The new initiatives sponsored by DOSECC and IRIS were pursued to the extent that funds were available, but along with COCORP the programmes were crippled by last-minute cuts in the 1988 budget for the National Science Foundation.

5. Earth materials research

COSEPUP recommendation (iii), "Physics and Chemistry of Geological Materials", led to a series of informal meetings of interested scientists and initiation of a workshop by the Board of Earth Science. The workshop of 44 scientists met in 1986, and prepared four documents on geochemistry, petrology, mineral physics and rock physics. The steering committee of 8 chaired by Charles Prewitt prepared a comprehensive report based on the four documents, which followed as Appendices in the 1987 publication on "Earth Materials Research".

This report is full of ideas and suggestions. The recommendations for "future directions" are:

(i) There should be increased support for: (a) upgrading and greater use of existing experimental and analytical facilities; (b) development and application of new experimental and analytical techniques; and (c) interpretation and modelling of experimental and analytical results.
(ii) A concerted programme is recommended for the collection, synthesis, and distribution of the samples required for earth materials research.

In connection with item (i), development of four technologies was recommended: synchrotron radiation facilities; high-pressure, high-temperature, large-volume experimentation; microscale, in-situ analytical instrumentation; and accelerator mass spectrometry.

The report noted that the study of earth materials was essential for the interpretation of the results of several other major scientific efforts, serving as an anchor to geological and geophysical theoretical and field measurements. Three major topics were identified where significant advances could be expected from research on earth materials:

(i) Mantle convection.
(ii) Material transport through fluid flow.
(iii) Evolution of continents.

All of these topics can be represented in figure 1. Systematic rock sampling and analysis would produce a global geochemical data base that could be integrated with the global geophysical data base to produce a complete view of the mechanisms of evolution on the Earth's interior. Only through global sampling of magmatic rocks from ocean floors and continents and a carefully designed isotopic study can we obtain information about the long-term variations of mantle convection.

6. Mission to planet Earth

In June of 1988, The Space Science Board of the National Research Council published the findings of a study initiated early in 1984, when NASA requested a study "to determine the principal scientific issues that the discipline and space science would face during the period from 1995 to 2015." There were six Task Groups, each preparing a volume under the umbrella title "Space Science in the Twenty-First Century". The volume "Mission to Planet Earth", prepared by an 11-person committee chaired by Don L Anderson, outlines a unified programme for studying the Earth, from its deep interior to its fluid envelopes.

The proposals emphasize the need for an integrated and inter-related set of satellite and surface observations. One central observing element for the mission consists of an "Earth Observing System" (EOS) of satellites collecting environmental data on a global, consistent, repetitive, and long-term basis. The space-based elements are considered in six broad areas: land, oceans, atmospheres, radiation budget, atmospheric chemistry, and geodynamics. Among the land-based elements of the mission is the "Permanent Large Array of Terrestrial Observatories" system (PLATO), which can be viewed as a gigantic terrestrial observatory that can be treated as a telescope or accelerator.

The measurement strategy proposed for implementation involves a total system composed of about 8 orbiters, 1000 ocean systems, 1000 automated land stations, thousands of simple surface stations, and itinerant ships, planes and balloons. Some of the subsystems are relatively new concepts: ocean drifting instruments, pop-up buoys, and smart ground stations. Smart ground stations with an on-site computer capable of "intelligent" decisions and built-in recording system could transmit data by satellite telemetry. Among the many simultaneous measurements possible, such instruments could measure the structure of the Earth's interior, earthquake source mechanisms, strong ground motion in seismic areas, volcanic activities, and tidal forces.

New management and organization systems will be required to handle the abundance of global data that would accrue from the space and ground stations. A programme is needed to yield a coherent integration of data analysis and data management technologies. The Task Force also recommended that state-of-art computing technology be utilized for theoretical modelling of the complex, turbulent earth processes.

The essence of the proposal is to set forth a concerted and integrated research programme on the origin, evolution, and nature of our planet and its place in the solar system. The primary research objectives are addressed by four themes that are developed at length:

(i) To determine the composition, structure, and dynamics of the Earth's interior and crust, and to understand the processes by which the Earth evolved to its present state.
(ii) To establish and understand the structure, dynamics, and chemistry of the oceans, atmosphere, and cryosphere and their interactions with the solid Earth, including the global hydrological cycle, weather, and climate.
(iii) To characterize the interactions of living organisms among themselves and with the physical environment, including their effects on the composition, dynamics, and evolution of the ocean, atmosphere, and crust.
(iv) To monitor and understand the interaction of human activities with the natural environment.

The geodetic and seismological systems proposed will provide significant advances in mapping crustal motions in zones of deformation, and in mapping density variations throughout the mantle. These results will determine the nature of mantle convection. The fundamental problem is the long-term evolution of the mantle, which depends on convection that was different earlier than the contemporary convection. The nature of the hydrosphere and atmosphere depends on the outgassing of the mantle, but the coupling of the solid earth processes and the oceanic and atmospheric processes occurs on time scales of millions of years. The wide disparity of time and space scales represented by geophysics, geochemistry, fluid dynamics, and biological processes on Earth can be addressed for the first time by the global data sets and modelling on high-speed supercomputers.

7. Future prospects

Future prospects begin with successful applications of the recent past. For me, the most exciting recent developments have involved the emerging picture of mantle convection, and the associated history of crustal formation via magmatic processes occurring in both mantle and crust, which are included in topic (i) of "Mission to Planet Earth". I find it particularly fascinating that the details of mantle convection are being revealed by two completely different approaches on very different scales: seismologists are defining density and anisotropic variations for the whole mantle, and geochemists are cataloging and interpreting trace element and isotope variations in the rocks from the Earth's magmatic effluent and small samples of mantle rocks. Vigorous pursuit of these two approaches promises much for future understanding of mantle convection.

Prospects are also very good for a multiplicity of approaches to the problems represented by the structures depicted in figure 1. The combination of gravity and geophysical fluid dynamic calculations with seismic tomography has been very fruitful. We can expect regular advances and occasional breakthroughs from parallel studies in laboratory experiments with minerals and rocks at increasingly high pressures in both large-volume massive presses, and in diamond-anvil apparatus. Experimental studies will include phase equilibrium studies, kinetic studies, and determination of the physical properties of minerals, rocks, melts, and vapours, all of which provide fundamental information for the interpretation of mantle processes, and the chemical differentiation which is ultimately controlled by the physical behaviour of solid-liquid-vapour systems. Small-scale details such as the degree of liquid-connectivity in a partially melted rock, which is controlled by surface tension and the dihedral angle between melt and minerals, have important consequences for the escape and migration of magmas from source rocks, one of the significant processes associated with convection in figure 1 (Watson and Brenan 1987). Computer modelling will obviously become increasingly important in dealing with large data sets and testing theoretical ideas.

If we can determine the internal dynamic state of the Earth, we are well on the way to understanding geology. It is mantle convection that generates and later redigests the lithosphere, causes the partial melting, magma migration, and the chemical differentiation that controls the formation and evolution of the continents, which are then modified by uplift, erosion, and further mass redistribution and geochemical differentiation associated with denudation and the formation of sedimentary blankets on land and beneath the oceans. It is mantle convection which welds parts of the sediments onto continents and returns part to the Earth's mantle. It is mantle convection which releases volatile components to the hydrosphere and atmosphere, and which recycles them through the crust and mantle.

The Earth is a system with all levels interconnected. We cannot understand the causes of short-term changes in the atmosphere such as the greenhouse effect and depletion of the ozone layer if we do not understand the causes of major climatic changes such as ice ages, and the consequences of volcanic eruptions or emergent continents for the CO_2 concentration in the atmosphere. If we understand mantle convection we have one most important key for understanding the interactions between fluid envelopes and the lithosphere. Sediments and other rocks of the crust contain the only historical record of these interactions. The International Geosphere-Biosphere Program (IGBP) is studying global change with focus on a time scale of decades to centuries. The success of the study of interacting ocean-atmosphere-biota system depends upon suitable consideration of the longer-term interactions of this system with the lithosphere.

The approaches advocated above are no longer novel but I would expect to be surprised from time to time by new dramatic discoveries from them. Any advancement in instrument design or a new instrument may reveal a deeper level of understanding. Futurists who make predictions in any field have a low chance of success because one discovery may change the situation completely. An important lesson was presented to our profession when paleomagnetism and magnetic anomalies turned out to be key studies in the recognition and formulation of plate tectonic theory. The lesson is that great discoveries may spring from anywhere. We must attract inquisitive and creative minds to explore the earth sciences.

Acknowledgements

This research was supported by the Earth Sciences section of the US National Science Foundation, Grant EAR87-19792.

References

Allègre C J 1987 Isotope geodynamics; *Earth Planet. Sci. Lett.* **86** 175–203

Anderson D L 1982 The chemical composition and evolution of the mantle: Advances in earth and planetary sciences; in *High-pressure research in geophysics* (eds) S Akimoto and M H Manghnani (Dordrecht: Reidel) pp. 301–318

Davies G F 1984 Geophysical and isotopic constraints on mantle convection: an interim synthesis; *J. Geophys. Res.* **89** 6017–6040

DePaolo D J and Wasserburg G J 1976 Inferences about mantle sources and mantle structure from variations of $^{143}Nd/^{144}Nd$; *Geophys. Res. Lett.* **3** 743–746

Earth Materials Research 1987 Report of a Workshop on Physics and Chemistry of earth materials. Board on Earth Sciences (Washington D.C.: National Academy Press)

Hager B H and Clayton R W 1989 Constraints on the structure of mantle convection using seismic observations, flow models and the geoid; in *Mantle convection* (ed.) W R Peltier (New York: Gordon and Breach) pp. 658–763

Hofmann A W and White W M 1982 Mantle plumes from ancient oceanic crust; *Earth Planet. Sci. Lett.* **57** 421–436

Jacobson S B and Wasserburg G J 1981 Transport models for crust and mantle evolution; *Tectonophys.* **75** 163–179

Oppertunities for Research in the Geological Sciences 1983 Board on Earth Sciences (Washington D.C.: National Academy of Sciences)

Research Briefings 1983 (Washington D.C.: National Academy Press)

Ringwood A E 1982 Phase transformations and differentiation subducted lithosphere: implications for mantle dynamics, basalt petrogenesis, and crustal evolution; *J. Geol.* **90** 611–643

Space Science in the 21st Century: Imperatives for the Decades 1995–2015 Mission to Planet Earth 1988 Task group on earth sciences, Space Science Board (Washington D.C.: National Academy Press)

Takahashi E 1986 Melting of dry peridotite KLB-1 up to 14 GPa: Implications on the origin of peridotitic upper mantle; *J. Geophys. Res.* **91** 9367–9382

Watson E B and Brenan J B 1987 Fluids in the lithosphere, I. Experimentally determined wetting characteristics of $CO_2 - H_2O$ fluids and their implications for fluid transport, host-rock physical properties, and fluid inclusion formation; *Earth Planet. Sci. Lett.* **85** 497–515

Wyllie P J 1988a Solidus curves, mantle plumes and magma generation beneath Hawaii; *J. Geophys. Res.* **93** 4171–4181

Wyllie P J 1988b Magma genesis, plate tectonics and chemical differentiation of the earth; *Rev. Geophys.* **26** 370–404

Application of diamond-anvil-cell technique to study f-element metals and some materials relevant to planetary interiors

JAGANNADHAM AKELLA*

Lawrence Livermore National Laboratory, University of California, Livermore, California 94550, USA

Abstract. The field of static high pressure research using the DAC has attained phenomenal growth during the last decade and is still growing. Recent advances in the diamond-anvil-cell (DAC) technology have made it possible to study the high pressure properties of various materials to 200 GPa and more. DAC experimental results from our work on actinides and lanthanides along with data from others for some materials that have planetary science relevance are briefly discussed.

Keywords. Multi-megabar pressure; diamond anvil cell; X-ray diffraction; actinides; lanthanides; planetary materials; equation-of-state (EOS).

1. Introduction

Ultra-high pressure and temperature studies to probe the properties of materials under extreme conditions were possible in the past only by shock wave and nuclear explosive techniques. However, in these techniques physical properties are transient and measurements must be made within an extremely short time. With the development of the diamond-anvil-cell (DAC) apparatus and multi-anvil solid media devices, very high static pressures and temperatures could be sustained over long periods of time. Of the two types of apparatus, DAC has become the most favoured device because a wide variety of measurements e.g. X-ray, spectroscopic, electrical etc could be made up to very high pressures (>200 GPa) and ~ 5000 K. The three most important areas of investigation with DAC are (i) equation of state studies, (ii) pressure-induced structural phase changes and (iii) metallization under pressure.

In this report an attempt will be made to describe some of the advances made in the apparatus design for generating ultra-high pressures and high temperatures. Results for some materials that are relevant to physics and planetary sciences will be briefly discussed.

2. Apparatus

The historical background for generating high pressures is an interesting one (see Bridgman 1931). Since the first reported experimental work in 1700's, the rise in experimental pressure has been exponential (figure 1). Bridgman, considered the

---

*Work performed under the auspices of the US Department of Energy by the Lawrence Livermore National Laboratory under contract number W-7405-ENG-48.

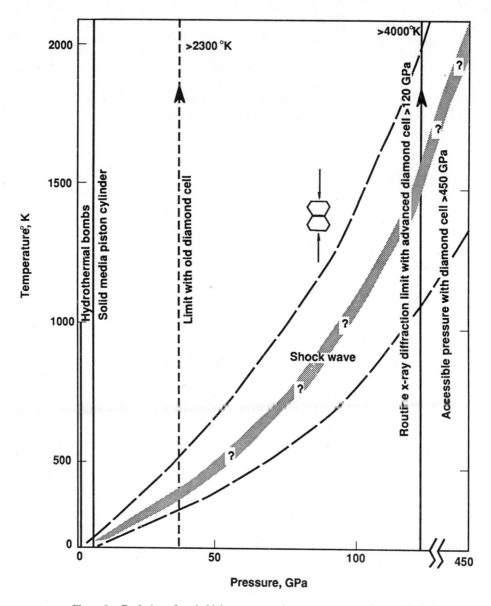

Figure 1. Evolution of static high pressure and temperature experimental limits.

father of modern high pressure research, developed solid-media apparatus (Bridgman's opposed anvil device, piston-cylinder apparatus) capable of generating ~ 10 GPa (1 GPa = 10 kilobars) static pressure and which could be heated to a few hundred degrees centigrade. He also developed techniques to measure the pressures and temperatures. Apparatus to generate still higher pressures were developed by Drickamer and Balchan (1962), the supported anvil device; the belt apparatus and the girdle-anvil device by Hall (1960), Bundy (1962) and others at General Electric.

The most advanced design among the solid-media devices is the multi-anvil apparatus. The apparatus that is gaining importance in the mineral physics research is the multi-anvil solid media high pressure device with X-ray capabilities or MAX (Shimomura et al 1984, 1985; Yagi et al 1987). The maximum attainable pressure and temperature in a MAX press depend upon the size of the anvils. Truncation of the anvils is from 12 and 3 mm an edge; which translates to a maximum pressure-generating capability of 3 to 12 GPa respectively in this apparatus. At the highest pressure (12 GPa) temperatures up to 800°C can be sustained, while at 6 GPa, 1600°C temperature can be sustained. Endo et al (1987) generated pressures in the range of 60 GPa at room temperature with this apparatus. A number of Japanese researchers are working on increasing the P-T range of this apparatus. The uniqueness of this kind of apparatus is the simultaneous generation of high pressures at high temperatures, large sample volumes (several cubic millimeters) and *in situ* X-raying capability of the experimental material.

Large sample volumes provide adequate material for X-ray diffraction of silicates which have relatively lighter atoms. Silicate melting, viscosity measurements, kinetics of phase transformations, equation of state measurements etc (relevant to earth sciences) can be carried out simultaneously at high pressures and temperatures with this gear.

The basic limitations in designing a static high pressure apparatus are the mechanical integrity of the pressure chamber (compressive strength of the sintered tungsten carbide) and plastic flow or mechanical failure of the apparatus confining the cell. It is also worth noting that higher pressures in these apparatus are achieved by integrating the total applied force onto a smaller sample volume (generally few cubic millimeters). This means that as the pressure capability increased the sample size decreased. For a detailed description of the various types of apparatus see the reviews by Spain (1977) and Sherman and Stadtmuller (1988).

In the mid-1950's another kind of high pressure revolution was taking place albeit at a slower pace. Lawson and Tang (1950) first came up with the idea of using a diamond, which is the hardest material known to man, for the containment of pressure. They designed a miniature piston-cylinder apparatus by drilling a hole in the centre of the diamond, pressing the material inside that hole with a piston and X-raying the sample *in situ*. This rudimentary design was independently improved upon by Jamieson et al (1959) and Weir et al (1959) to do X-ray diffraction and infra-red absorption measurements respectively. Several other crucial and important discoveries at the National Bureau of Standards have made the diamond-anvil cell (DAC) an extremely valuable research tool in earth sciences, materials science, physics, chemistry and even biology. Van Valkenburg (1965) introduced the use of a metal gasket between the diamonds for generating hydrostatic pressure. For *in situ* pressure calibration Forman et al (1972) proposed the ruby fluorescence technique and Piermarini et al (1975) introduced the 1:4 ethanol-methanol mixture as a pressure medium for the generation of hydrostatic pressures around the sample.

Jayaraman (1983, 1986) gave detailed description of various DAC designs. In this report the DAC design of Mao and Bell (1975) will be discussed as it is the one that most investigators use to generate pressure in the range of 100 GPa (megabar and 1000 kilobars) and above.

The diamond-anvil cell consists essentially of two ~ 0.3 carat, brilliant cut, flawless, type 1 diamonds in an opposed-anvil configuration (figure 2). The stones are modified to have 16 facets with the sharp tip (culet) polished to a flat face. This culet face could be modified to a single bevel or double-bevelled tip with bevel angle 5–10° and a

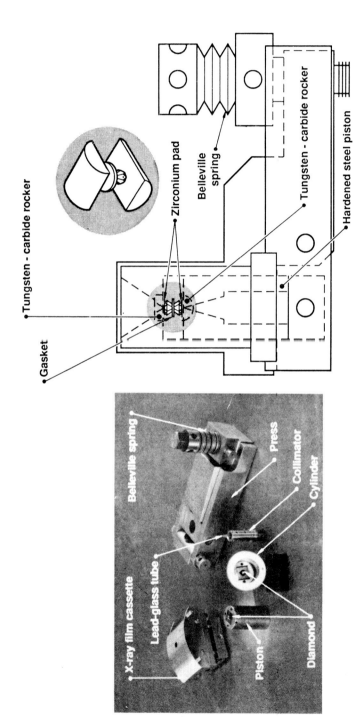

Figure 2. Diamond-anvil pressure cell cutaway view showing various components.

central flat of $\sim 25\,\mu$m to $100\,\mu$m (figure 3). The size of the central flat and bevel angle is varied depending upon the type of experiments to be conducted and the maximum pressure to be generated (Moss et al 1986; Moss and Goettel 1987).

In a diamond-anvil cell, the two diamonds are mounted over zirconium pads placed between the diamonds and carbide rockers and these pads act as an intermediate support and also to distribute any concentrated forces caused by surface imperfections in the tungsten-carbide rockers. Zirconium can also act as a β filter for Mo X-radiation. The tungsten carbide rockers are hemicylinders with their axes at right angles so that they can be tilted and moved along their long axis to align the anvils perfectly parallel. These tungsten-carbide rockers have proper apertures built in them to allow X-ray and other kinds of radiations to enter and exit through the diamond anvils.

A gasket, which is either a strip or a circular metal disc made out of Rene' 41, Inconel, work hardened stainless steel or rhenium, is placed between the culets of the anvils and force is applied which indents (prepresses) the surface. A hole of $50-100\,\mu$m is drilled at the centre of the indented area of the gasket for loading the sample together with small ruby chips for pressure calibration, and the pressure medium. The latter may be a liquid or a gas (table 1). In the case of a gas, loading is accomplished by remotely assembling the apparatus in a pressure vessel filled with the gas to be used.

When the ruby chip is excited by a helium-cadmium laser light, it emits fluorescence at two characteristic frequencies. The wavelengths of these two sharp spectral lines (R_1 and R_2) vary as a well-calibrated function of pressure. When the pressure ceases to be hydrostatic because the fluid pressure medium transforms into a solid or becomes very viscous on compression, the non-uniformity of the resulting stresses could broaden the fluorescent peaks. Pressure can also be determined by mixing the sample

Table 1. Pressure media and the limiting pressures below which they remain fluid and continue to exert hydrostatic pressure.*

Material	Hydrostatic pressure limit (GPa)
1:4 ethanol/methanol	~ 20
Neon[a]	~ 20
Argon[a]	~ 9.0
Helium[a]	~ 60
Hydrogen[a]	> 60

*Each of these media may be used in quasi-hydrostatic experiments at somewhat higher pressures. The quasi-hydrostatic range merges gradually into the non-hydrostatic range as fluids become more viscous without any definite line of demarcation.

[a] The gases are injected into the diamond cell under high pressure (about 200 MPa); they liquefy when the applied pressure exceeds their vapour pressure; or loaded into DAC in a specially built cryostat.

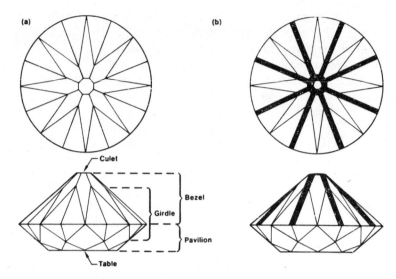

Figure 3. Brilliant cut diamond with (**a**) flat culet, (**b**) single bevel (coloured) and a central flat.

with a marker material whose volume under pressure could be calculated from the X-ray lattice parameters and compared with the known pressure-volume relationship [equation-of-state EOS] from shock-wave data.

Diamond-anvil cells are well suited for X-ray diffraction, spectroscopy and electrical measurements. To determine the crystal structure of the material, a well-collimated 75–100 μm beam of monochromatized X-rays are passed through the sample (and through both diamonds) and the resulting diffraction pattern from the sample under pressure is recorded on a film. Using these data, the crystal structure can be determined and the lattice parameters calculated.

With the advances made in diamond-anvil-cell technology, it is now possible to generate pressures in the multi-megabar (> 100 GPa) range. However, in order to achieve such high pressures, diamond-anvils are designed to have single bevel or double bevel angles (Moss *et al* 1986, 1987) with a central flat of 50 μm or less in diameter for the sample. Because of the prevailing non-hydrostaticity at these pressures (it is 5–10 GPa over 75 μm (Vohra *et al* 1986) depending upon the type of pressure medium used), conventional X-ray sources are not suited to obtain accurate data. With conventional X-ray sources, the X-ray spot cannot be collimated to less than 50 μm. Also by collimating the X-ray beam to a smaller size the net intensity will be so low that it is not possible to get X-ray data within a reasonable time (not even tens of days, Smith and Akella 1987). Thus, it is necessary to use X-rays from a synchrotron source and solid-state detectors (energy dispersive X-ray diffraction EDX) to collect data (Smith and Akella 1987). Synchrotron sources are approximately eight orders of magnitude more brilliant than a rotating anode type, and the highly coherent X-ray beam can be collimated down to 5–10 μm. A combination of high beam intensity and excellent collimation is essential to eliminate the pressure gradient problem (over 10 μm the gradient is < 0·5 GPa) and the time required for data collection (10–30 minutes at each pressure).

Even though our primary focus is on obtaining crystal-structure data as a function of pressure at room temperature (Akella 1983), diamond-anvil studies have many other uses. With the addition of a laser or resistance heater or with cryogenic cooling, we can explore the pressure-volume-temperature relationship of any material. With appropriate modifications, the diamond-anvil technique can be used to make conductivity and resistance measurements. Infra-red and Raman spectroscopy studies are routinely done with DAC. It is even possible to study the change in valence as a function of pressure in transition metals with Mössbauer spectroscopy. Jayaraman (1985, 1986) reviewed in detail the use of DAC.

3. Experimental phase transformations and equation-of-state studies

As was mentioned earlier the focus of our (LLNL) DAC programme is to study phase changes under pressure and static EOS. In this report a brief review of some of our data on lanthanide and actinide metals will be presented along with the studies made by others on some gases that have planetary importance.

3.1 Rare-earths and actinides

The rare-earth series of elements (lanthanum through lutetium) are characterized by a gradual filling of the $4f$ shell in the atom. This shell is located deep within the atom and thus the associated magnetic moments are highly localized (Johansson and Rosengren 1975). The lanthanide elements form a long series whose physical properties vary smoothly with $4f$ occupancy. Since the outer regions of the different lanthanide atoms are very similar to each other one may expect a broadly related behaviour in their solid modifications (Gschneidner 1961; Spedding and Daane 1961; Taylor and Darby 1972). If so, the lanthanide phase diagrams may be fitted into a single universal phase diagram in which the same sequence of changes could be represented either with decreasing atomic number or with increasing pressure. This possibility has stimulated both theoretical and experimental studies on lanthanides.

In the actinide series (actinium through lawrencium) the $5f$ shell is gradually filled and according to Johansson and Rosengren, although the $5f$ wave function is fairly localized, it is more extended than the lanthanide $4f$. Kmetko and Hill (1970) have shown, from energy-band calculations on actinides, that the earlier actinide metals (Pa-Pu) give good evidence for an itinerant character of the $5f$ electron. On the other hand, the $5f$ electrons in heavy actinides (Am and above) are localized and their properties resemble those of the rare-earth elements which have localized $4f$ electrons. So another important motivation to study both lanthanides and actinides is to understand any similarities between these two groups of elements. This has led experimentalists to examine the high pressure structural changes in actinides and lanthanides (Akella et al 1979, 1980, 1981; Smith et al 1981; Roof 1980, 1981; Smith and Akella 1982, 1984, 1986; Mao et al 1982; Benedict 1984; Benedict et al 1985; Benedict 1986; Akella and Smith 1986a; Akella et al 1986b; Grosshans 1987).

After a systematic study of some rare earth elements, Jayaraman (1978) proposed a generalized high-pressure phase transformation sequence, hcp-Sm type-dhcp-fcc. This sequence shows a general increase in the cubicity for the structures as the pressure is increased. According to Jayaraman the structural sequence in the rare-earth elements progresses towards lower Z numbers as the pressure is increased. The data that were collected over the years at our laboratory both on lanthanides and actinides are presented briefly in this report. Grosshans, Holzapfel and Benedict (see Grosshans 1987) have also worked on the rare-earth elements using a DAC. Benedict et al (1986) modified the original generalized phase diagram for the lanthanides given by Gschneidner (1985). A short review of the differences between our data and that of Grosshans, Benedict and Holzapfel is appropriate.

Benedict et al proposed that the $4f$ electron delocalization is accompanied by a large volume change and the structure of the new phase at that pressure will have a lower symmetry (monoclinic); which they called the "collapsed phase". For Sm they proposed that this phase could appear at about 36·0 GPa and our data do not substantiate their pressure and the predicted large volume change. Secondly, we differ from them in that we find the post-fcc phase in the lanthanides to be a distorted form of the fcc phase (Vohra et al 1984). We argue on theoretical grounds (McMahan and Young 1984) and on the interpretation of our X-ray diffraction patterns collected by angular dispersive technique on film, that it could be a six-layered hexagonal packing of ABCACB repeat called triple-hexagonal-close packed (thcp) structure (Smith and Akella 1984). It is important to note that both interpretations for the post-fcc phase have some drawbacks, and much more precise X-ray diffraction data are required to resolve this issue. Benedict et al (1985), also suggest in their generalized phase diagram that in the heavy rare-earths holmium through lutetium, the high pressure dhcp phase changes directly to the d-fcc phase without the intermediate fcc phase [the general sequence is hcp-sm type-dhcp-fcc-dfcc-collapsed (?) phase (monoclinic)]. Our recent studies on Gd (Akella et al 1988) and Tb (unpublished) suggest that at high pressure there could be a stability field for fcc holmium contrary to Benedict et al's (1985) observation. A detailed account of all our data on the rare-earth elements will be presented elsewhere.

Among actinides under pressure we investigated Th, U, Pu, Am and Cm while Roof (1980, 1981) studied Pu and Am. Benedict (1987) recently published a review of the experimental work done on the actinides at Karlsruhe, West Germany.

Thorium which has an fcc structure at ambient conditions does not undergo a structural change even up to 100 GPa (Akella 1988b). However, a break in the slope of the P-V curve (compressibility) was detected above 20 GPa, which has been attributed to an electronic s-d transformation in the metal. This was not reported in the earlier DAC studies on Th by Bellusi et al (1981) and Benjamin et al (1981).

A possible structural change in uranium was suggested at about 70·0 GPa (Akella et al 1985). However, an unequivocal confirmation is still needed. On the other hand, alpha plutonium which has a complex monoclinic crystal structure with sixteen atoms in the unit cell at one atmosphere changes to a simple hexagonal-close-packed (hcp) structure at elevated pressure (Akella 1983). Recently Dabos (1987) in her doctoral thesis work on plutonium reported a high pressure structural transformation for this metal to an orthorhombic structure. If her identification of the structure is correct then the high pressure behaviour of plutonium is reminiscent of the trivalent rare-earth elements (as pressure increases the structures of lower Z elements appear). On the contrary, if the high pressure structural form for plutonium is hcp then one is tempted to presume that it may behave like the localized f-electron metals of heavy actinides as pressure is increased. More detailed experimental and theoretical work is needed to understand the high pressure behaviour of this complex metal.

At room pressure americium has a double hexagonal close-packed (dhcp) structure.

As pressure is increased it goes through three structural transformations (Akella et al 1979, 1980; Smith et al 1981), AmI dhcp-AmII fcc-AmIII monoclinic (α'' uranium like) – AmIV orthorhombic (?). The transformation pressures were 5.0 ± 1.0, $\sim 11.0 \pm 1.0$ and ~ 15.0 GPa. Roof et al (1981) have also observed structural changes at similar pressures but they assigned AmIII to have a body-centered monoclinic structure and AmIV as an orthorhombic structure. Benedict assigned different structures for AmIII and AmIV phases in his earlier reports, but recently (review paper) he proposed AmIII to have trigonal and AmIV orthorhombic structures. According to him, the AmIII to IV transformation occurs at a slightly higher pressure than that reported by Smith et al (1981); Roof et al (1981).

Akella et al (1981) reported data on Cm up to 20.0 GPa in which they mentioned a possible structural change around 18.0 GPa from ambient dhcp phase. No further details of this work were published. Benedict also worked on Cm and could publish his data to 50.0 GPa and the structures are CmI dhcp-CmII fcc-CmIII orthorhombic at 23.0 and 43.0 GPa respectively. Benedict and his associates also investigated the high pressure properties of californium and berkelium. Their data are given in detail in a recent publication by Benedict (1987).

3.2 High pressure DAC studies relevant to planetary interiors

The internal structures of the outer giant planets of our solar system, Jupiter, Saturn, Uranus and Neptune are known to differ from that of the inner or terrestrial planets. The recent flights of Pioneer and Voyager to some of these planets have stimulated renewed interest in their internal constitution and structure. Slattery (1977) and Hubbard and MacFarlane (1980) have predicted that the outer planets could essentially have three layers, a top layer of molecular hydrogen and helium; a middle layer of metallized hydrogen and helium (Jupiter and Saturn) or "ices" of ammonia, methane and water (Uranus and Neptune) and a rock core of iron, nickel, magnesium and silicon oxides. Recently Hubbard and Marley (1988) proposed revised models for these planets. It is beyond the scope of this report to discuss various models.

In order to construct structural models that predict correctly the observed characteristics of these planets, one must have accurate equations of state, melting curves, intermolecular potentials, and electrical conductivities for materials that are relevant to planets e.g. H_2, He, NH_3, CH_4 and N_2 etc. over a wide range of physical conditions (Ross and Nellis 1981). Also some gases like hydrogen may become a superconducting metal in the 2–3 megabar (200–300 GPa) pressure range (Ashcroft 1968; Ross and Shishkevish 1977; Ross and McMahan 1976, 1981). It is also important to understand the planetary and the high explosives physics characteristics of another diatomic gas viz. nitrogen. Insulator-to-metallic transition in nitrogen is also predicted at about 70 GPa (McMahan and LeSar 1985).

Ross and Nellis (1981) and Nellis (1987) have discussed the shock-wave studies (EOS) on some of these liquified gases and application of the data in understanding the giant planets. Recently Nellis et al (1988) reported electrical conductivity measurements and shock-compression equation of state for the planetary "ices" ammonia, methane, and "synthetic Uranus" composition at shock pressures and temperature up to 75 GPa and 5000 K.

With the modified diamond-anvil design, pressures up to 200 GPa are readily accessible in the DAC apparatus. This capability has stimulated researchers in the

DAC area to pursue the static equation-of-state studies and phase transformations in H_2 and N_2 at room temperature (Reichlin *et al* 1985; Bell *et al* 1986). Even though pressure-volume data were obtained, none of these workers have so far observed metallization of H_2 and N_2 in their studies; however, evidence was found for the weakening of the molecular bond which is a precursor of metallization. Further, nitrogen at 180 GPa became virtually opaque suggesting an incipient transformation to an electrically conducting state and hydrogen remained a clear and transparent solid suggesting a good insulator at this pressure (Bell *et al* 1986). Polian *et al* (1985) suggested that H_2O "ice" may transform to a dense phase with the oxygen forming a face-centered cubic lattice at about 100 GPa. Recent investigation of "ice" by Hemley *et al* (1987) to 128 GPa indicated no metallization and the structure is body-centered cubic. They also pointed out that ice samples in their experiments remained transparent up to 128 GPa with no indication of band gap closure. For further details the reader is referred to the original papers by these authors.

From the standpoint of planetary interiors of the large outer planets, diamond-anvil cell studies so far on H_2, N_2 and H_2O have not shown metallization of these species up to the pressures mentioned above. H_2 and H_2O ice remained transparent insulating materials, while N_2 may have an incipient transformation to an electrically conducting state. If the present pace of development in DAC technology is sustained, metallization of N_2 and H_2 will, in the near future, be achieved as in xenon (Moss *et al* 1986; Reichlin *et al* 1985), and this eventually will help in the fine tuning of the structural models of the planets.

4. Conclusion

For conducting scientific inquiry into condensed matter physics and planetary interior (geochemical) problems, under static pressure conditions, DAC is a very valuable and versatile research tool. Small samples are sufficient and the data obtained are very accurate.

Interfacing the DAC with the synchrotron X-radiation has further made it possible to study (i) ultra-high pressure and temperature X-ray crystallography, (ii) kinetic studies and transient phenomena, (iii) phase diagrams in the P-T field even when the temperature is of a transient nature (e.g. simultaneous laser heating and X-raying) and (iv) EXAFS. Results obtained from these investigations could provide an excellent data base to understand the physics of matter (developing theories of bonding and electronic states of matter etc.) useful for geophysical and planetary sciences and perhaps modelling of systems that are too complex for present-day theory to handle. The field of static high pressure research using DAC is still expanding and there seems to be no end to it in the near future.

Acknowledgements

The experimental work performed at the Lawrence Livermore National Laboratory on the actinides and lanthanides was supported by the Physics and Chemistry Departments. Thanks are due to Drs A K McMahan, D Young, R Grover, N Holmes and M Ross of the H-Division with whom I had many technical discussions. I am personally appreciative of the association of Dr G S Smith with me on the "DAC project" all these years. I appreciate the support and encouragement received from Dr H Graboske.

I have known Dr D Lal from the time I was at UCLA in 1967, and I am very happy to dedicate this paper to him on his sixtieth birthday.

References

Akella J 1983a Diamonds: Powerful tools for high pressure research; *Energy Technol. Rev.* 11–19
Akella J 1983b Ultra-high pressure studies on uranium and plutonium, Internal Publication, Lawrence Livermore Laboratory, Berkeley
Akella J, Johnson Q and Schock R N 1980 Phase transformations in Am at high pressure: Relation to rare-earth elements; *J. Geophys. Res.* **85** 7056–7058
Akella J, Johnson Q, Thayer W and Schock R N 1979 Crystal structure of the high pressure form of Am; *J. Less-Common Met.* **68** 95–97
Akella J, Reichlin R L, Smith G S and Schwab M 1981 High pressure studies on curium, Actinides 81; *Extended Abstract* 216–217
Akella J, Smith G S and Weed H 1985 Static high pressure diamond-anvil studies on uranium to 50 GPa; *J. Phys. Chem. Solids* **46** 399–400
Akella J and Smith G S 1986a High pressure diamond-anvil studies on neodymium to 40 GPa; *J. Less-Common Met.* **116** 313–316
Akella J, Xu J and Smith G S 1986b Static high pressure studies on Nd and Sc; *Physica* **B139–B140** 285–288
Akella J, Smith G S and Jephcoat A P 1988a High pressure phase transformation studies in gadolinium to 106 GPa; *J. Phys. Chem. Solids* **49** 573–576
Akella J, Johnson Q, Smith G S and Ming L C 1988b Diamond-anvil cell high pressure X-ray studies on thorium to 100 GPa; *High Pressure Res.* **1** 91–95
Ashcroft N W 1968 Metallic hydrogen: A high temperature superconductor? *Phys. Rev. Lett.* **21** 1748
Bell P M, Mao H K and Hemley R J 1986 Observations of solid H_2, D_2, and N_2 at pressures around 1·5 Mbar at 25°C; *Physica* **B139–B140** 16–20
Bellusi G, Benedict U and Holzapfel W B 1981 High pressure X-ray diffraction of thorium to 30 GPa; *J. Less-Common Met.* **78** 147–153
Benedict U 1984 Properties of actinide metals under pressure; *J. Phys.* **45** 145–148
Benedict U 1986 Systematics of f electron delocalization in lanthanide and actinide elements under pressure; *Physica* **B&C144** 14–18
Benedict U 1987 The effect of high pressure on actinide metals; in *Handbook on the physics and chemistry of the actinides* (eds) A J Freedmand and G H Lander (Amsterdam: Elsevier) and references mentioned therein
Benedict U, Itie J P, Dufour C, Dabos S and Spirlet J C 1985 Delocalization of $5f$ electrons in amerecium metal under pressure: Recent results and comparison with other actinides; *Americium and curium technology* (eds) N M Edelstein, J D Navratil and W W Schulz (Dordrecht: D Reidel) 213–224
Benjamin T M, Zou G, Mao H K and Bell P M 1981 Equation of state for thorium metal UO_2 and a high pressure phase of UO_2 to 650 Kbar; *Carnegie Institution Geophysical Laboratory Year Book* **80** 280–283
Bridgman P W 1931 *The physics of high pressure* (New York: Dover)
Bundy F P 1962 General principles of high pressure apparatus design; in *Modern high pressure research* (ed.) R H Wentorf Jr (Washington D.C.: Butterworths) Vol. 1 pp. 1–24
Dabos-Seignon S 1987 Etudes par diffraction X du comportement sous pression d'actinides (N_p, Pu, Am-Cm) et de composes d'actinides AnX2, et An X, Doctoral thesis, University of Paris, Paris
Drickamer H G and Balchan A S 1962 High pressure optical and electrical measurements; in *Modern very high pressure techniques* (ed.) R H Wentorf Jr (Washington D.C.: Butterworths) Vol. 1, pp. 25–50
Endo S, Toyama N, Ishibashi A, Chino T, Fujita E, Shimomura O, Sumiyama K and Tomli Y 1987 Determination of alpha-epsilon transition pressure in Fe-V alloy, in *High pressure research in mineral physics* (eds) M H Manghnani and Y Syono Vol. **39** pp. 29–33
Forman R A, Piermarini G J, Barnett J D and Block S 1972 Pressure measurement made by the utilization of ruby sharp line luminescence; *Science* **176** 284–285

Grosshans W A 1987 *Rontengenbeugung an enigen seltenen erden unter druck*, Ph.D. thesis, University of Paderborn, West Germany

Gschneider K A Jr 1961 *Rare earth alloys* (Princeton N.J.: Van Nostrand)

Gschneider K A Jr 1985 Pressure dependence of the intra rare earth generalized binary phase diagram; *J. Less-Common Met.* **110** 1

Hall H T 1960 Ultra high pressure high temperature apparatus; *Rev. Sci. Instrum.* **31** 125–131

Hemley R J, Jephcoat A P, Mao H F, Zha C S, Finger L W and Cox D E 1987 Static compression of H_2O-ice to 128 GPa (1·28 Mbar); *Nature (London)* **330** 737–739

Hubbard W B and MacFarlane J J 1980 Structure and evolution of Uranus and Neptune; *J. Geophys. Res.* **85** 225–234

Hubbard W B and Marley M S Optimized Jupiter, Saturn and Uranus interior models (preprint)

Jamieson J C, Lawson A W and Nachtrieb N D 1959 Device for obtaining X-ray diffraction patterns from substances exposed to high pressure; *Rev. Sci. Instrum.* **30** 1016–1019

Jayaraman A 1978 High pressure systematics in rare earth; in *Handbook on the Physics and Chemistry of the rare earths* (eds) K A Gschneider and L Eyring (Amsterdam: North Holland) Vol. 1 ch. 9

Jayaraman A 1983 Diamond-anvil cell and high pressure physical investigations; *Rev. Mod. Phys.* **55** 65–108

Jayaraman A 1985 Recent developments in static high pressure research; in *Shock waves in condensed matter* (ed.) Y M Gupta (New York: Plenum Press) 13–36

Jayaraman A 1986 Ultra-high pressures; *Rev. Sci. Instrum.* **57** 1013–1031

Johansson B and Rosengren A 1975 Generalized phase diagram for the rare-earth elements: Calculations and correlations of bulk properties; *Phys. Rev.* **B11** 2836

Kmetko E A and Hill H H 1970 in the *Proceedings of the Fourth International Conference on Plutonium and other Actinides* (ed.) W N Miner (New York: The Metallurgical Society of AIME)

Lawson A W and Tang T Y 1950 Diamond bomb for obtaining powder pictures at high pressures; *Rev. Sci. Instrum.* **21** 815

Mao H K and Bell P M 1975 Design of a diamond-windowed high pressure cell for hydrostatic pressure in the range 1 bar to 0·5 Mbar; *Geophys. Lab. Year Book* **74** 402–405

Mao H K, Hazen R M, Bell P M and Wittig J 1981 Evidence for 4f-shell delocalization in praseodymium under pressure; *J. Appl. Phys.* **52** 4572–4574

McMahan A K and LeSar R 1985 Pressure dissociation of solid nitrogen under one Mbar; *Phys. Rev. Lett.* **A105** 129

Moss W C and Goettel K A 1987 Finite element design of diamond anvials; *Appl. Phys. Lett.* **50** 25

Moss W C, Hallquist J O, Reichlin R, Goettel K A and Martin S 1986 Finite element analysis of DAC: Achieving 4·6 Mbar; *Appl. Phys. Lett.* **48** 1258

Moss W (Private communication)

Nellis W J 1987 Nitrogen at high pressure, *APS Topical Conference on Shock waves in condensed matter, Monterey, CA. UCRL*-96768

Nellis W J, Hamilton D C, Holmes N C, Radousky H B, Ree F H, Mitchell A C and Nicol M 1988 The nature of the interior of Uranus based on studies of planetary ices at high dynamic pressure; *Science* **240** 779–786

Piermarini G J, Block S and Barnett J D 1973 Hydrostatic limits in liquids and solids to 100 Kbar; *J. Appl. Phys.* **44** 5377–5382

Polian A, Besson J M and Grimsditch M 1985 in *Solid state physics under pressure* (ed.) S Minomura (Tokyo: Terra Scientific) p. 93

Reichlin R 1988 (Private communication)

Reichlin R, Schiferl D, Martin S, Vanderborg C and Mills R L 1985 Optical studies on nitrogen to 130 GPa; *Phys. Rev. Lett.* **55** 1464–1467

Roof R B 1980 Compression and compressibility studies of plutonium-gallium alloy. Advances in X-ray analysis, (eds) D K Smith, C S Barrett, D E Layden and P K Predecki *29th Annual Conf. Application of X-ray Analysis* (Denver: Plenum Press) 221–230

Roof R B 1981 Structure relationships in americium metal, Actinides 81; *Extended Abstract* 213–215

Ross M and Shishkevish C 1977 Molecular and metallic hydrogen, *ARPA Report No.* R-2056

Ross M and McMahan A K 1976 Comparison of theoretical models for metallic H_2; *Phys. Rev.* **B13** 5154–5157

Ross M and McMahan A K 1981 The metallization of some simple systems; in *Physics of solids under high pressure* (eds) J S Schilling and R N Shelton (Amsterdam: North Holland) 161–168

Ross M and Nellis W 1981 Shock wave studies modelling the giant planets; *Energy Technol. Rev.* 1–10

Sherman W F and Stadtmuller A A 1988 *Experimental techniques in high pressure research* (New York: John Wiley)

Shimomura O, Yamoka S, Yagi T, Wakatsuki M, Tsuji K, Fukunaga O, Kawamura H, Aoki K and Akimoto S 1984 *Mater. Res. Soc. Symp. Proc.* **22** 17

Shimomura O, Yamoka S, Yagi T, Wakatsuki M, Tsuji K, Fukunaga O, Kuwamura H, Aoki K and Akimoto S 1985 Multi-anvil type X-ray system for synchrotron radiation; in *Solid state physics under pressure* (ed.) S Minomura (Tokyo/London: KTK/Reidel) 351–356

Slattery W L 1977 The structure of the planets Jupiter and Saturn; *Icarus* **32** 58–72

Smith G S, Akella J, Reichlin R, Johnson Q, Schock R N and Schwab M 1981 Crystal structure of americium in the pressure range, 11–13 GPa. Actinides 81; *Extended Abstract* 218–220

Smith G S and Akella J 1982 Reexamination of the crystal structure of a high pressure phase in praseodymium metal; *J. Appl. Phys.* **53** 9212–9213

Smith G S and Akella J 1984 On the possibility of Pr III having a thcp structure; *Phys. Lett.* **A105** 132–133

Smith G S and Akella J 1986 Crystallographic calculations for possible high-pressure lanthanide structure models; *Phys. Lett.* **A118** 136–138

Smith G S and Akella J 1987 X-ray diffraction applications; *Synchrotron Radiation Volume, Energy Technol. Rev.* 23–28

Spain I L 1977 *High pressure technology* (eds) I L Spain and J Paawwe (New York: Dekker) Vol. 1

Spedding F H and Daane A H 1961 *The rare earths* (New York: Wiley)

Taylor K N R and Darby M I 1972 *Physics of rare earth solids* (London: Chapman and Hall)

Van Valkenberg A 1965 Visual observations of single crystal transitions under true hydrostatic pressures up to 40 Kbars, *Conference Internationale sur-les-Hautes Pressions, Le Creusot, Saone-et-Loire, France*

Vohra Y K, Brister K E, Weir S T, Duclos S J and Ruoff A L 1986 Crystal structures at megabar pressure determined by the use of the Cornell synchrotron source; *Science* **231** 1136–1138

Vohra Y K, Vijayakumar V, Godwaland B K and Sikka S K 1984 Structure of the distorted fcc high pressure phase of the trivalent rare-earth metals; *Phys. Rev.* **B30** 6205–6207

Weir C E, Lippincott E R, Van Valenburg A and Bunting E N 1960 Studies of infrared absorption spectra of solids at high pressures; *Spectrochim Acta* **16** 58–73

Yagi T, Akaogi M, Shimoumura O, Suzuki T and Akimoto S 1987 In situ observations of the olivine-spinel phase transformation in Fe_2SiO_4 using synchrotron radiation; *J. Geophys. Res.* **92** 6207–6213

Applications of Raman spectroscopy in earth and planetary sciences

SHIV K SHARMA

Hawaii Institute of Geophysics and Hawaii Natural Energy Institute, University of Hawaii, 2525 Correa Road, Honolulu, HI 96822, USA

Abstract. The present status of the applications of Raman spectroscopy for investigating earth and planetary materials under high pressures and temperatures is reviewed. The diamond anvil cell (DAC), which can be used for generating pressures in the megabar range, is briefly introduced. Following this is a presentation of high-pressure Raman techniques used in conjunction with the DAC, and time-resolved Raman techniques for use at high temperatures. Results of recent high-pressure and high-temperature Raman measurements on a number of metal oxides and silicates have demonstrated that, with this technique, it is feasible to investigate pressure- and temperature-induced phase transitions and map the P-T diagrams of these materials in the laboratory. A whole section is devoted to high-pressure Raman studies of condensed gases (including H_2, D_2, H_2-He mixtures, H_2O, CH_4, NH_3 and NH_3-H_2O mixtures) of interest to planetary scientists. The high-pressure Raman measurements on hydrogen, which have been extended to ~ 2.5 Mb at 77 K, have demonstrated that hydrogen remains a molecular solid to the highest pressure. Single-pulse multichannel Raman measurements of shocked nitrogen and water have demonstrated that Raman techniques can be utilized for investigating structures and properties of planetary materials under dynamic high pressures and temperatures. It is speculated that a combination of the time-resolved multichannel Raman technique along with the technique of laser heating in the DAC will allow measurements of the structures and vibrational properties of materials to very high pressures ($\leqslant 2$ Mb) and temperatures ($\leqslant 2000$ K). The availability of static high-pressure and high-temperature Raman data of planetary materials along with the Raman data on shocked materials under dynamic pressure will help to improve our modelling and understanding of planetary interiors.

Keywords. Raman spectroscopy; diamond anvil cell; time-resolved Raman spectrometry; high-pressure Raman spectroscopy.

1. Introduction

A complete knowledge of the chemical composition and phases of materials present in planetary interiors can help us to understand various phenomena observed on planetary surfaces. Most of the direct information about the interior of the Earth has come from seismic data. Complementary information has also come from rock composition, magnetic variation and laboratory experiments on rocks at ambient and high pressures. The surface compositions of the other planets in the solar system are more poorly known, and seismic information is non-existent except for the Moon. As the exact composition and structures of materials in the planetary interiors are unknown, scientists must investigate the structure and properties of planetary materials, such as metal oxides, silicates, H_2, CH_4, ammonia, ices, etc under high pressures and temperatures. During the past several years, advances in micro-Raman and time-resolved Raman techniques, as well as advances in high-pressure diamond anvil cell (DAC) designs have made it possible to investigate the structure and vibrational properties of these materials at very high pressures and temperatures relevant to planetary interiors. In this article high-pressure Raman techniques and recent applications of high-pressure Raman spectroscopy and time-resolved Raman spectroscopy in earth and planetary sciences are discussed.

2. Experimental methods

The details of DAC design and its usage have been reviewed in several recent publications (Jayaraman 1983, 1986; Ferraro 1984). The basic idea of generating high-pressure within a DAC is illustrated in figure 1. A sample is placed in a punctured metal gasket mounted between the flat parallel faces of two opposing anvils, and is subjected to pressure when the anvils are brought closer together. Variations in DAC designs result from different ways of applying force and maintaining alignment under pressure. Five different types of DACs have evolved over the past several years. Jayaraman (1983) summarized the various design features of these cells. All of these cells can be used for measuring high-pressure Raman spectra of materials, provided the conical opening behind the exit diamond is 40° or larger. Hirsch and Holzapfel (1981) described a cell design in which a single crystal sapphire is used for the backing of one anvil to create an enlarged aperture for Raman spectroscopy.

Most of the DACs can be used for generating pressure $\leqslant 350$ kbar. Figure 2 shows the DAC developed by Mao and Bell (1978a, b) for generating pressures exceeding

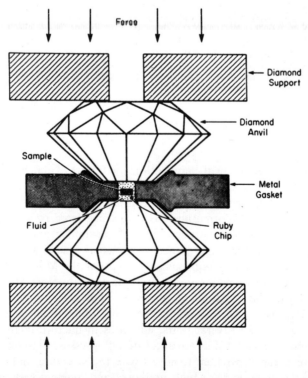

Figure 1. Basic configuration of the diamond anvil cell.

Raman spectroscopy in earth and planetary sciences 265

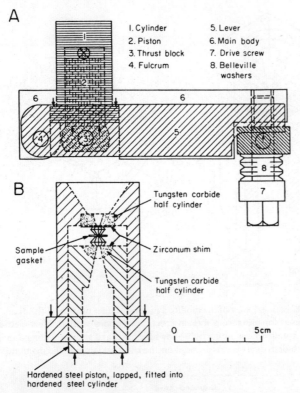

Figure 2. Schematic diagram of (**A**) Mao-Bell (M-B) diamond anvil cell (Mao and Bell 1978a, b) and (**B**) M-B cell piston-cylinder details.

1 megabar (Mb). In a later design (see figure 3), the cell was modified so that scattered radiation can also be measured in 90° and 135° geometries (Mao and Bell 1980) in addition to the more-common 180° geometry. The use of bevelled diamond anvil geometry has made it possible to generate pressures in excess of 2·5 Mb (Bell *et al* 1984; Goettel *et al* 1985). Mao and coworkers have carried out a number of experiments using bevelled diamond anvils with various ratios for A and B, and different bevel angles, where A is the diameter of the bevelled face and B is the diameter of the flat part of the bevelled face (see figure 4). In the 2·75 Mb experiments, the A dimension was 300 μm, B was 50 μm, and the bevel angle was 5°. More recently pressures as high as 5·5 Mb have been generated with the bevelled anvils (Xu *et al* 1986). Moss *et al* (1986) also reported ultra-high pressures of 4·6 Mb in a similar cell. The Mao-Bell cell has been extensively used for Raman spectroscopic studies, and has enabled the measurement of H_2 Raman spectra from pressures as high as 2·5 Mb (Hemley and Mao 1988).

2.1 *Diamond anvil cell*

2.1a *Diamond anvils*: An important first step for high-pressure Raman spectroscopy in the DAC is the selection of diamonds that exhibit low fluorescence in the laser beam

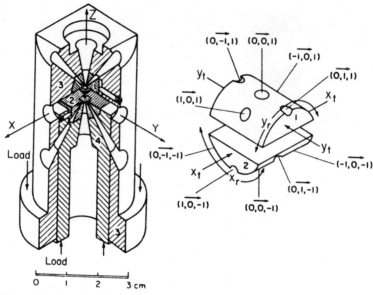

Figure 3. A sectional view of the modified piston-cylinder portion of the diamond anvil cell on the left used for measuring Raman spectra in various scattering geometries (Mao and Bell 1980). Exploded view on the right shows the ports and geometric direction indices in the tungsten carbide rockers (parts 1 and 2 on the left) that support the diamond anvils.

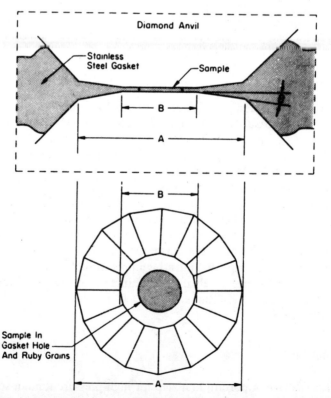

Figure 4. Schematic diagram of bevelled diamond anvils.

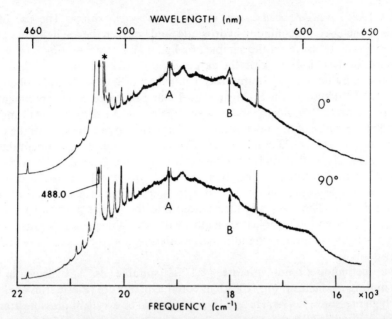

Figure 5. Raman/fluorescence spectra of n-D_2 and type-I diamond anvils in $0°$ and $90°$ scattering geometries stimulated with 200 mW of 488·0 nm line of an Ar^+ laser. Sample pressure 25 kb. A and B indicate regions of first- and second-order Raman scatter of diamond anvils.

(figure 5). Sharma (1979) quantified fluorescence intensities in different types of diamonds by comparing the relative intensities of the 1333 cm^{-1} first-order diamond line (C_{1333}) with those of the fluorescence maximum ($C_{f.max}$) and minimum ($C_{f.max}$) respectively. The resulting ratios, N_1 and N_2, are defined by $N_1 = C_{1333}/C_{f.max}$ and $N_2 = C_{1333}/C_{f.min}$, where C_{1333}, $C_{f.max}$ and $C_{f.min}$, respectively, refer to the number of counts per second at the peak position of the first-order Raman diamond line, at the fluorescence (or second-order phonon) maximum and at the fluorescence minimum, i.e. minimum background. It is found that some type-IIb diamonds with $N_1 =$ 448–305 and $N_2 = 4050$–2050 usually have a fluorescence background lower in intensity than the second-order diamond Raman spectrum. These diamonds are most suitable for studying weak Raman scatterers (Adams and Sharma 1977). According to Hirsch and Holzapfel (1981) an acceptable diamond for Raman spectroscopy should have a second-order band at 2500 cm^{-1} at least three times stronger than the flat luminescence background.

In general, most of the yellowish type-I diamonds exhibit very strong fluorescence ($N_1 \simeq 4·0$, $N_2 \simeq 30$). Colourless type-I and type-IIa diamonds with $N_1 = 384$–242 and $N_2 = 1500$–2180 can be used with strong Raman scatterers. In addition to low fluorescence, only those anvils that are free from micro-cracks and inclusions, and have low birefringence patterns due to strain should be used for high-pressure work. Wong and Klug (1983) reported that it is possible to use type-I diamonds for Raman measurements with UV-laser radiation. Although such anvils can be used only with a limited range of laser excitation, these could be useful for making Raman measurements of materials simultaneously maintained at high pressures and temperatures.

2.1b *Gaskets, pressure media, and in-situ pressure calibration*: The use of a metal gasket in the DAC for confining hydrostatic fluid between the anvils allows generation of hydrostatic pressure on the sample (see figure 1). For generating moderately high pressures ($\leqslant 100$ kilobars (kb)), the gasket is prepared by drilling a hole $\sim 300\,\mu$m in diameter in a thin (0·1–0·25 mm thick) metal sheet of Inconel X750, molybdenum, or tempered T301 stainless steel (Jayaraman 1986). For conducting experiments at very high pressures (>100 kb), the metal gasket is first indented in the DAC and a hole is drilled at the centre of the indentation. All the burrs around the sides of the hole in the gasket must be removed before use. The cylindrical sample compartment thus created in the gasket is loaded under the binocular microscope in the following way: the gasket is seated on one of the diamond flats with the hole in the centre of the indentation. The gasket must be seated in the same orientation as it had when making the indentation. The sample is then placed into the gasket hole along with a small (5–20 μm) ruby chip. The hole is then filled with a suitable pressure-transmitting fluid, often a 4:1 (by volume) methanol:ethanol mixture, and the gasket is sealed between the two anvils.

The methanol:ethanol mixture (4:1) is suitable as a hydrostatic pressure transmitting medium up to ~ 100 kb at room temperature (Piermarini *et al* 1973). At very high pressures, Xe, Ar, He and H_2 are found to be excellent pressure-transmitting media. A number of techniques have been developed for filling the gasket hole with H_2, He or noble gases either by cryogenic filling (Mao and Bell 1979; Liebenberg 1979; Silvera and Wijngaarden 1985) or by a high-pressure gas-loading system (Besson and Pinceaux 1979; Mills *et al* 1980). Argon is found to be a convenient pressure-transmitting medium because a droplet of Ar can be trapped in the gasket hole at ~ 86 K, and does not have the more severe problems associated with loading H_2 and He.

In-situ pressure in the DAC can be determined by measuring the red shift of the R_1 fluorescence line of the ruby crystal (Barnett *et al* 1973; Piermarini and Block 1975; Adams *et al* 1976). The shift is almost linear up to 300 kb at a rate of 0·365 Å/kb or 0·753 cm^{-1} kb^{-1}. The following equation proposed by Mao *et al* (1978) can be used for estimating pressure from the R_1 ruby line shift over an extended pressure range:

$$P(\text{kb}) = 3808[(\lambda/\lambda_0)^5 - 1], \qquad (1)$$

where λ_0(nm) is the wavelength of R_1 line at 1 bar and λ(nm) is the wavelength at pressure P.

2.2 *High-pressure Raman spectroscopy*

Most of the early high-pressure Raman studies with HgI_2, a very strong Raman scatterer, showed the feasibility of high-pressure Raman spectroscopy (Brasch *et al* 1968; Postmus *et al* 1968; Adams *et al* 1973). Lack of an efficient optical coupling between the DAC and Raman spectrometer, however, made it difficult to measure the high-pressure Raman spectra of materials having medium-to-weak Raman scattering efficiency. Adams *et al* (1977) who analysed the optical path of scattered light from the sample in DAC concluded that a high numerical aperture system is essential for efficient collection of scattered light from the sample in DAC. These workers used a 90° off-axis ellipsoidal mirror both in 0° and 180° scattering geometries (figure 6) to couple the DAC with a CODER triple monochromator (Adams *et al* 1977). Later

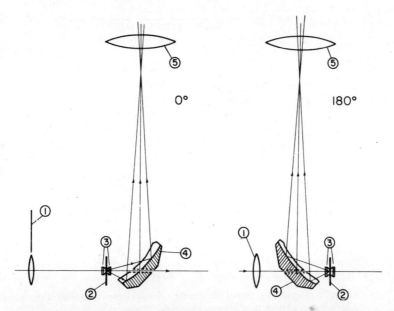

Figure 6. Schematic diagram showing the use of 90° off-axis ellipsoidal mirror in 0° and 180° high-pressure experiments (Adams *et al* 1977; Sharma 1979): 1, laser focusing lens; 2, metal gasket; 3, diamond anvils; 4, off-axis ellipsoidal mirrors; and 5, collecting lens.

Sharma (1979) used a 90° off-axis ellipsoidal mirror to couple the Mao-Bell DAC (figure 6) with a Jobin-Yvon double monochromator (model HG-2S) and successfully measured the Raman spectra of H_2 to pressures as high as 630 kb (Sharma *et al* 1980a, b). A commercially available sample compartment from Spex Industries (UVISIR Illuminator, model 1459), consisting of a 90° off-axis ellipsoidal mirror can also be used successfully in the forward scattering geometry. An optical diagram of this system is shown in figure 7. The laser beam is focused to a fine spot (30–40 μm) by the lens, and light scattered in the forward direction is collected by the ellipsoidal mirror and focused on the slit of the Spex double monochromator (model 1403).

Weinstein and Piermarini (1975) were the first to carry out Raman measurements in the 180° scattering geometry on Si and GaP to pressures over 100 kb using conventional lens collection optics and a double monochromator. A lens-based optical system used for 180° Raman scattering measurements of the interaction of H_2 and D_2 with a metal sample in the DAC is shown in figure 8 (Takahashi and Neill 1985). A small mirror deflects the focused beam ($\sim$ 30–40 μm diameter spot) onto the sample placed inside the DAC. The scattered light is focused onto the slit of the Raman spectrometer by the collecting lens, whose focal length is between 75 and 130 mm and whose f value ranges from f/1 to f/2 (Nakamura *et al* 1979; Takahashi and Neill 1985).

Although the optical coupling systems described in previous paragraphs have been used successfully for high-pressure Raman spectroscopy in DAC, alignment is often difficult and one is not certain where the laser beam is hitting the sample. These drawbacks have now been overcome by adapting a micro-Raman system so as to optically couple the DAC with a single channel or a multichannel Raman spectrometer.

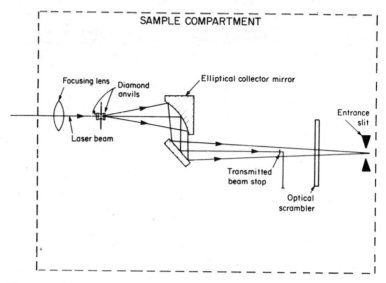

Figure 7. The diamond anvil cell and 90° off-axis ellipsoidal mirror in the modified sample compartment of Spex double monochromator (Spex 1403).

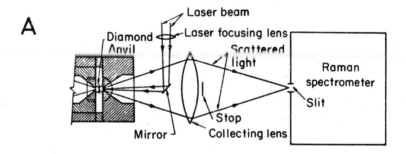

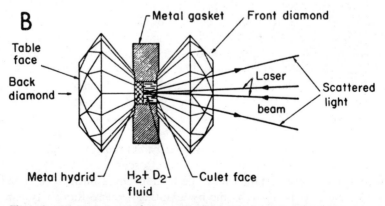

Figure 8. Schematic for Raman scattering measurements with DAC in the backscattering geometry with conventional lens collecting system (Takahashi and Neill 1985).

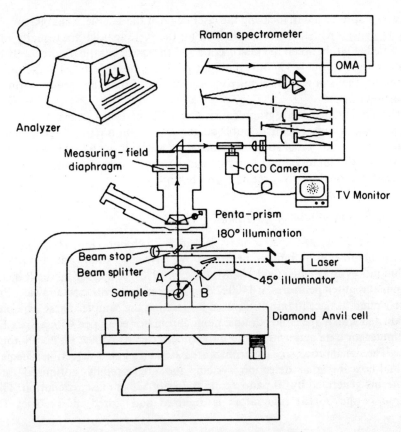

Figure 9. Details of a micro-Raman multichannel spectrograph.

Figure 9 shows a schematic diagram of a multichannel micro-Raman system developed at the University of Hawaii for high-pressure Raman spectroscopy with the DAC. The system is built around a modified Leitz Ortholux-I microscope optically coupled with a Spex Triplemate spectrograph (model 1877), and an optical multichannel detector (OMA III EG&G Princeton Applied Research) for use in either a 180° or 135° scattering geometry (Mao *et al* 1985; Sharma *et al* 1985b). The microscope is equipped with a variable aperture for selecting the excited image area of the sample, a penta-prism for simultaneous viewing of the sample, and a back-lighted spectrometer slit. The microscope was modified to allow sufficient working distance between the microscope objective and microscope stage in order to accommodate an 8·25-cm high diamond anvil cell (Mao and Bell 1978a, b). Samples are typically excited with the 488·0 nm line of a continuous wave (CW) Ar^+ ion laser. A long-working-distance 16 × objective (A in figure 9) focuses the laser beam and also serves to collect the scattered light in the 180° scattering geometry. In the 135° scattering geometry, the laser beam is focused by a long-focal-length 16 × objective (B in figure 9) mounted at 45° to the optical axis of the microscope. Objectives A and B are mounted so that their foci coincide at a point along the optic axis of the microscope. Radiation collected by the microscope objective (A) is focused on the adjustable aperture, then reflected by a 90° prism through an achromatic doublet, and focused on the entrance slit of the spectrograph. To aid in focusing the laser beam at the sample, a charge-coupled device (CCD) camera with an attenuation filter for the exciting laser line (usually 488·0 nm) is mounted at 90° between the microscope and the spectrograph (figure 9). By rotating a plane mirror through 45° into the light path it is possible to view the focused laser beam at the sample. The mirror is removed from the beam during spectral measurements.

The Triplemate spectrograph consists of a 0·22 m double-stage system and a 0·6 m single-stage system. The first double-stage provides subtractive dispersion and is used primarily for rejection of stray light, including Rayleigh-scattered light. The dispersion viewed by the multichannel detector can be changed by selecting one of the three gratings mounted on the grating turret assembly.

2.3 *Time-resolved Raman spectrometry*

Time-resolved Raman spectrometry is a useful technique for measuring the Raman spectra of minerals at high temperatures. We have successfully adopted time-resolved techniques (Van Duyne *et al* 1974; Mulac *et al* 1978) for measuring the Raman spectra of minerals at temperatures < 1400 K. For these experiments, a windowless Pt-10% Rh wire-wound furnace (figure 10) is used for heating the sample. These experiments are conducted with a Spex 1403 double monochromator equipped with a Spex Datamate computer for data acquisition and a thermoelectrically cooled GaAs photomultiplier tube. The signal processing electronics are based on a gated-detection scheme utilizing a dual-boxcar signal detection system, and are carefully optimized for Raman scattering generated by 10-nanosecond pulses of 532 nm laser excitation. The analog output of the boxcar integrators is digitized and stored in the Spex Datamate

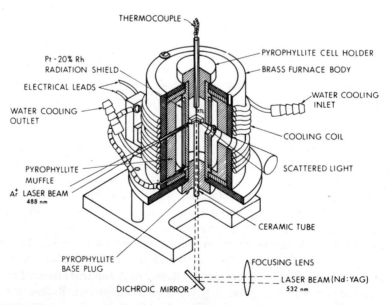

Figure 10. A windowless Pt-10% Rh wire-wound furnace for studying Raman spectra of minerals at high temperatures and ambient pressure.

computer. With this system it is possible to record the Raman spectra of silicate minerals at high temperatures free from interference of blackbody radiation.

The problems associated with interference from blackbody radiation at $T \leqslant 1073$ K can be largely overcome by proper selection of the excitation radiation (e.g. the 488·0 nm line of a CW Ar^+ ion laser; Bates 1972; Bates and Quist 1972; Gilbert et al 1975a, b; Mitchell and Raptis 1983). For materials such as GeO_2 quartz, which is a strong Raman scatterer, it is possible to measure the Raman spectra with 488·0 nm excitation at temperatures up to 1400 K. Multiple scans are, however, necessary at temperatures greater than 1200 K to obtain a good signal-to-noise ratio in the spectrum.

3. Applications

Both conventional and advanced Raman spectroscopic techniques are being increasingly used to examine a wide variety of earth and planetary materials under high pressures and temperatures (e.g. Hemley et al 1987; Sharma 1989). These studies are of considerable interest because the results of these investigations can yield information about phase transitions in materials under high P and T, and the vibrational frequencies can be used for calculating thermodynamic properties as well as for checking theoretical models. These data will help us in improving our understanding of the structure of planetary interiors. Following is a brief review of some of these applications.

3.1 *Raman spectroscopic studies of minerals at high pressures and high temperatures*

It is of great significance to earth scientists to investigate the structures and thermodynamic properties of minerals at high pressures and temperatures. The structures of a number of minerals have been studied with X-ray diffraction techniques at high pressures and temperatures as well as under the simultaneous application of high pressures and high temperatures (for a recent review see Hazen and Finger 1982). Raman spectroscopic investigation of minerals at high P and T has been rather limited (Ferraro 1984). The notable exception is α-quartz which has been investigated with Raman spectroscopy by a number of workers at high pressures (Asell and Nicol 1968; Hemley 1987; Jayaraman et al 1987) and high temperatures (Raman and Nedungadi 1940; Nedungadi 1940; Narayanaswamy 1948; Bates and Quist 1972) as well as under uniaxial stress (Tekippe et al 1973). In fact high-temperature investigations on α-quartz date back to 1940 when Raman and his coworkers investigated the Raman spectrum and identified a soft mode responsible for the α- to β-quartz phase transition at ~ 846 K (Raman and Nedungadi 1940; Nedungadi 1940).

3.1a *Quartz form of SiO_2 and GeO_2, and high-pressure silica polymorphs*: With recent advances in multichannel micro-Raman spectroscopic instrumentation, high-pressure Raman spectra of α-quartz, coesite and stishovite have been measured in the diamond anvil cell (Hemley 1987). Very recently, the high-pressure Raman spectra of α-quartz and isostructural $AlPO_4$ (berlinite) have also been studied in the DAC with a conventional single-channel laser Raman spectrophotometer equipped with a

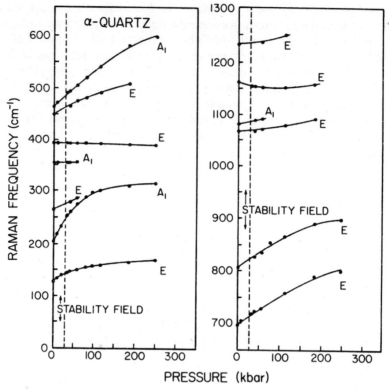

Figure 11. Pressure dependence of the frequency of the Raman active modes of α-quartz form of SiO$_2$ (Hemley 1987).

photon-counting detection system (Jayaraman et al 1987). The pressure dependencies of the phonon frequencies of the α-quartz form of SiO$_2$ and GeO$_2$ are shown in figures 11 and 12 respectively.

Bates (1972) and Bates and Quist (1972) have used laser-Raman spectroscopy to investigate α- to β-phase transitions in quartz and cristobalite. Both α- and β-quartz forms of SiO$_2$ are hexagonal and have the same number of atoms in the unit cell ($Z = 3$), but belong to two different space groups (α-quartz: $D_3^4(P3_12)$, β-quartz: $D_6^4(P6_222)$, and Si atoms at C_2 and D_2 sites, respectively). According to factor group analysis, the following modes are expected to occur in these two polymorphs:

α-quartz:
$$\Gamma(D_3^4) = 4A_1(\text{R}) + 4A_2(\text{IR}) + 8E(\text{R, IR});$$

β-quartz:
$$\Gamma(D_6^4) = A_1(\text{R}) + 2A_2(\text{IR}) + 2B_2(\text{i.a.}) + 3B_1(\text{i.a.}) + 4E_1(\text{R, IR}) + 4E_2(\text{R});$$

where R is Raman-active; IR, infrared-active; and i.a., inactive.

Scott and Porto (1967) reinvestigated the polarized Raman spectrum of an oriented single crystal of the α-quartz form of SiO$_2$ excited with a laser source, and have identified all the modes predicted on the basis of group theory. The results of these

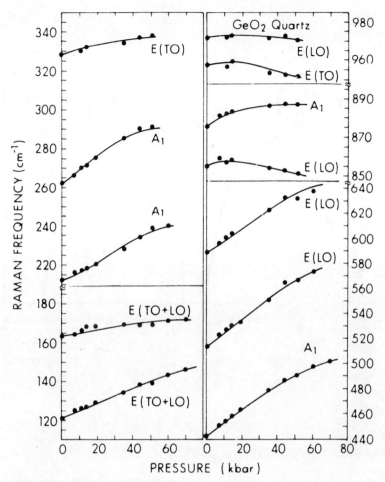

Figure 12. Pressure dependence of the frequencies of the Raman-active modes of α-quartz form of GeO$_2$.

Raman studies of the quartz polymorphs of SiO$_2$ are reviewed by Scott (1974).

The metastable α-quartz form of GeO$_2$ is isostructural with that of SiO$_2$; there is, therefore, a close similarity between the Raman spectra of these materials. The Raman bands in the spectrum of GeO$_2$ appear at a lower frequency because of the heavier mass of Ge atoms relative to the mass of Si atoms and also because of the lower value of the Ge-O force constant relative to that of Si-O in these otherwise isostructural materials. Sarver and Hummel (1960) placed the α- to β-transition in GeO$_2$ at 1311 K on the basis of a small discontinuity at 1273 K in the thermal expansion of the quartz form of GeO$_2$ along the *a* and *c* directions.

Recently the effects of high temperature on the Raman spectra of the α-quartz forms of GeO$_2$ and SiO$_2$ were investigated by the present author (figures 13–15). In this study, polycrystalline samples of the quartz form of GeO$_2$ and SiO$_2$ were used because a single crystal of the α-quartz form of GeO$_2$ was not available. The comparative study has shown that the α- to β-quartz transition does not take place in GeO$_2$ quartz even if the material is superheated to 10°C above the melting point (figure 14). The

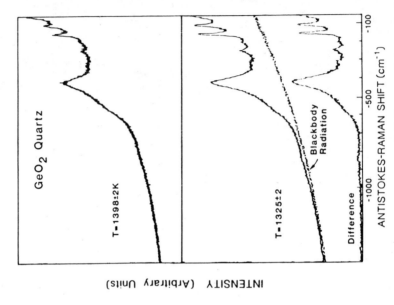

Figure 14. AntiStokes-Raman spectra of GeO_2-quartz below and above the melting temperature ($T_M = 1387$ K) at ambient pressure.

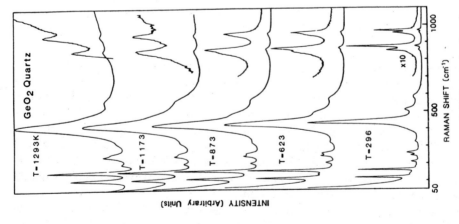

Figure 13. Stokes-Raman spectra of GeO_2-quartz at various temperatures and ambient pressure.

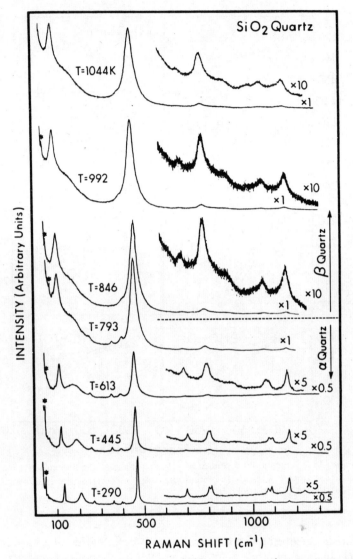

Figure 15. Stokes-Raman spectra of SiO_2-quartz at various temperatures and ambient pressures.

present investigation has also shown that the frequencies of a number of phonon bands of the α-quartz form of SiO_2 show considerable nonlinear temperature variation beginning several hundred degrees below the transition temperature (figure 16). In the case of the α-quartz form of GeO_2, the frequencies of Raman-active phonons vary linearly with temperature (figure 17).

The α- to β-phase transition, even in the polycrystalline sample of SiO_2 quartz, is clearly indicated by the softening, or rapid decrease towards low frequency, of the 207 cm^{-1} band and the nonlinear temperature variation of several phonon bands (figure 16). The dependencies of the Raman frequencies of SiO_2 quartz on temperature reported by various workers (Nedungadi 1940; Narayanaswamy 1948; Bates and

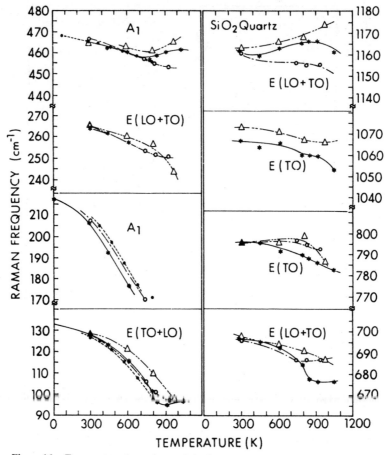

Figure 16. Temperature dependence of the frequencies of the Raman active modes of SiO_2 quartz at ambient pressure (°, Nedungadi, 1940; ○, Narayanaswamy, 1948; △, Bates and Quist 1972; *, present study).

Quist 1972) are in agreement with the present study as far as the trends are concerned. The differences in the magnitude of frequencies of optic phonons of quartz as reported by various workers (figure 16) are most pronounced for high-frequency modes associated with the E-type LO and TO phonons. These differences may be attributable to the fact that the frequencies of these modes in quartz are strongly dependent on the orientation of the crystal. The observed inflection near 847 K (the $\alpha - \beta$ transition temperature) in the frequency versus T curves of the 125 and 464 cm^{-1} bands probably indicates the presence of a third phase between α- and β-quartz. Recent neutron, X-ray and electron diffraction studies of SiO_2 quartz have shown the existence of an incommensurate phase within a small temperature range (~ 1.3 K) around the usual $\alpha - \beta$ transition temperature (Gouhara et al 1983; Ghose 1985; Dolino 1986).

By combining the high-pressure and -temperature data of minerals and isostructural materials, one can separate the pure volume dependencies of phonon frequencies at 296 K. If the phonon frequency, v_i, is treated as a function of volume

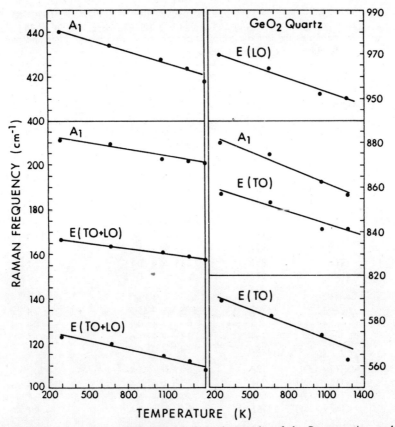

Figure 17. Temperature dependence of the frequencies of the Raman-active modes of quartz-form of GeO_2 at ambient pressure.

Table 1. Thermal expansion and bulk modulus data for α-quartz form of GeO_2 and SiO_2

	$\beta/(10^{-5} K^{-1})$	K_0/(Mb)	K_0'
GeO_2	$2{\cdot}99^a$	$0{\cdot}391(4)^b$	$2{\cdot}2(5)$
SiO_2	$4{\cdot}17^c$	$0{\cdot}371(2)$ Mb^d	$6{\cdot}2(1)$

Data from Ref. ^{a}Murthy (1962), ^{b}Jorgensen (1978), ^{c}Shafer and Roy (1956) and ^{d}Levien et al (1980).

and temperature, then according to Peercy and Morosin (1973):

$$(\partial \ln v_i / \partial T)_P = (\partial \ln V / \partial T)_P \cdot (\partial \ln v_i / \partial \ln V)_T + (\partial \ln v_i / \partial T)_V \tag{1}$$

$$= -\beta K_T (\partial \ln v_i / \partial T)_T + (\partial \ln v_i / \partial T)_V, \tag{2}$$

where β refers to the volume thermal expansion coefficient $(\partial \ln V / \partial T)$; and K_T refers to the isothermal bulk modulus, $(\partial P / \partial \ln V)_T$. Values of K_T and β for the α-quartz form of SiO_2 and GeO_2 are summarized in table 1. The volume dependencies of the α-quartz forms of SiO_2 and GeO_2 are given in table 2.

Table 2. Logarithmic pressure and temperature derivatives of frequencies of the Raman-active phonons for α-quartz form of GeO_2 and SiO_2 and the separation of the isobaric temperature derivatives into pure-volume and pure-temperature contributions at 296 K.

Modes	v_i(cm^{-1})	$(\partial \ln v_i/\partial T)_P = -\beta K_T(\partial \ln v_i/\partial P)_T + (\partial \ln v_i/\partial T)_V$		
GeO_2				
A_1	212	-5.3^a	$= -2.8$	-2.5
	262	-3.0	$= -2.5$	-0.5
	440	-4.4	$= -2.3$	-2.11
	880	-2.9	$= -0.3$	-2.6
E(TO + LO)	121	-6.1	$= -3.6$	-2.5
(TO + LO)	163	-5.1	$= -0.9$	-4.2
(TO + LO)	957	-2.9	$= 1.1$	-4.0
E(TO)	326	-2.45	$= -0.7$	-1.75
(LO)	514	-5.35	$= -2.7$	-2.7
(TO)	583	-4.1	$= -1.6$	-2.5
(TO)	857	-2.15	$= +0.1$	-2.25
(LO)	972	-2.2	$= -0.02$	-2.18
SiO_2				
A_1	207^a	-17.9	$= -15.5^c$	2.4
	355	0.3	$= +0.7$	-0.4
	464	-3.1	$= -1.6$	-1.5
	1082	-1.4	$= -0.17$	-1.2
E(TO + LO)	128	-26.0	$= -5.8$	-20.2
(TO + LO)	264	-3.2	$= -2.8$	-0.4
(TO + LO)	696	-1.8	$= -1.3$	-0.5
(TO + LO)	1161	-0.5	$= -0.4$	-0.1
E(TO)	394	-1.5	$= +0.05$	-1.55
(LO)	401	0	$= 0.05$	-0.05
(TO)	450	-4.3^b	$= -1.5$	-2.8
(LO)	511	N.D.*		
(TO)	795	-1.5	$= -0.9$	-0.6
(LO)	307	-1.6^b	$= -0.9$	-0.8
(TO)	1066	-1.3	$= -0.15$	-1.15
(LO)	1231	-1.0^b	$= -0.1$	-0.9

^{a} In units of $10^{-5} K^{-1}$.
Data from Refs. ^{b} Bates and Quist (1972) and ^{c} Hemley (1987).
* N.D., not determined.

In GeO_2, the temperature dependencies of most of the Raman bands result about equally from thermal expansion and the pure temperature term owing to higher anharmonicities (table 2) (Maradudin and Fein 1962; Cowley 1963). In α-quartz SiO_2, for most of the bands the self-energy shifts are dominated by the pure-volume term due to thermal expansion. The lower degree of anharmonicity of SiO_2-quartz phonons as compared with the corresponding GeO_2-quartz phonons is also indicated by the lesser degree of broadening of the v_s(Si-O-Si) band with temperature relative to the v_s(Ge-O-Ge) band (figures 13, 15).

3.1b *Rutile forms of GeO_2, TiO_2, SnO_2 and SiO_2, and high-pressure polymorphs of*

TiO$_2$: The high-pressure Raman spectra of the rutile forms of GeO$_2$, TiO$_2$ and SnO$_2$, which are isostructural with stishovite and have cations in six-fold coordination, have been investigated at ambient temperature (Mammone and Sharma 1979, 1980a, b; Mammone *et al* 1980, 1981). According to factor group analysis of rutile structure oxides (space group $P4_1/nmm(D_{4h}^7)$ with $z = 2$), the modes are expected to apportion as follows:

$$\Gamma(D_{4h}^7) = A_{1g}(\text{R}) + A_{2g}(\text{i.a.}) + B_{1g}(\text{R}) + B_{2g}(\text{R}) + 2B_{1u}(\text{IR}) + E_g(\text{R}) + 3E_u(\text{IR})$$

where R is Raman-active; IR, infrared-active; and i.a., inactive.

One of the Raman-active modes (the B_{1g} mode) of the rutile forms of GeO$_2$ and TiO$_2$ shows softening at high pressures (figures 18–21), indicating instability of rutile-type structures at very high pressures. Recent Raman studies of stishovite (figures 22 and 23) have shown that this phase of SiO$_2$ also shows softening of the B_{1g} mode at high pressures (Hemley 1987).

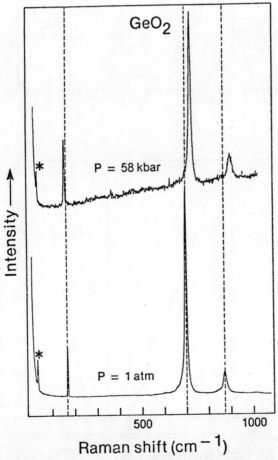

Figure 18. Raman spectra of tetragonal GeO$_2$ at 1 atm and 58 kb. Asterisk (*) marks the plasma line (Mammone *et al* 1981).

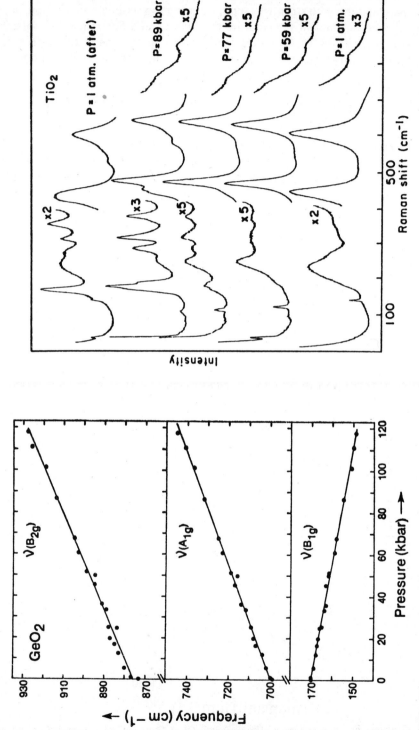

Figure 20. Raman spectra of rutile (TiO$_2$) at room temperature and various pressures (Mammone et al 1980).

Figure 19. Pressure dependence of the frequencies of the Raman active modes in tetragonal GeO$_2$ with symmetries A_{1g}, B_{1g} and B_{2g} (Mammone et al 1981).

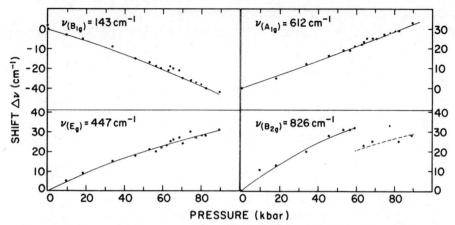

Figure 21. Pressure dependence of various Raman-active mode frequencies of rutile TiO_2 at room temperature. Circles, rutile TiO_2; squares, TiO_2-II; triangles, intermediate between rutile TiO_2 and TiO_2-II (new bands beginning to appear) (Mammone et al 1980).

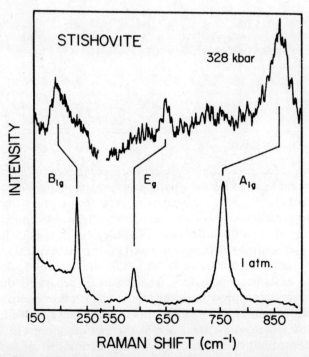

Figure 22. Raman spectra of stishovite at 1 atm and 328 kb ($T = 298$ K) (Hemley 1987).

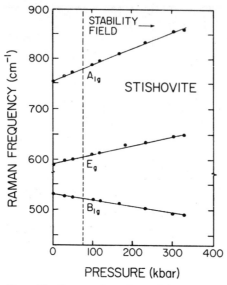

Figure 23. Pressure dependence of the B_{1g}, E_g and A_{1g} Raman bands of stishovite at room temperature (Hemley 1987).

Table 3. Raman frequencies (cm^{-1}) and mode-Grüneisen parameters (0.1 MPa) of stishovite and other rutile-type oxides[a].

Sym.	SiO$_2^a$		GeO$_2$		SnO$_2$		TiO$_2$	
	v_i	γ_i	v_i	γ_i	v_i	γ_i	v_i	γ_i
A_{1g}	753	1·38 ± 0·04	700	1·4	613	3·6	612	1·6
						1·8		1·5
B_{1g}	231	−1·58 ± 0·06	170	−2·8	121	−10·4	143	−5·0
								−5·4
B_{2g}	967	—	873	1·3	776	2·6	826	—
						1·6		
E_g	589	1·00 ± 0·03	—	—	476	3·2	447	2·4
						1·6		2·2

Data from [a] Hemley (1987).

Hemley (1987) has calculated the mode-Grüneisen parameters of stishovite and other oxides with the rutile structure; these parameters are summarized in table 3. The smaller value of v_i for the B_{1g} mode in stishovite indicates that SiO$_2$ in the rutile structure appears to remain stable with respect to the B_{1g} displacement to much higher compression relative to the other oxides (Hemley 1987). Using equations (1) and (2), Mammone and Sharma (1980a, b) combined the thermal expansion and bulk modulus data for rutile oxides (table 4) with the pressure and temperature dependencies of the v_i Raman frequencies. These workers separated the isobaric temperature derivatives into pure-volume and pure-temperature contributions at 298 K. These results are summarized in table 5. In GeO$_2$ and SnO$_2$, self-energy shifts are dominated by the pure-volume term due to thermal expansion. In TiO$_2$, the temperature dependence results about equally from the thermal expansion and pure-temperature terms associated with higher order anharmonicities.

Table 4. Thermal expansion and bulk modulus data for rutile-structure oxides[a].

	$\beta/(10^{-5}\,K^{-1})$	K_0/(Mb)	K'_0
TiO_2	2·35	2·2	6·84
GeO_2	2·027	2·65	6·16
SnO_2	1·17[a]	2·03	
SiO_2		2·88	

[a] Data from Mammone et al (1981).

Table 5. Logarithmic pressure and temperature derivatives of the frequencies of the Raman-active phonons for TiO_2, GeO_2 and SnO_2 and the separation of the isobaric temperature derivatives into pure-volume and pure-temperature contributions at 298 K[a].

Modes	$v_i(cm^{-1})$	$(\partial \ln v_i/\partial T)_P =$	$-\beta K_T(\partial \ln v_i/\partial P)_T +$	$(\partial \ln v_i/\partial T)_V$
TiO_2				
B_{1g}	143	0·4[b] =	12·3	−11·9
E_g	447	−13 = −	4·9	−8·1
A_{1g}	612	−1·8 = −	3·0	+1·2
B_{2g}	826	3 = −	4	+7
GeO_2				
B_{1g}	170	2·6 =	5·7	−3·1
A_{1g}	700	−3·1 = −	2·9	−0·2
B_{2g}	873	−3·0 = −	2·7	−0·3
SnO_2				
E_g	473	−1·8 = −	1·7	−0·1
A_{1g}	633	−4·3 = −	2·0	−2·3
B_{2g}	775	−4·4 = −	1·7	−2·7

[a] Data from Samara and Peercy (1973) and Mammone et al (1981).
[b] In units of $10^{-5}\,K^{-1}$.

The effect of high pressures on the TiO_2 rutile to TiO_2-II phase transition has also been reinvestigated with Raman spectroscopy in a diamond anvil cell (Mammone and Sharma 1979; 1980a, b; Mammone et al 1980). It was found that phase transition is very sensitive to the presence of shear stresses. For example, in a Drickamer-type cell using NaCl as the pressure-transmitting medium, the onset of the transition was observed at $P > 30$ kb (Nicol and Fong 1971). In a diamond anvil cell with a 4:1 methanol:ethanol mixture as the pressure-transmitting medium, the onset of rutile to the TiO_2-II phase transition was detected by Raman spectroscopy at $P > 70$ kb (Mammone and Sharma 1979a; Mammone et al 1980) (figures 20, 21). The high-pressure Raman spectra of TiO_2-II have been studied to 372 kb, and a reversible but sluggish phase transition from TiO_2-II to TiO_2-III was detected in the pressure range 200–300 kb (figure 24). In the 200–300 kb range, metastable coexistence of TiO_2-II and TiO_2-III was observed. At 372 kb, TiO_2-II appears to be completely transformed into TiO_2-III (figure 23). In many respects, the Raman spectrum at 372 kb has similar

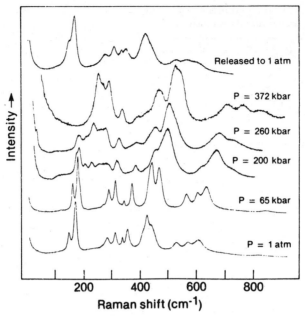

Figure 24. Raman spectra of TiO_2-II at room temperature and various pressures (Mammone et al 1981).

spectral characteristics to those of BaF_2 and PbF_2 (Kessler et al 1974). The Raman spectrum of TiO_2-III further implies that the primitive cell contains at least four formula units (Mammone et al 1981).

3.1c *Enstatite ($MgSiO_3$) and its high-temperature polymorphs*: The effect of high temperature on the structures of a number of polymorphs of enstatite ($MgSiO_3$) has been investigated with time-resolved Raman spectroscopy. The high-temperature Raman spectra of ortho-enstatite (figure 25) show that the ortho- to proto-enstatite transition takes place at 1320 K (Sharma et al 1987). Although no Raman-active soft mode is observed in the high-temperature Raman spectra of ortho-enstatite, the frequencies of a number of phonons show nonlinear variation with temperature before the transition (figure 26). The ortho- to proto-enstatite transition at 1320 K is marked by the appearance of a new and strong lattice mode at $98\,cm^{-1}$ (figure 25). The internal modes of $[Si_2O_6]$ pyroxene chains show rather small variations at the phase transition. On quenching the proto-enstatite to room temperature it transforms irreversibly to low clinoenstatite (figure 27).

Figure 27 shows the effect of temperature on clinoenstatite which undergoes the low-to-high clinoenstatite phase transition at 1273 K (Sharma et al 1987). The frequencies of a number of Raman-active phonons of low-clinoenstatite show nonlinear variation with temperature below the transition temperature (figure 28). The transition is marked by the appearance of new lattice modes at 131 and $155\,cm^{-1}$. The various polymorphs of enstatite give rise to distinct lattice modes in the Raman spectra (figure 27). Raman spectroscopy can be used, therefore, to identify $MgSiO_3$ polymorphs in natural as well as synthetic minerals.

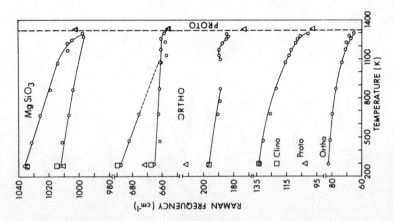

Figure 26. Temperature dependence of the frequency of the Raman active modes of orthoenstatite ($MgSiO_3$) at 1 atm pressure.

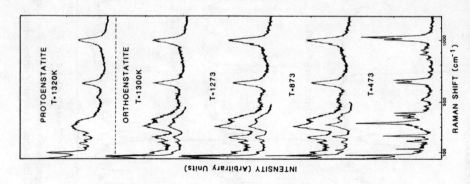

Figure 25. Time-resolved Raman spectra of orthoenstatite ($MgSiO_3$) at various temperatures and ambient pressure (excitation 533 nm Nd:YAG laser system, slit 2 cm^{-1}; gated detection).

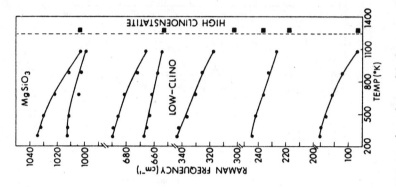

Figure 28. Temperature dependence of the frequency of the Raman active modes of low clinoenstatite.

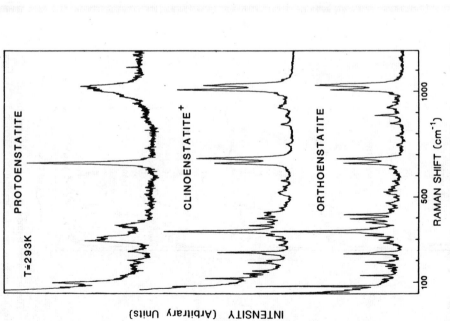

Figure 27. Raman spectra of various polymorphs of orthoenstatite. The sample of clinoenstatite contains a small amount of orthoenstatite.

3.1d *ZrO_2, Mg_2SiO_4 and other minerals*: The high-pressure Raman spectra of ZrO_2 have been reported recently (Arashi and Ishigame 1982; Arashi 1987). Using a modified DAC, the Raman spectra of ZrO_2 were measured up to 350°C and 70 kb (Arashi 1987). The Raman data below 350°C and at various pressures have been used to investigate the monoclinic to orthorhombic phase transition in ZrO_2, and for constructing a tentative P–T diagram. Arashi (1987) has demonstrated the feasibility of making Raman measurements under simultaneous application of high pressures and of temperatures higher (350°C vs 130°C) than used in the study by Adams and Sharma (1976).

Most of the silicate minerals are weak Raman scatterers. With a single-channel Raman spectrometer it is difficult to measure the effect of high-pressures on all Raman-active modes of these minerals. The effect of high pressure on the most intense bands of Mg_2SiO_4, that appear at 854 and 920 cm^{-1} at ambient conditions, has been reported up to 65 kb (Besson *et al* 1982).

The Raman spectrum of forsterite (Mg_2SiO_4) as a function of temperature has been measured to 1409 K (Sharma *et al* 1985a). The frequencies of the Raman bands of Mg_2SiO_4 decrease linearly with increasing temperature. On the basis of calculations from the high-temperature X-ray diffraction measurements of Mg_2SiO_4 and olivine crystals, it had been suggested that there may be a slight shortening of Si-O bonds at high temperatures (Hazen and Finger 1982). The results of high-temperature Raman studies, however, show that the frequency of the $v_s(Si-O^-)$ symmetric stretching mode of SiO_4^{4-} ions decreases with increasing temperature, indicating that at high temperature there is an elongation of the Si–O$^-$ bonds in forsterite. The results of the high-temperature Raman study of Ni_2SiO_4 polymorphs also indicate that there is elongation of Si–O bonds in Ni-olivine at high temperature. The Ni_2SiO_4-spinel to Ni-olivine transition at ambient pressure has been investigated and a weakening of Si–O bonds in the spinel at high temperatures has been observed (Yamanaka and Ishii 1986). The phonon density-of-states and specific heat of forsterite have been calculated with a rigid-ion model (Rao *et al* 1987) and good agreement is found with the experimental measurement by inelastic neutron scattering. Refinement of the theoretical model will become possible with more extensive high pressure and temperature Raman data for Mg_2SiO_4.

Salje and Viswanathan (1976) and Salje and Werneke (1982) have investigated the Raman spectra of a number of minerals at moderately high pressures and temperatures. These workers have used the pressure and temperature dependencies of phonon frequencies of minerals and crystal structure data to calculate thermodynamic properties and P-T phase diagrams. In a series of papers, Kieffer (1979a, b, 1982, 1985) calculated the thermodynamic properties of minerals from the vibrational spectra. Most of these calculations are based on vibrational spectroscopic data at ambient conditions. It is anticipated that with the availability of high-pressure and -temperature Raman and infrared spectroscopic data, scientists will be able to model the thermodynamic properties of minerals at elevated pressures and temperatures corresponding to conditions existing in the interiors of the planets.

3.2 *Raman spectroscopic study of condensed fluids at high pressures*

Hydrogen is the most abundant element in the universe and is a major constituent of the giant planets. In the mantles of the Jovian planets, which contain mixtures of helium and hydrogen, hydrogen is believed to be mostly in a metallic form under these extreme P-T conditions. The structure and properties of hydrogen in the megabar pressure range have been of interest for a number of years both from a fundamental viewpoint and for testing models of the interiors of these planets (Stevenson 1982, 1983; Hubbard 1984; Cole 1984). Theoretical studies of solid hydrogen at 0 K have predicted that, at sufficiently high pressure, hydrogen is expected to become a metal (Wigner and Huntington 1935; Schneider 1969). Ross and Shishkevish (1977) reviewed numerous calculations of the properties of hydrogen at high density, including the metallic form, over the past 50 years. In addition to the prediction of an insulator-to-metal transition associated with the destruction of molecular bonding, metallization by band overlap in the molecular solid has been predicted (Ramaker et al 1975; Friedli and Ashcroft 1977).

It has been proposed that metallic hydrogen may be a high-temperature superconductor (Ashcroft 1968). The quest for metallic hydrogen led to the development of cryogenic and other filling techniques to trap condensed gases in the DAC for pressurization. As a result, the high-pressure behaviour of H_2, D_2, CH_4, N_2, O_2 and rare gas solids has been investigated with Raman, Brillouin scattering, optical microscopy and X-ray structure analysis (Jayaraman 1983; Hemley et al 1987). In several cases their equations of state have been determined and compared with theory. In the following sections, results of high-pressure Raman measurements of planetary gases are reviewed and discussed.

3.2a *Hydrogen and deuterium*: Raman spectroscopy has provided the most successful probe of the structure of compressed hydrogen and deuterium at ultra-high pressures. The first high-pressure Raman studies of normal (n)-hydrogen at room temperature were carried out in the pressure range 20–630 kb (Sharma et al 1979, 1980a, b). Later, high-pressure Raman spectra of n-D_2 were measured in the pressure range 8–537 kb (Sharma et al 1980c). In the Raman spectra of n-H_2 and n-D_2, both the rotational and the H–H and D–D bond stretching vibrational modes (vibron) were measured as a function of pressure. In the n-H_2 and n-D_2 fluid phase, the rotational lines broaden, become diffuse and shift slightly towards higher frequencies with increasing pressure. The rotational lines disappear above the solidification pressure (55 kb) in the spectrum of n-H_2 but are observable in the spectrum of n-D_2 up to 146 kb (figure 29). The H_2 vibron band, $Q_1(1)$, at $4155 \cdot 2 \text{ cm}^{-1}$ at ambient temperature was followed up to 626 kb (figure 30), and the n-D_2 vibron band, $Q_1(2)$, at $2987 \cdot 23 \text{ cm}^{-1}$, to 537 kb (figure 31). In the spectra of H_2 and D_2, the vibron band broadens with pressure in the fluid phase and suddenly sharpens at the solidification pressure. At the solidification point, the decrease in the half-width of the $Q_1(1)$ band of H_2 from 15 to $5 \cdot 6 \text{ cm}^{-1}$, and that of the $Q_1(2)$ band of D_2 from 8 to 5 cm^{-1} is believed to be due to a reduction in the collision-induced broadening in the solid phase. At the freezing pressure, the frequencies of H_2 and D_2 vibron bands decrease by $1 \cdot 8 \text{ cm}^{-1}$ and 1 cm^{-1}, respectively (Sharma et al 1980b, c). At the ambient-pressure freezing point, the frequencies of the $Q_1(1)$ band of n-H_2 and the $Q_1(0)$ and $Q_1(1)$ lines of n-D_2 decrease by $2 \cdot 53$ and $2 \cdot 51 \text{ cm}^{-1}$ respectively (Bhatnagar et al 1962). The small decreases in the frequencies of n-H_2 and n-D_2 vibrons at the freezing pressure indicate that repulsive forces are significant in these compressed solids at room temperature (Sharma et al 1979; 1980a, b). The large decrease in the linewidths of the vibron bands of n-H_2 and n-D_2 at the freezing pressures can

Raman spectroscopy in earth and planetary sciences 291

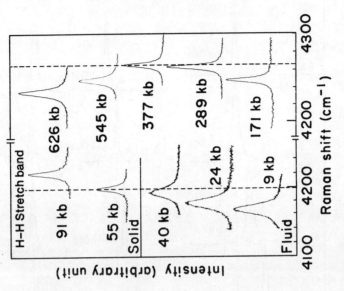

Figure 30. The $Q_1(1)$ Raman active vibron of n-H$_2$ at room temperature and various pressures (Sharma et al 1980b).

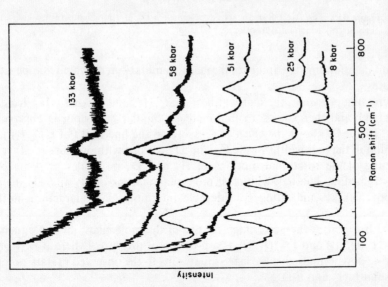

Figure 29. Raman spectra of n-D$_2$ in the region of rotational modes at room temperature and various pressures (Sharma et al 1980c).

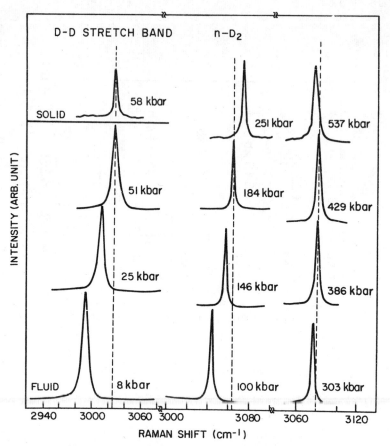

Figure 31. The $Q_1(2)$ Raman active vibron of n-D_2 at room temperature and various pressures (Sharma *et al* 1980c).

be used for determining the fluid-solid phase boundary at elevated temperatures and pressures.

With increasing pressures, the half-width of the $Q_1(1)$ band of solid n-H_2 decreases to about 175 kb and then shows a small increase. Similar behaviour is observed in solid n-D_2. Figure 32 shows the effect of pressure on the line width of n-D_2 (vibron). The reduction in the half-widths of the H_2 and D_2 vibrons with pressure is either due to a lessening of the rotational motions of H_2 and D_2 molecules or some other secondary transition. The half-widths of molecular vibrons of H_2 and D_2 increase rapidly above 300 kb, indicating considerable intermolecular interaction at these pressures.

Figure 33 is a plot of the percentage change of the molecular vibron frequencies, $\Delta v_R/v_0$, of H_2 ($Q_1(1)$) and D_2 ($Q_1(2)$) as a function of pressure, (where R = H or D). The following polynomials give the least-squares fits to the observed v_R data points in fluid and solid n-H_2 and n-D_2.

Fluid phase:

$$v_H = 1\cdot553P - 0\cdot0115P^2, \qquad (3)$$

Raman spectroscopy in earth and planetary sciences 293

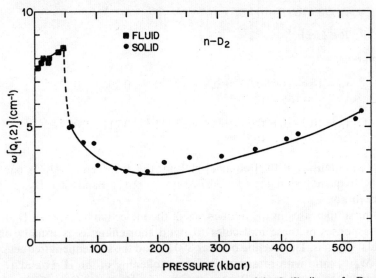

Figure 32. Pressure dependence of the halfwidth (ω) of the $Q_2(2)$ vibron of n-D_2.

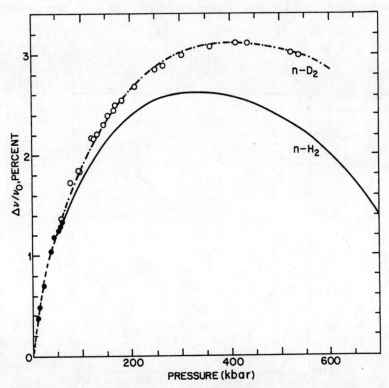

Figure 33. The shift of $Q_1(1)$ and $Q_1(2)$ vibrons ($\Delta\nu/\nu_0$ in percent) of n-H_2 and n-D_2, respectively, as a function of pressure (Sharma *et al* 1980c).

$$v_D = 1{\cdot}1168P - 0{\cdot}827 \times 10^{-3}P^2$$
$$P \leqslant 55\,\text{kb}. \tag{4}$$

Solid phase:
$$v_H = 14{\cdot}6 + 0{\cdot}780P - 0{\cdot}229 \times 10^{-2}P^2 + 0{\cdot}296 \times 10^{-5}P^3 \\ - 0{\cdot}165 \times 10^{-8}P^4, \tag{5}$$

$$v_D = 10{\cdot}2 + 0{\cdot}600P - 0{\cdot}164 \times 10^{-2}P^2 + 0{\cdot}211 \times 10^{-5}P^3 \\ - 0{\cdot}116 \times 10^{-8}P^4. \tag{6}$$

In these equations, v_H is the frequency shift for the $4155{\cdot}2\,\text{cm}^{-1}$, $Q_1(1)$, band of $n\text{-}H_2$; v_D is the frequency shift for the $2987{\cdot}23\,\text{cm}^{-1}$, $Q_1(2)$, band of $n\text{-}D_2$; and P is the pressure in kb.

The initial increases in the frequencies of the molecular vibrons of H_2 and D_2 are direct responses of these molecules to bond shortening as is usually observed in molecular crystals. The leveling off near 300 kb and the subsequent frequency decrease of the $\Delta v_R/v_0$ ratio with pressure indicate a softening of the H–H and D–D bonds. The changes observed in the Raman spectra of $n\text{-}H_2$ and $n\text{-}D_2$ with pressure closely resemble each other with two significant differences: (i) In both the fluid and solid phases, the D–D stretching band is much narrower than that for H–H stretching at corresponding pressures; this may result from the relatively heavy mass of the deuterium nuclei; and (ii) the frequency ratio, v_D/v_H, increases linearly with increasing pressure in both fluid and solid phases (figure 34). The v_D/v_H data can be fit to the

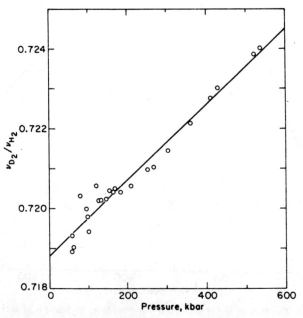

Figure 34. Plot of pressure dependence of v_{H_2}/v_{D_2}.

straight line:

$$\nu_D/\nu_H = 0.719 + 0.945 \times 10^{-5} P, \qquad (7)$$

where P is the pressure in kb. The changes in the above ratio have been recently attributed to the large ratio zero-point motion of hydrogen atoms (Hemley et al 1987).

Silvera and Wijngaarden (1981) carried out Raman scattering studies of ortho(o)-D_2 and para(p)-H_2 in the DAC to 540 kb at liquid-He temperature. These workers found a second-order phase transition only in o-D_2. The $J = 2$ rotational band undergoes a remarkable broadening in the pressure range 200–278 kb and then splits at high pressures. The splitting of the rotational band is attributed to a phase transition in which spherically symmetric molecules of o-D_2 go into an orientationally ordered state at low temperature; the initial broadening is a precursor effect. Silvera and Wijngaarden (1981), and Wijngaarden et al (1982) have also observed turnover in the vibron frequency of p-H_2 and o-D_2 at low temperatures.

Mao et al (1985) have extended the Raman measurements on the vibron band of n-H_2 and n-D_2 to 1.47 and 1.26 Mb, respectively, using a DAC, with bevelled diamond anvils, and a multichannel micro-Raman spectrometer. Both H_2 and D_2 remained transparent, and therefore non-metallic, to the highest pressures reached in this experiment. Figure 35 shows the Raman spectra of n-H_2 and n-D_2 under these ultra-

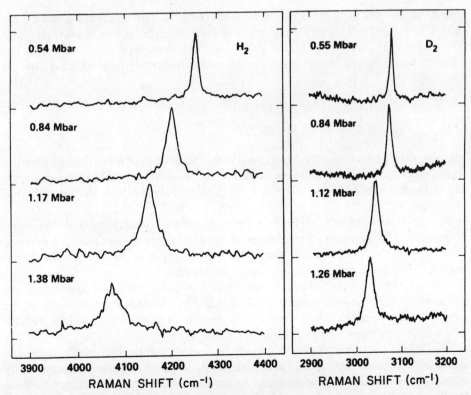

Figure 35. Raman spectra of solid normal hydrogen and deuterium at room temperature and high pressures (Mao et al 1985).

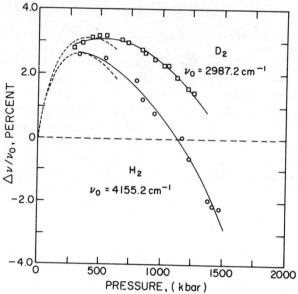

Figure 36. Pressure dependence of the frequency of the intermolecular stretching mode in solid hydrogen and deuterium. The zero pressure frequencies are indicated. The dashed lines are fits to the data of Sharma *et al* (1980c) and Mao *et al* (1985).

high pressures. The percentage frequency shifts of the Raman bands of H_2 and D_2 are shown in figure 36. The zero pressure frequency (horizontal line, $\Delta v = 0\,cm^{-1}$) in figure 35 is $4155 \cdot 2\,cm^{-1}$ for H_2 and $2987 \cdot 2\,cm^{-1}$ for deuterium (Stoicheff 1957). The solid curves in figure 36 are quadratic least-square fits to the data given by the functions

$$v_H = 105 \cdot 78 + 55 \cdot 66 P - 135 \cdot 35 P^2, \tag{8}$$

$$v_D = 65 \cdot 72 + 100 \cdot 77 P - 95 \cdot 70 P^2, \tag{9}$$

where P is the pressure in Mb (Mao *et al* 1985). The dotted curves in the low pressure region are polynomial fits to the previous data for the fluid (to 0·055 Mb; equations (3) and (4)) and the solid (0·055–0·63 Mb, hydrogen; 0·055–0·54 Mb, deuterium) (Sharma *et al* 1980a, b).

The new data of Mao *et al* (1985) in figure 36 follow a similar trend of decreasing frequencies. The frequency of the vibron of solid hydrogen crosses the zero-pressure value and the shift becomes negative on further compression. At 1·47 Mb the vibron band in hydrogen is $90\,cm^{-1}$ below the initial value.

The bandwidth of the vibron in the spectra of solid hydrogen and deuterium increases with increasing pressure above 175 kb. At 1·26 Mb, the half-width of the vibron reaches $41\,cm^{-1}$ in hydrogen and $20\,cm^{-1}$ in deuterium. At 1·47 Mb the half-width of the hydrogen vibron is estimated to be $80\,cm^{-1}$.

The shifts of the vibrons to lower frequencies suggest that the intramolecular bonds continue to weaken up to 1·47 Mb. The solid, however, remains predominantly molecular even at the highest pressure. As mentioned before, the difference between the pressure shifts of the vibron in hydrogen and deuterium is significant. Reference equations of state calculated for the two isotopes do not indicate large differences in compressibility (Ross *et al* 1983), and therefore the large shift in frequency of the hydrogen stretching band may imply that dissociation is more advanced than in the case of deuterium under equivalent pressure (Mao *et al* 1985).

Recently, Hemley and Mao (1988) have further extended Raman measurements on $n\text{-}H_2$ to 2·0 Mb at 77 K. The Raman spectra of $n\text{-}H_2$ at 77 K and various pressures are shown in figure 37. The pressure dependence of the vibron of solid hydrogen at 77 K is shown in figure 38. As in the previous studies outlined above, a single, well-resolved

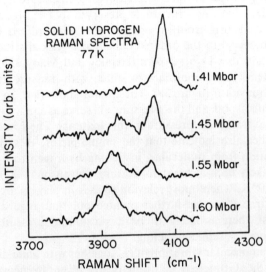

Figure 37. High-pressure Raman spectra of solid normal hydrogen at 77 K (Hemley and Mao 1988).

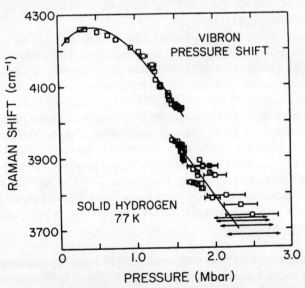

Figure 38. Pressure dependence of the vibron of solid hydrogen in the low- and high-pressure phases at 77 K (Hemley and Mao 1988).

band that decreases in frequency with pressure above 300 kb has been observed. At 1·45 Mb, however, a second broad band is observed 103 ($\pm$ 5) cm^{-1} below that of the first. The two bands coexist over a small pressure interval, with the intensity of the new band growing at the expense of the first, with increasing pressure. At 1·6 Mb, the first peak completely disappears. This discontinuous change in the vibron frequency of hydrogen with pressure indicates a phase transition.

According to single-crystal X-ray diffraction data, solid n-H$_2$ exists in a disordered hexagonal-close-packed structure to 265 kb (Hazen et al 1987; Mao et al 1988). The continuity of the Raman spectrum with a decrease in temperature to 77 K and increase in pressure indicates that this structure is preserved to 1·45 Mb. It has been proposed that n-H$_2$ may transform from an orientationally disordered hexagonal-close-packed structure ($P6_3/mmc$) to the ordered cubic-close-packed structure ($Pa3$) similar to the transformation in o-D$_2$ at 280 kb (Hemley and Mao 1988). The new vibron retains its band identity and continues to soften with pressure at pressures greater than that of the transition (figure 38). These data show that, to $\sim$ 2·5 Mb, hydrogen remains a molecular solid, and this pressure will serve as a new lower bound on the molecular-to-atomic transition pressure in this material. The results of these high-pressure Raman studies also indicate that the simple picture of a single, well-defined and abrupt transition from molecular, insulating hydrogen to monatomic, metallic hydrogen is not likely even at low temperature. Stevenson (1983) suggested that at a temperature $\sim 10^4$ K, where most relevant phases in planets are fluid, one would suspect a gradual transition. He further pointed out that, regardless of the nature of the transition(s), there is likely to be a substantial pressure range at $T \sim 10^4$ K for which hydrogen is semiconducting.

The softening of the intramolecular vibration of H$_2$ together with finite dissociation (H$_2 \to$ 2H) could reduce the Gruneisen v_i dramatically, even to negative value (Stevenson 1983). This could have profound implications for the dynamics of the interiors of the giant planets. The specific heat could also be anomalous and is important because it affects estimates of planetary thermal evolution and interpretations of observed heat flux.

Spectroscopic study of other diatomic molecules at high pressure can be used to gain some insight into the behaviour of hydrogen. High-pressure Raman measurements of solid nitrogen to 1·40 Mb (Reichlin et al 1985) and 1·7 Mb (Bell et al 1986) indicate weakening of the N $\equiv$ N bond on compression, but also show evidence of several phase transitions in the molecular solid. High-pressure Raman studies at both high and low temperatures have been useful for determining the phase diagram of solid molecular nitrogen at lower pressures (Buchsbaum et al 1984; Schiferl et al 1985; Zinn et al 1987). Recently, single-pulse multiplex coherent anti-Stokes Raman scattering (CARS) has been used to observe the vibrational spectra of fluid N$_2$ at several pressures and temperatures (up to 350 kb and 4400 K) (Schmidt et al 1987). The results of CARS measurements show the effect of very high temperatures on the frequency of the N $\equiv$ N vibron on compression. At pressures up to 340 kb and 4400 K, nitrogen exists as molecular fluid.

3.2b *Hydrogen-helium mixtures*: Schouten et al (1985) made measurements on a H$_2$-He mixture with initial helium concentration $x = 0.58$ in a DAC up to a pressure of 50 kb, at which fluid-fluid and solid-fluid separation of phases occurs as a function of T. Loubeyre et al (1987) investigated the binary phase diagram of H$_2$-He mixtures up to 100 kb at $T = 295$ and 373 K. These authors calibrated the shift of the Q_1 Raman-active vibron H_2 molecules as a gauge of helium concentration at a given pressure. They found that the richer the surrounding medium in helium, the larger the shift of the Q_1 vibron. The measurements of optically observed phase transitions in mixtures of known initial concentrations were accomplished by Raman measurements of the shift of the Q_1 vibron of H_2. In this way it was possible to determine detailed isothermal phase relations for fluid H_2-He mixtures, an important system for interior models of the giant planets.

3.3 *Water, methane and ammonia*

In order to understand the geology and history of icy bodies in the solar system, it is important to know the phase relations of their constituent ices. The principal ice-forming molecular compounds are water (H_2O), methane (CH_4) and ammonia (NH_3). These components are expected to occur in the solar system with abundance ratios for O, C, and N of 100:25:18 (Hunten *et al* 1984). Because water is the most abundant solid ice phase, methane and ammonia are most likely to be stable as hydrates. Although the properties of icy bodies with core pressures of less than 5 kb (e.g. Enceladus and Mimas) can be predicted based on atmospheric phase diagrams, understanding the large and more tectonically active icy satellites such as Titan and Triton requires knowledge of phase relations to at least 50 kb (Johnson and Nicol 1987).

3.3a *Water*: The PT-phase diagram for water has been determined by several investigators and is reviewed by Pistorius (1976). Mishima and Endo (1980) re-examined the phase relations among the nine crystalline polymorphs of ice at low temperature and moderately high pressures ($P \leqslant 40$ kb). As the various polymorphs of ice are metastable at low temperature, the Raman and infrared spectra of various polymorphs (Ic, Ih, II, III, V, VII and VIII) have been studied at ambient pressures and low temperature (77 K) using precompressed samples (Taylor and Whalley 1964; Marckmann and Whalley 1964; Wong and Whalley 1976). Hawke *et al* (1974) and Adams *et al* (1977) demonstrated that it is feasible to detect various phases of ice with Raman techniques in a sapphire anvil and a diamond anvil cell, respectively. The Raman spectral measurements by Walrafen *et al* (1982) on ice-VII to 360 kb indicate a lengthening of the molecular O–H bond on compression. Hirsch and Holzapfel (1986) observed a discontinuity in the Raman spectrum of ice VIII at 100 K between 300 and 400 kb, and have suggested that this is indicative of the transition to the ice-X phase in which symmetric hydrogen bonds are formed. The force constants in the high-pressure ices are related to a first-order phase transition from the "molecular" phase VIII to "ionic" phase X (Holzapfel *et al* 1987).

In addition to the large number of crystalline polymorphs of ice, there has been significant interest in the transformation and physical properties of amorphous forms of ice produced under pressure at low temperature (Mishima *et al* 1984; Klug *et al* 1986). Recently, *in situ* Raman spectroscopy with the diamond anvil cell has been used for studying water at pressures to 300 kb and temperatures from 77 to 300 K (Hemley *et al* 1988). It has been found that ice-I transforms to high-density amorphous (hda)-ice at 10 kb and 77 K in the DAC. The transformation has been observed at temperatures as high as 155 K without being occluded by crystallization of I to II (or

IX) phase. The hda-ice transformed at 45 kb and 77 K to ice-VII. It has also been found that temperature-induced amorphization occurs in ice-VII at 170 K and ~ 1 atm pressure. On the basis of these Raman data it has been suggested that hda-ice may have a limited role during accretion of icy bodies because of its instability under high pressure. Amorphization of high-pressure ice phases may occur, however, in a convecting body, if temperatures remain low enough to inhibit crystallization (Hemley et al 1988).

With a new single-pulse time-resolved spontaneous Raman technique, Holmes et al (1985) succeeded in measuring the Raman spectra of fluid water to 260 kb and 1700 K. These Raman data indicate that intermolecular hydrogen bonding, dominant at 120 kb, is nearly absent at 260 kb. No evidence for the presence of hydrated protons was observed. It has been suggested that the most probable mechanism for the high conductivity of shocked water is the dissociation of water molecules into H^+ and OH^- ions (Holmes et al 1985). These measurements are very significant because of the potential offered by Raman spectroscopy for the micro-scale experimental investigation of molecular and ionic processes in shocked fluids.

3.3b *Methane*: As pointed out above, methane is an important constituent of our solar system, and is the simplest of saturated hydrocarbon compounds. In phases without orientational order, tetrahedral CH_4 molecules may be considered to be spherical and the anisotropy of their intermolecular interactions should be quite small. Raman spectroscopy has been used extensively for investigating the six solid phases of methane at low temperature ($T < 80$ K) and moderate pressures ($P < 10$ kb) (Stevenson 1957; Fabre et al 1979, 1982; Thiery et al 1985). The melting curve of methane has been measured up to 387 K (Grace and Kennedy 1967); at 293 K, methane crystallizes near 1·3 kb into phase I (Hazen et al 1980). The phases of solid methane in the pressure range 16–187 kb at room temperature have been investigated *in situ* in the DAC with high-pressure Raman spectroscopy (Sharma et al 1980d). Hebert et al (1987) have measured the methane internal vibrational modes up to 200 kb by Raman scattering, the refractive index up to 120 kb, and the elastic constant up to 320 kb. On increasing the pressure from 50 to 51 kb there are discontinuous changes in the Raman spectra of the internal modes of CH_4, indicating the presence of a new phase (referred to as phase V by Sharma et al 1980d and phase IV by Hebert et al 1987). Another high-pressure phase transition was discovered by Sharma et al (1980d) at 118 kb and room temperature by the appearance of a shoulder in the v_1 region of the Raman spectrum. This transition, of which the high-pressure phase was referred to as VI by Sharma et al (1980d), is sluggish and incomplete even at 187 kb. It should be pointed out that a low-pressure phase at $P = 9$ kb and $T = 4·2$ K, discovered by Thiery et al (1985) and referred to as phase VI, was not known in 1980, and is different from the high-pressure phase above 118 kb. Raman studies of Hebert et al (1987) have confirmed the presence of the 118 kb phase of methane. These workers have, however, assigned this as phase VII. Despite these literature inconsistencies in the assignment of the names to the high-pressure phases of methane, such studies have, however, demonstrated that Raman spectroscopy is a valuable technique in mapping the P-T phase boundaries of methane.

3.3c *Ammonia and ammonia-water system*: The phase diagram and transition properties of ammonia from 200 to 305 K and 0·9–10·5 kb have been measured by

Mills et al (1982). Raman spectra of solid phases I, II and III of ammonia have been reported in the range 214 to 280 K and up to 8·7 kb (Gauthier et al 1987). Recently, high-pressure Raman measurements on ammonia have been extended to 750 kb at room temperature (Gauthier et al 1986). Around 3·8 kb, phase III transforms to a birefringent phase IV in which the spectrum of the vibron modes is different from that observed for phase III. The spectrum of phase IV consists of at least five overlapping bands, as compared to three bands observed in the spectrum of phase III. The strongest peak in the spectrum of phase IV has been assigned to the v_1 band and the two components v_{3a} and v_{3b} to the degenerate v_3 mode of ammonia. The frequency of the v_1 band decreases further with increasing pressure whereas those of v_{3a} and v_{3b} are observed to increase slightly (Gauthier et al 1986).

The transformation of phase IV to phase V near 140 kb is indicated by a change of the slopes, dv/dP, of the three vibration modes of ammonia, as well as a change in the shape of the spectrum in the region of vibron modes, indicating the presence of six overlapping bands. On the basis of Brillouin scattering, it is suggested that the IV-V transition is a first-order transition (Gauthier et al 1986). Raman evidence has been found for the presence of another high-pressure phase (phase VI) of ammonia close to 600 kb. Gauthier et al (1986) have speculated on the formation of a symmetrical hydrogen-bonded solid at 750 kb, as proposed for the phase X of ice (Hirsch and Holzapfel 1986).

The atmospheric-pressure phase relationships of the ammonia-water system in the liquidus region were investigated early in this century by Postma (1920), Elliott (1924) and others using optical observation of crystallization and melting. At that time, two stoichiometric compounds were identified, ammonia hemihydrate, $2NH_3 \cdot H_2O$ and ammonia monohydrate $NH_3 \cdot H_2O(M)$. Discovery of a third compound, ammonia dihydrate, $NH_3 \cdot H_2O(D)$, was reported by Rollet and Vuillard (1956). The ammonia-water phase diagram and its implications for icy satellites have been examined with high-pressure Raman spectroscopy by Nicol and coworkers in a series of papers (Johnson et al 1984, 1985; Johnson and Nicol 1987; Cynn et al 1989). The melting curve of monohydrate was reported by Johnson et al (1985) for pressures from 0 to 60 kb. Johnson and Nicol (1987) described parts of the ammonia-water phase diagram of $(NH_3)_x(H_2O)_{1-x}$ mixtures for $0 \leqslant x \leqslant 0.34$ on the basis of experimental observation in the DAC at $P = 0$ to 50 kb and $T = 240$ to 370 K. Cynn et al (1989) found that because of reaction between caustic samples and metal gaskets, there were inconsistencies in the melting curve for the dihydrate when extrapolated from the data of Johnson and Nicol (1987). Cynn et al (1989) have overcome these problems by employing Inconel 718, stainless steel 301 or work-hardened stainless steel 316 gaskets, electroplated with gold according to a procedure outlined by Parker (1963). These workers were thus able to identify phases in the ammonia-water system at various compositions and pressures to 80 kb and temperatures from 125 to 400 K. Phase assignments were made by optical characterization and confirmed by Raman spectroscopy. The formation of a stoichiometric solid with a composition of $(NH_3)_x(H_2O)_{1-x}$, where $x = 0.33$ was observed in a rather narrow range of conditions: from 15 to 20 kb and 193 to 208 K. At room temperature, the stable phases observed were fluid, high-pressure ices (VI and VII) and ammonia monohydrate. Near $x = 0.33$ the stable phases observed were high-pressure ices (VI, VII, and VIII), $NH_3 \cdot H_2O$, and another phase tentatively identified as ammonia dihydrate. Cynn et al (1989) have discussed the implications of these data on the geophysics of large icy satellites of the outer solar system. According to these workers the planetary application of the P-T phase data is heavily dependent upon the projected temperature of the mantle region of the icy satellite. These workers have further pointed out that the reactivity of concentrated ammonia in the liquid phase toward N_2 (Hemley and Mao 1988; Young *et al* 1987). More work is, however, needed if the diamond line shift is to be used as a secondary pressure standard.

Recent shock wave experiments on a mixture of 36 mol% anorthite and 64 mol% diopside analog for natural basalt have shown that the silicate melt undergoes relatively smooth compression at ~ 2000 K to 250 kb; above this pressure a marked lowering of compressibility occurs (Rigden *et al* 1988). It has been suggested that the gradual structural changes characteristic of the low-pressure regime, such as changes of Al^{3+} and Si^{4+} coordination by oxygen from four-fold to six-fold are essentially complete by ~ 250 kb. The single-pulse Raman measurements of the shocked silicate liquids could provide valuable insights about the structural transitions with pressure at high temperatures. Evidently static high-pressure and high-temperature Raman data in the DAC combined with the Raman data under dynamic pressures of shocked materials will help to improve our modelling and understanding of planetary interiors.

Acknowledgements

I am grateful to several colleagues for many helpful discussions and for providing me reprints and preprints of their work. They are in alphabetical order: Dr D M Adams, Dr W A Bassett, Dr R J Hemley, Dr A Jayaraman, Dr R Jeanloz, Dr H K Mao, Dr M Nicol, Dr D Heinz, Dr D J Stevenson and Dr E M Stolper. I wish to thank Dr T F Cooney and J P Urmos, for critically reading the manuscript. Special thanks to Carol Koyanagi for expertly handling the job of typing the manuscript. Financial support from the National Science Foundation (grants EAR 85-08971 and EAR 85-17888) is gratefully acknowledged. This is the Hawaii Institute of Geophysics contribution No. 2110.

References

Adams D M, Appleby R and Sharma S K 1976 Spectroscopy at very high pressure: Part X. Use of ruby R-lines in the estimation of pressure at ambient and low temperatures; *J. Phys.* **E9** 1140–1144

Adams D M, Payne S J and Martin K M 1973 The fluorescence of diamonds and Raman spectroscopy at high pressures using a new design of diamond anvil cell; *Appl. Spectrosc.* **27** 377–381

Adams D M and Sharma S K 1976 Spectroscopy at very high pressure: Part XI. Raman study of polymorphs of carbon tetrachloride and carbon tetrabromide and their photo-decomposition products; *J. Chem. Soc. Dalton Trans.* 2424–2429

Adams D M and Sharma S K 1977 Selection of diamonds for infrared and Raman spectroscopy; *J. Phys.* **E10** 680–682

Adams D M, Sharma S K and Appleby R 1977 Spectroscopy at very high pressures: Part 14. Laser Raman scattering in ultra-small samples in the diamond anvil cell; *Appl. Opt.* **16** 2572–2575

Arashi H 1987 Raman spectroscopic studies at high temperatures and high pressures: Application to determination of P-T diagram of ZrO_2; in *High pressure research in mineral physics* (eds) M H Manghnani and Y Syono (Tokyo: Terra Scientific) pp. 335–340

Arashi H and Ishigame M 1982 Raman spectroscopic studies of the polymorphism in ZrO_2 at high pressures; *Phys. Status Solidi* **A71** 313–321

Asell J F and Nicol M 1968 Raman spectrum of a quartz at high pressures; *J. Chem. Phys.* **49** 5395–5399
Ashcroft N W 1968 Metallic hydrogen: A high-temperature superconductor?; *Phys. Rev. Lett.* **21** 1748–1749
Barnett J D, Block S and Piermarini G J 1973 An optical fluorescence system for quantitative pressure measurement in the diamond anvil cell; *Rev. Sci. Instrum.* **44** 1–9
Bassett W A and Ming L C 1972 Disproportionation of Fe_2SiO_4 to $2FeO + SiO_2$ at pressures up to 250 kb and temperatures up to 3000°C; *Phys. Earth Planet. Inter.* **6** 154–160
Bassett W A and Weathers M A 1987 Temperature measurement in a laser-heated diamond cell; in *High pressure research in mineral physics* (eds) M H Manghnani and Y Syono (Tokyo: Terra Scientific) pp. 129–134
Bates J B 1972 Raman spectra of α and β cristobalite; *J. Chem. Phys.* **57** 4042–4047
Bates J B and Quist A S 1972 Polarized Raman spectra of β-quartz; *J. Chem. Phys.* **56** 1528–1533
Bell P M, Mao H K and Goettel K 1984 Ultrahigh pressure: Beyond 2 Mb and the ruby fluorescence scale; *Science* **226** 542–544
Bell P M, Mao H K and Hemley R J 1986 Observations of solid H_2, D_2 and N_2 at pressures around 1·5 Mb at 25°C; *Physica* **B139&140** 16–20
Besson J M and Pinceaux J P 1979 Melting of helium at room temperature and high-pressure; *Science* **206** 1073–1075
Besson J M, Pinceaux J P, Anastaopoulos C and Velde B 1982 Raman spectra of olivine up to 65 kb; *J. Geophys. Res.* **87** 773–775
Bhatnagar S S, Allin E J and Welsh H L 1962 The Raman spectra of liquid and solid H_2, D_2 and HD at high resolution; *Can. J. Chem.* **40** 9–23
Boppart H, van Straaten J and Silvera I F 1985 Raman spectra of diamond at high pressures; *Phys. Rev.* **B32** 1423–1425
Brasch J W, Melveger A J and Lippincott E R 1968 Laser excited Raman spectra of samples under very high pressures; *Chem. Phys. Lett.* **2** 99–100
Buchsbaum S, Mills R L and Schiferl D 1984 Phase diagram of N_2 determined by Raman spectroscopy from 15 to 300 K at pressures to 52 GPa; *J. Chem. Phys.* **88** 2522–2525
Cole G H A 1984 *Physics of planetary interiors* (Bristol, UK: Hilger) 152 pp
Cowley R A 1963 Lattice dynamics of anharmonic crystals; *Adv. Phys.* **12** 421–480
Cynn H C, Boone S, Koumvakalis A and Nicol M 1989 Phase diagram for ammonia-water mixtures at high pressures: Applications to icy satellites; in *Proc. Lunar Planet. Sci. Conf. XIX* (eds) G Ryder and V L Sharpton (Houston USA: Lunar and Planetary Inst.) pp. 433–441
Dolino G 1986 The incommensurate phase of quartz; in *Incommensurate phases in dielectrics: 2, Materials* (eds) R Blinc and A P Levanyuk (Amsterdam: North-Holland) pp. 205–232
Elliott L D 1924 The freezing point cure of the system water-ammonia; *J. Phys. Chem.* **28** 887–889
Fabre D, Thiery M M and Kobashi K 1982 Raman spectra of solid CH_4 under high pressure. II. New phases below 9 kb at 4·2 K; *J. Chem. Phys.* **76** 4817–4827
Fabre D, Thiery M M, Vu H and Kobashi K 1979 Raman spectra of solid CH_4 under pressure. I. Phase transition between phases II and III; *J. Chem. Phys.* **71** 3081–3088
Ferraro J R 1984 *Vibrational spectroscopy at high external pressures: The Diamond Anvil Cell* (New York: Academic Press) 264 pp.
Friedli C and Ashcroft N W 1977 Combined representation method for use in band-structure calculations. Applications to highly compressed hydrogen; *Phys. Rev.* **B16** 662–672
Gauthier M, Pruzan Ph, Besson J M, Hamel G and Syfosse G 1986 Investigation of the phase diagram of ammonia by Raman scattering; *Physica* **B139&140** 218–220
Gauthier M, Pruzan Ph, Chervin J C, Tanguy L and Besson J M 1987 Dynamical properties of ammonia from Raman scattering up to 75 GPa; in *Dynamics of Molecular Crystals* (Amsterdam: Elsevier) pp. 141–146
Ghose S 1985 Lattice dynamics, phase transitions and soft modes; in *Microscopic to macroscopic; Reviews in Mineralogy* (eds) S W Kieffer and A Navrotsky (Washington D.C.: Miner. Soc. Amer.) **14** 127–164
Gilbert B, Mamantov G and Begun G M 1975a A simple Raman cell and furnace usable at temperatures higher than 1000° for corrosive melts; *Appl. Spectrosc.* **29** 276–278
Gilbert B, Mamantov G and Begun G M 1975b Raman spectra of aluminum fluoride containing melts and the ionic equilibrium in molten cryolite type mixtures; *J. Chem. Phys.* **62** 950–955
Goettel K A, Mao H K and Bell P M 1985 Generation of static pressures above 2·5 megabars in a diamond-anvil pressure cell; *Rev. Sci. Instrum.* **56** 1420–1427
Goncharov A F, Makarenko I N and Stichov S M 1985 Raman scattering from a diamond at pressures up to 72 GPa; *JETP Lett.* **41** 184–187 (English translation)

Gouhara K, Li Y H and Kato N 1983 Studies of the α-β transition of quartz by means of *in situ* X-ray topography; *J. Phys. Soc. Jpn* **52** 3821–3828

Grace J D and Kennedy G C 1967 The melting curve of five gases to 30 kb; *J. Phys. Chem. Solids* **28** 977–982

Hanfland M, Syassan K, Fahy S, Louie S G and Cohen M L 1985 Pressure dependence of the first-order Raman mode in diamond; *Phys. Rev.* **B31** 6896–6899

Hawke R S, Syassen K and Holzapfel W B 1974 An apparatus for high pressure Raman spectroscopy; *Rev. Sci. Instrum.* **45** 1598–1601

Hazen R M and Finger L W 1982 *Comparative crystal chemistry* (New York: John Wiley) 231 pp.

Hazen R M, Mao H K, Finger L W and Bell P M 1980 Structure and compression of crystalline methane at high pressure and room temperature; *Appl. Phys. Lett.* **37** 288–289

Hazen R M, Mao H K, Finger L W and Hemley R J 1987 Single-crystal X-ray diffraction of n-H_2 at high pressure; *Phys. Rev.* **B36** 3944–3947

Hebert P, Polian A, Loubeyre P and Le Tollec R 1987 Optical studies of methane under high pressure; *Phys. Rev.* **36** 9196–9201

Heinz D L and Jeanloz R 1987 Temperature measurements in the laser-heated diamond cell, in *High-pressure research in mineral physics* (eds) M H Manghnani and Y Syono (Tokyo: Terra Scientific) pp. 113–128

Hemley R J 1987 Pressure dependence of Raman spectra of SiO_2 polymorphs: α-quartz, coesite, and stishovite; in *High-pressure research in mineral physics* (eds) M H Manghnani and Y Syono (Tokyo: Terra Scientific) pp. 347–360

Hemley R J, Bell P M and Mao H K 1987 Laser techniques in high-pressure geophysics; *Science* **237** 605–612

Hemley R J, Chen L and Mao H K 1988 New transitions between crystalline and amorphous ice (abst.); *Trans. Am. Geophys. Union (Eos)* **69** 1049

Hemley R J and Mao H K 1988 Phase transition in solid molecular hydrogen at ultrahigh pressures; *Phys. Rev. Lett.* **61** 857–860

Hirsch K V and Holzapfel W B 1981 Diamond anvil high-pressure cell for Raman spectroscopy; *Rev. Sci. Instrum.* **52** 52–55

Hirsch K R and Holzapfel W B 1986 Effect of high pressure on the Raman spectra of ice VIII and evidence for ice X; *J. Chem. Phys.* **84** 2771–2775

Holmes N C, Nellis W J and Graham W B 1985 Spontaneous Raman scattering from shocked water; *Phys. Rev. Lett.* **55** 2433–2436

Holzapfel W B, Johnson P G and Grosshans W A 1987 Molecular dissociation in solids under pressure; in *Dynamics of molecular crystals* (ed) Lascombe Jean (Amsterdam: Elsevier) pp. 133–140

Hubbard W B 1984 *Planetary interiors* (New York: Van Nostrand Reinhold) 334 pp.

Hunten D M, Tomasko M G, Flasar F M, Samuelson R E, Strobel D F and Stevenson D J 1984 Titan; in *Saturn* (eds) T Gehrels and M S Matthews (Tucson, Arizona: University of Arizona Press) pp. 671–787

Jayaraman A 1983 Diamond anvil cell and high-pressure physical investigations; *Rev. Mod. Phys.* **55** 65–108

Jayaraman A 1986 Ultrahigh pressures; *Rev. Sci. Instrum.* **57** 1013–1031

Jayaraman A, Wood D L and Maines R G Sr 1987 High-pressure Raman study of the vibrational modes in $AlPO_4$ and SiO_2 (α-quartz); *Phys. Rev.* **B35** 8316–8321

Jeanloz E and Heinz D L 1984 Experiments at high temperature and pressure: Laser heating through the diamond cell; *J. Phys. (Paris)* **45** Colloque C8 83–92

Johnson M L and Nicol M 1987 The ammonia-water phase diagram and its implications for icy satellites; *J. Geophys. Res.* **92** 6339–6349

Johnson M L, Schwake A and Nicol M 1984 The ammonia-water phase diagram, I, Water-rich region at low (< 4 GPa) pressure (abst.); in *Fifteenth Lunar and Planetary Science Conference Abstracts*, Lunar and Planet. Sci. Inst., Houston, Texas, pp. 405–406

Johnson M L, Schwake A and Nicol M 1985 Partial phase diagram for the system NH_3-H_2O: The water rich region; in *Ices in the solar system* (ed.) J Klinger (Hingham, Massachusetts: Reidel) pp. 39–47

Jorgensen J D 1978 Compression mechanisms in α-quartz structures—SiO_2 and GeO_2; *J. Appl. Phys.* **49** 5473–5478

Kessler J R, Monberg E and Nicol M 1974 Studies of fluorite and related divalent fluoride systems at high pressure by Raman spectroscopy; *J. Chem. Phys.* **60** 5057–5065

Kieffer S W 1979a Thermodynamics and lattice vibrations of minerals: 1. Mineral heat capacities and their relationships to simple lattice vibrational models; *Rev. Geophys. Space Phys.* **17** 1–19

Kieffer S W 1979b Thermodynamics and lattice vibrations of minerals: 3. Lattice dynamics and an approximation for minerals with application to simple substances and framework silicates; *Rev. Geophys. Space Phys.* **17** 35–59

Kieffer S W 1982 Thermodynamics and lattice vibrations of minerals: 5. Applications to phase equilibria, isotopic fractionation, and high-pressure thermodynamic properties; *Rev. Geophys. Space Phys.* **20** 827–849

Kieffer S W 1985 Heat capacity and entropy: Systematic relations to lattice vibrations; in *Microscopic to macroscopic, Rev. Mineral.* (eds) S W Kieffer and A Navrotsky **14** 65–126

Klug D D, Mishima O and Whalley E 1986 Raman spectrum of high-density amorphous ice; *Physica* **B139&140** 475–478

Levien L, Prewitt C T and Weidner D J 1980 Structure and elastic properties of quartz at pressure; *Am. Mineral.* **65** 920–930

Liebenberg D H 1979 A new hydrostatic medium for diamond anvil cells to 300 kb pressure; *Phys. Lett.* **A73** 74076

Loubeyre P, Le Toullec R and Pinceaux J P 1987 Binary phase diagram of H_2-He mixtures at high pressures; *Phys. Rev.* **B36** 3723–3730

Mammone J F and Sharma S K 1979 Raman study of TiO_2 under high pressures at room temperature; *Carnegie Inst. Washington Year Book* **78** 636–640

Mammone J F and Sharma S K 1980a Pressure and temperature dependence of Raman spectra of rutile-structure oxides; *Carnegie Inst. Washington Year Book* **79** 369–374

Mammone J F and Sharma S K 1980b Raman evidence for a new high-pressure room temperature polymorph of TiO_2; *Carnegie Inst. Washington Year Book* **79** 367–369

Mammone J F, Sharma S K and Nicol M 1980 Raman study of rutile (TiO_2) at high pressure; *Solid State Commun.* **34** 799–802

Mammone J F, Nicol M F and Sharma S K 1981 Raman spectra of TiO_2-II, TiO_2-III, SnO_2 and GeO_2, at high pressure; *J. Phys. Chem. Solids* **42** 379–384

Mao H K and Bell P M 1978a Design and varieties of the Mb cell; *Carnegie Inst. Washington Year Book* **77** 904–908

Mao H K and Bell P M 1978b High-pressure physics: Sustain static generation of 1·36 to 1·72 Mb; *Science* **200** 1145–1147

Mao H K and Bell P M 1979 Design of the diamond-window high-pressure apparatus for cryogenic experiments; *Carnegie Inst. Washington Year Book* **78** 659–663

Mao H K and Bell P M 1980 Experiment for in-situ high-pressure light-scattering measurements; *Carnegie Inst. Washington Year Book* **79** 411–415

Mao H K, Bell P M and Hemley R J 1985 Ultrahigh pressures: Optical observations and Raman measurements of hydrogen and deuterium to 1·47 Mb; *Phys. Rev. Lett.* **55** 99–102

Mao H K, Bell P M, Shaner J and Steinberg D 1978 Specific volume measurements of Cu, Mo, Pd and Ag and calibration of the ruby R_1 fluorescence pressure gauge from 0·06 to 1 Mb; *J. Appl. Phys.* **49** 3276–3283

Mao H K, Jephcoat A P, Hemley R J, Finger L W, Zha C S, Hazen R M and Cox D E 1988 Synchrotron X-ray diffraction measurements of single-crystal hydrogen to 26·5 GPa; *Science* **239** 1131–1134

Maradudin A A and Fein A F 1962 Scattering of neutrons by an anharmonic crystal; *Phys. Rev.* **128** 2589–2608

Marckmann J P and Whalley E 1964 Vibrational spectra of the ices. Raman spectra of ice VI and ice VII; *J. Chem. Phys.* **41** 1450–1453

Mills R L, Liebenberg D H, Bronson J C and Schmidt L C 1980 Procedure for loading diamond cells with high pressure gas; *Rev. Sci. Instrum.* **51** 881–895

Mills R L, Liebenberg D H and Pruzan Ph 1982 Phase diagram and transition properties of condensed ammonia to 10 kb; *J. Phys. Chem.* **86** 5219–5222

Ming L C and Bassett W A 1974 Laser heating in the diamond anvil press up to 2000°C sustained and 3000°C pulsed at pressures up to 260 kb; *Rev. Sci. Instrum.* **45** 1115–1118

Mishima O, Calvert L D and Whaley E 1984 'Melting' ice I at 77 K and 10 kb: A new method of making amorphous solids; *Nature (London)* **310** 393–395

Mishima O and Endo S 1980 Phase relations of ice under pressure; *J. Chem. Phys.* **73** 2454–2456

Mitchell E W J and Raptis C 1983 Raman scattering from molten alkali halides; *J. Phys.* **C16** 2973–2985

Moss W C, Hallquist J O, Reichlin R, Goettel K A and Martin S 1986 Finite element analysis of the diamond anvil cell: achieving 4·6 Mb; *Appl. Phys. Lett.* **48** 1258–1260

Mulac A J, Flower W L, Hill R A and Aeschliman D P 1978 Pulsed spontaneous Raman scattering technique for luminous environments; *Appl. Opt.* **17** 2695–2699

Murthy M K 1962 Thermal expansion properties of vitreous and crystalline germania; *J. Am. Cer. Soc.* **45** 616–617

Nakamura T, Tominaga Y, Udagawa M, Kojima S and Takashige M 1979 Light scattering studies on structural phase transition; *Solid State Commun.* **32** 95–101

Narayanaswamy P K 1948 The α-β transformation in quartz; *Proc. Indian Acad. Sci.* **A28** 417–422

Nedungadi T M K 1940 Effect of temperature on the Raman spectrum of quartz; *Proc. Indian Acad. Sci.* **A11** 86–95

Nicol M and Fong M Y 1971 Raman spectrum and polymorphism of titanium dioxide at high pressures; *J. Chem. Phys.* **48** 2240–2248

Parker E A 1963 Gold; in *Modern electroplating* (ed.) F A Lowenhein (New York: John Wiley) pp. 207–277

Peercy P S and Morosin B 1973 Pressure and temperature dependences of the Raman-active phonons in SnO_2; *Phys. Rev.* **B7** 2779–2786

Piermarini G J and Block S 1975 Ultrahigh pressure diamond anvil cell and several semiconductor phase transition pressures in relation to the fixed point pressure scale; *Rev. Sci. Instrum.* **46** 973–979

Piermarini G J, Block S and Barnett J S 1973 Hydrostatic limit in liquids and solids to 100 kb; *J. Appl. Phys.* **44** 5377–5382

Pistorius C W F T 1976 Phase relations and structures of solids at high pressures; *Prog. Solid State Chem.* **11** 1–151

Postma S 1920 Le systeme ammonique-eau; *Rec. Trav. Chim.* **39** 515–536

Postmus C, Maroni V A, Ferraro J R and Mitra S S 1968 High-pressure laser Raman spectra; *Inorg. Nucl. Chem. Lett.* **5** 269–274

Ramaker D E, Kumar L and Harris F E 1975 Exact-exchange crystal Hartree-Fock calculations of molecular and metallic hydrogen and their transition; *Phys. Rev. Lett.* **34** 812–814

Raman C V and Nedungadi T M K 1940 The α-β transformation of quartz; *Nature (London)* **145** 147

Rao K R, Chaplot S L, Choudhury N, Ghose S and Price D L 1987 Phonon density of states and specific heat of forsterite, Mg_2SiO_4; *Science* **236** 64–65

Reichlin R, Schiferl D, Martin S, Vanderborgh C and Mills R L 1985 Optical studies of nitrogen to 130 GPa; *Phys. Rev. Lett.* **55** 1464–1467

Rigden S M, Ahrens T J and Stolper E M 1988 Shock compression of molten silicate: Results for a model basaltic composition; *J. Geophys. Res.* **93** 367–382

Rollet A-P and Vuillard G 1956 Sur un nouvel hydrate de l'ammoniac; *C.R. Acad. Sci. Paris* **243** 383–386

Ross M and Shishkevish C 1977 *Molecular and metallic hydrogen*; (Santa Monica, Calif: Rand Corporation) 111 pp.

Ross M, Ree R H and Young D A 1983 The equation of state of molecular hydrogen at very high density; *J. Chem. Phys.* **79** 1487–1494

Salje E and Viswanathan K 1976 The phase diagram calcite-aragonite as derived from the crystallographic properties; *Cont. Mineral. Petrol.* **55** 55–67

Salje E and Werneke Ch 1982 How to determine phase stabilities from lattice vibrations; in *High pressure researches in geoscience* (ed.) W Schreyer (Stuttgart, West Germany: E Schweizerbast sche Verlagsbuchhandlung) pp. 321–348

Samara G A and Peercy P S 1973 Pressure and temperature dependence of the static dielectric constants and Raman spectra of TiO_2 (rutile); *Phys. Rev.* **B71** 1131–1148

Sarver J F and Hummel F A 1960 Alpha to beta transition in germania quartz and a pressure-temperature diagram for GeO_2; *J. Am. Ceram. Soc.* **43** 336

Schiferl D, Buchsbaum S and Mills R L 1985 Phase transition in nitrogen observed by Raman spectroscopy from 0·4 to 27·4 GPa at 15 K; *J. Phys. Chem.* **89** 2324–2330

Schmidt S C, Moore D S and Shaw M S 1987 Vibrational spectroscopy of fluid N_2 up to 34 GPa and 4400 K; *Phys. Rev.* **B35** 493–496

Schneider T 1969 Metallic hydrogen I; *Helv. Physica Acta* **42** 957–989

Schouten J A, Van den Bergh L C and Trappeniers N J 1985 Demixing in a molecular hydrogen-helium mixture up to 50 kb; *Chem. Phys. Lett.* **114** 401–404

Scott J F 1974 Soft-mode spectroscopy: Experimental studies of structural phase transitions; *Rev. Mod. Phys.* **46** 83–128

Scott J F and Porto S P S 1967 Longitudinal and transverse optical lattice vibrations in quartz; *Phys. Rev.* **161** 903–910

Shafer E C and Roy R 1956 Study of silica structure phases: I. GaPO$_4$, GaAsO$_4$ and GaSbO$_4$; *J. Am. Ceram. Soc.* **39** 330–336

Sharma S K 1979 Raman spectroscopy at very high pressure; *Carnegie Inst. Washington Year Book* **78** 660–665

Sharma S K 1989 Applications of advanced Raman spectroscopic techniques in earth sciences, in *Vibrational spectra and structure 17B* (Amsterdam: Elsevier) pp. 513–568

Sharma S K, Byahut S P and Lam P K 1985a Raman spectroscopic study of forsterite (Mg$_2$SiO$_4$) at high temperatures (abst.); *Trans. Am. Geophys. Union (Eos)* **66** 1058

Sharma S K, Mao S K, Bell P M and Xu J A 1985b Measurement of stress in diamond anvils with micro-Raman spectroscopy; *J. Raman Spectrosc.* **16** 350–352

Sharma S K, Mao H K and Bell P M 1979 Raman study of n-H$_2$ under very high pressures at room temperature; *Carnegie Inst. Washington Year Book* **78** 645–649

Sharma S K, Mao H K and Bell P M 1980a Raman spectra of supercritical phases and the crystalline solids of n-H$_2$ and n-D$_2$ under very high pressures; in *High pressure science and technology* (eds) B Vodar and Ph Marteau (New York: Pergamon Press) Vol. 2, pp. 1101–1103

Sharma S K, Mao H K and Bell P M 1980b Raman measurements of hydrogen in the pressure range 0·2–630 kb at room temperature; *Phys. Rev. Lett.* **44** 886–888

Sharma S K, Mao H K and Bell P M 1980c Raman measurements of deuterium in the pressure range 8–537 kb at room temperature; *Carnegie Inst. Washington Year Book* **79** 358–364

Sharma S K, Mao H K and Bell P M 1980d Phase transitions in methane under high pressures at room temperature—a Raman spectral study; *Carnegie Inst. Washington Year Book* **79** 351–355

Sharma S K, Ghose S and Urmos J P 1987 Raman study of high-temperature phase transition in ortho- and clinoenstatite (abst.); *Trans. Am. Geophys. Union (EOS)* **68** 433

Silvera I F and Wijngaarden R J 1981 New low-temperature phase of molecular deuterium at ultrahigh pressure; *Phys. Rev. Lett.* **47** 39–43

Silvera I F and Wijngaarden R J 1985 Diamond anvil cell and cryostat for low-temperature optical studies; *Rev. Sci. Instrum.* **56** 121–124

Stevenson D J 1982 Interiors of the giant planets; *Annu. Rev. Earth Planet. Sci.* **10** 257–295

Stevenson D J 1983 Condensed matter physics of planets: Puzzles, progress and predictions; in *High pressure in science and technology* (eds) C Homan, R K MacCrone and E Whalley (New York: North Holland) pp. 357–368

Stevenson R 1957 Solid methane-change in phase under pressure; *J. Chem. Phys.* **27** 656–658

Stoicheff B P 1957 High resolution Raman spectra of gases. IX. Spectra of H$_2$, HD and D$_2$; *Can. J. Phys.* **35** 730–741

Takahashi P K and Neill D R 1985 The Hawaii hydrogen R & D program; in *Proc. Second Symp. on Hydrogen Produced from Renewable Energy* Oct. 22–24, 1985 (eds) O G Hancock and K G Sheinkof, Cocoa Beach, Florida, Florida Solar Energy Center, Florida, pp. 389–396

Taylor M J and Whalley E 1964 Raman spectra of ices Ih, Ic, II, III and V; *J. Chem. Phys.* **40** 1660–1664

Tekippe V J, Ramdas A K and Rodriguez S 1973 Piezospectroscopic study of the Raman spectrum of α-quartz; *Phys. Rev.* **B8** 706–717

Thiéry M M, Fabre D and Kobashi K 1985 Raman spectra of solid CH$_4$ and CD$_4$; *J. Chem. Phys.* **83** 6165–6172

Van Duyne R P, Jeanmaire D L and Shriver D F 1974 Mode-locked laser Raman spectroscopy—a new technique for the rejection of interfering background luminescence signals; *Anal. Chem.* **46** 213–222

Walrafen G E, Abebe M, Mauer F A, Block S, Piermarini G J and Munro R 1982 Raman and X-ray investigations of ice VII to 36·0 GPa; *J. Chem. Phys.* **77** 2166–2174

Weinstein B A and Piermarini G J 1975 Raman scattering and phonon dispersion in Si and GaP at very high pressure; *Phys. Rev.* **B12** 1172–1186

Wigner E and Huntington H B 1935 On the possibility of a metallic modification of hydrogen; *J. Chem. Phys.* **3** 764–770

Wijngaarden R J, Lagendijk A and Silvera I F 1982 Pressure dependence of the vibron in solid hydrogen and deuterium up to 600 kb; *Phys. Rev.* **B26** 4957–4967

Wong P T T and Klug D D 1983 Re-evaluation of type I diamonds for infrared and Raman spectroscopy in high-pressure diamond anvil cells; *Appl. Spectrosc.* **37** 285–286

Wong P T T and Whalley E 1976 Raman spectrum of ice VIII; *J. Chem. Phys.* **64** 2359–2366

Xu J A, Mao H K and Bell P M 1986 High-pressure ruby and diamond fluorescence: Observations at 0·21 to 0·55 terapascal; *Science* **232** 1404–1406

Yamanaka T and Ishii M 1986 Raman scattering and lattice vibrations of Ni_2SiO_4 spinel at elevated temperature; *Phys. Chem. Miner.* **13** 156–160

Young D A, Zha C-S, Boehler R, Yen J, Nicol M, Zinn A S, Schiferl D, Kinhead S, Hanson R C and Pinnick D A 1987 Diatomic melting curve to very high pressures; *Phys. Rev.* **B35** 5353–5356

Zinn A S, Schiferl D and Nicol M F 1987 Raman spectroscopy and melting of nitrogen between 290 and 900 K and 2·3 and 18 GPa; *J. Chem. Phys.* **87** 1267–1271

Subject Index

Ablation 87
Ablation constraints 87, 113
Acceleration 1
Accelerator mass spectrometry 61
Accretion 29
Acetaldehyde 195
Achondrites 180
Actinide metals 255
Actinides 249, 256
Actinium 255
AF demagnetization 231
Age-dating 216
Airy's method 1, 15
Alpha particle 173
Alpha plutonium 256
Alpha-quartz 273
Americium 256
Ammonia 263, 299
Ammonia ices 257
Ammonia-water system 300
Amorphization 300
Andradite 154
Anharmonicities 280
Anharmonicity 47
Anhysteretic remanent magnetization 231
Anorthite 302
Antarctica 88
Antarctica micrometeorites 87
Anti-Stokes, coherent scattering 298
Antiparticles 9
Antiprotons 9
Aquifers 214
^{26}Al cosmogenic nuclides 94
Ar^+ 268
^{40}Ar degassing equation 147
Ar^+ ion laser 271
Archaean rocks 178
Argon 153, 157
Asteroids 180
Asymptotes 205
Aurorae 127
Autochthonous phenomenon 189
Autotrophic carbon 192
Autotrophy 193

Bacillus subtilis 196
Baryon number 3
Basalt 239
Basin's response 212
^{10}Be and ^{9}Be 201
Belt apparatus 250
Berkelium 257
Berlinite 273
Beryllium concentration 201

Beryllium distribution 201
Biogenic microlithography 87
Biogeochemical evidence 189
Black smokers 65
Blackbody radiation 273
Blue Ice-I expedition 87
Blue-ice sediments 87
Börn-Oppenheimer approximation 39
Bouguer anomalies 24
Boxcar integrators 272

C_3 photosynthesis 193
CFB 242
CH_4 263
COCORP 243, 244
CODER triple monochromator 268
COKE computation 87
Californium 257
Calvin cycle 193
Carbon cycle 190
Carbon flux 147, 151
Carbonaceous chondrites 179, 180
Carbonaceous meteorites 87
Causality 144
Causality and uniqueness 137
^{14}C dating 61
Cerenkov radiation 69
Chaos theory 142
Charge-conjugation 4
Charge-coupled devices 272
Chemical evolution 189
Chondrites 46
Chondritic composition 87
Chondritic evolution 178
Chondrules 179
Chromite 65
Clinoenstatite 286
Clinopyroxenes 129
Coastal upwelling 205
Collapsed phase 256
Composite-pyrolyzable micrometeorite 87
Corals 127
Cosmic clays 116
Cosmic dust 87
Cosmic radiation 69
Cosmic spherules 97
Cosmogenic 61
Cosmogenic radionuclides 61
Cristabolite 274
Cryoconite 91
Crystalline unmelted micrometeorites 99
Cyanbacteria 192
Cyanoacetylene 195
Cyclic behaviour 127

Cyclotron resonance 75

D/H ratio 159
DOSECC 244
Darcy's law 222
Dark side 139
Data kernels 144
Deep time 138
Deep-sea sediment 64
Deformational phenomena 213
Degassing 147, 177
Degradation 142
Density of states 289
Deuterium 290
Diamond 65, 251, 265
Diamond anvil cells 249, 263, 265
Diatomic molecules 298
Dicke method 1, 20
Diffusion coefficient 173
Diopside 154, 302
Dipolar reversal 229
Dissociation processes 39
Double monochromator 269
Dynamic nuclear polarization 51

Earth structure 137
Eclipses 127
Eclogite 242
Effective stress 220
Electrical measurements 254
Electromagnetic interaction 69
Electromagnetism 1
Electron delocalization 256
Electrostatic force 4
Elementary particles 9
Ellipsoidal mirror 268
Emanation coefficient 174
Emanation from rocks 176
Enstatite 286
Eötvös experiments 13
Eötvös method 1
Equation of state 249, 251
Equilibrium isotope exchange 56
Equivalence principle 1, 6
1:4 Ethanol-methanol mixture 251
Etch canals 87
Experimental artifacts 153

FTMS 71
Faller and Kuroda experiments 1
Fayalite 154
Fe/Ni alloys 99
Fifth force 143
First static mass spectrometer 79
Fluid flow 218
Fluid inclusions 72
Fluid-rock interactions 213
Fluid-rock systems 224

Formaldehyde 195
Formic acid 195
Forsterite 154, 289
Fourier transform ion cyclotron resonance 71
Fourier transform mass spectrometry 71
Fourier transforms 75
Fractal 105, 142
Frictional heating 87

GPS 244
Gabbro-anorthositic crust 179
Gabbros 177
Galileo's method 1
Garnet 65
Gas extraction procedures 153
Gas phase decomposition 35
Gd 256
Gemstones 65
General relativity 68
Geochemical model 202
Geochemical ocean box model 201
Geodesy 67
Geodetic triangulations 67
Geodynamic satellites 14
Geodynamics 137
Geodynamo, models 230
Geoid 239
Geomagnetic field 229
Geomagnetic reversal 140
Geomorphology 213, 217
Geophysical 258
Geophysical time 137
Geophysical window 24
Geophysics 1
Giant thermal flux 87
Girdle-anvil device 250
Gneisses 177
Granites 177
Gravitation 1, 67
Gravitational constant 68
Gravitational instability 32
Gravitational redshift 68
Gravito-electrodynamic effect 31
Gravito-electrodynamic processes 32
Gravity measurements 1
Greenland 88
Greenland collections 93
Greenland ice cap 112
Greenland micrometeorites 87
Groundwater contamination 213, 215
Groundwater hydrology 213, 214
Gruneisen 298
Gruneisen parameters 284
Gyre 205

Half-widths 292
H_2 263, 268
Harmonic oscillator 40

Subject index

He 268
Heat flux 114
Helium 153, 257
Helium isotope ratios 165
Helium-cadmium laser light 253
High temperatures 249
High-pressure Raman spectroscopy 263
High-temperature superconductor 290
Holmium 256
Holocene period 233
Hotspot basalts 153
Hydrodynamic dispersion 222
Hydrogen 153, 257, 289
Hydrogen cyanide 195
Hydrogen-helium mixtures 298
Hydrogeology 213
Hydromagnetic engine 140
Hydrosphere 179
Hyperfine nuclear spin effects 35, 51

IRIS 244
Ices 263
Incompatible elements 179
Indian summer monsoon 205
Induced seismicity 219
Inertial mass 2
Infrared spectroscopy 255
Instrumentation 213
Internal area 173
Interplanetary dust particles 87
Interplanetary medium 110
Interstellar dust 179
Interstellar space 35
Inverse theory 144
Ionian Sea 127, 129
Iron meteorites 180
Irreversible processes 142
Isothermal remanent magnetization 231
Isotope effects 35
Isotope geochemistry 239
Isotopic analyses 153
Isotopic anomalies 71
Isotopic exchange reactions 35
Isotopic fractionation 35
Isotopic symmetry factors 39

Johnson noise 76
Josephinite 153, 155
Jovian planets 289
Jupiter 257

Kerogen 192, 193
Kinetic isotope effect 192
Kinetics 223
Lanthanide metals 255
Lanthanides 249, 256
Lanthanum 255
Laplace transforms 205

Laplacian detector 14
Laser decrepitation 72
Lawrencium 255
Life on Earth 189
Lighthill's equation 205
Local time 138
Low level counting 61
Lower peridotitic layer 179
Lunar laser ranging 13
Lutetium 255, 256

M regions 128
MORB 148, 162, 240
MORB basalt formation 64
Magmatism 177
Magnetic moments 255
Magnetic monopoles 69
Magnetite 65
Mantle 177
Mantle convection 239, 242
Mantle tomography 242
Mass and energy transport 213, 222
Mass spectrometry 71
Mass-independent fractionation 36
Meridional currents 206
Meteorites 180
Methane 299
Michelson-Morley experiment 67
Micro-Raman system 271
Micro-Raman techniques 263
Microbial mats 190
Microbialites 190
Microfossils 190
Micrometeorite flux 87, 110
Micrometeorites 87, 88
Mid-ocean ridge 153
Migration of fluids 213, 221
Mining of cosmic dust 87
Molecular diffusion 222
Molecular partition functions 41
Molecular seeding 190
Molecular symmetry effect 44
Molecular vibrons 292
Monsoon current 212
Monte Carlo methods 144
Monte Carlo technique 145
Moon 179
Mössbauer spectroscopy 255
Multi-anvil apparatus 250
Multi-megabar pressure 249

Natural remanent magnetization 230
Navier-Stokes equation 141
Neon 153, 157
Neptune 257
Neutrinos 70
New forces in nature 1
Newton's method 1, 12

Newtonian mechanics 143
Nitrogen 258
Noble gases 71, 153
Nodules 127
Non-Newtonian world 137
Non-equilibrium interactions 223
Nonlinear behaviour 139
Nonlinear inversion 144
Nonlinear processes 140
Nonlinearity 141
Nordvett-effect 13
Nuclear spin 39
Nuclear wastes 87, 120
Nutrient constituents 202

OIB 242
Ocean Island 153
Ocean floor 140
Ocean mixing time 202
Oceanic crust 213, 220
Olivine 179
Opposed anvil device 250
Organic evolution 189
Osmotic drive 223
Oxygen isotope 185

Palaeo-climatology 35
Palaeo-oceanography 35
Paleo-secular variation 229
Paleointensity 229
Paleomagnetic field 229
Paleontological evidence 189
Panspermia 189, 195
Parity 4
Partial melting 177
Particle physics 67, 69
Particles 9
Particulate cycling 203
Particulate removal 203
Pedology 213, 217
Penta-prism 271
Peridotite 240
Periodicity 229
Perovskite 240
Phase transformations 251
Phase transitions 273
Phonons, optic 278
Photochemical-electrodissociative reactions 36
Phreatic zone 218
Piclogite 242
Pioneer 257
Piston-cylinder apparatus 250
Planetary atmospheres 35
Planetary degassing 147
Planetary evolution 137
Planetary gases 290
Planetary materials 249
Planetary sciences 258

Planetesimal 29
Plasma 29
Plate margins 213, 220
Plate tectonics 64, 239
Plume 148
Porous media 214
Pound-Rebka experiment 6
Pre-solar nebula 35, 44
Prebiotic evolution 87, 115
Precambrian rocks 177
Prediction 137
Pressure transient analysis 215
Primitive mantle 154
Primordial 153
Pristine mantle rocks 64
Prokaryotic 189
Protostars 29
Pseudo-single domain grainsize range 231

Quaternary period 229
Quaternary sediments 229

Radiocarbon distribution in sea 201
Radiogenic 61
Radionuclides 61
Radon emanation 173
Raman diamond line 267
Raman scatterers 267
Raman spectroscopy 255, 263, 264
Rare gases 153
Rare nuclides 61
Recoil 174
Reduced partition functions 42
Regional groundwater systems 213, 218
Relativity 67
Residence times 97, 202, 203
Resonance ionization mass spectrometry 72
Rn 173
Ròzsa-Sélenyi floating accelerometer 1, 19
Ruby 253
Ruby fluorescence technique 251
Rutile 281

SMOW 159
Saturn 30, 257
Saturnian ring system 30
Screening 88
Secular variation 229
Sediment matrix control 229
Sedimentation 177
Sediments 127
Seismic discontinuity 240
Seismic tomography 239
Seismology 140
Self-shielding 54
Serpentine 154
Shock-wave data 254
Sialic lithosphere 177

Subject index

Silicate melting 251
Sm/Nd technique 177
Solar activity 127
Solar corona 127
Solar cycles 127
Solar system 29
Solar-terrestrial relationships 127
Solar-terrestrial system 127
Solid-state detectors 254
Somali current 205
Specific heat 289
Spectroscopy 254
Spinels 65
Stalagtites 127
Standardized earth-model 145
Static balance method 19
Static xenon analysis 71
Stirred remanent magnetization 232
Stishovite 281
Stromatolites 190
Subducted sediments 64
Sulphur isotopic fractionations 47
Summation dial 128
Sunspots 127
Symmetry selective fractionation 50
Synchrotron source 254

T-Tauri stars 29
Tb 256
Technetium 63
Teller-Redlich approximation 41
Temporal constraints 189
Terrestrial contamination 87, 118
Terrestrial engine 137
Terrestrial minerals 173
Terrestrial weathering 87
The atmosphere 177
The ocean 177
Thermodynamic 137
Thermoelastic theory 143
Thermoluminescence 128
Thermoluminescence in sea core 127
Thorium 256
Tidal friction 140
Time warps 138
Time's arrow 138
Time's cycle 138
Time-resolved Raman spectrometry 263
Tomographic method 142

Total partition function 41
Toxic wastes 216
Treanor pumping 39, 48
Triggered earthquakes 219
Tritium 161
Tungsten carbide 251
Turbulence 137, 138
Two-gyre system 205

Ultra-high pressures 249
Underground caves 127
Unimolecular dissociation 53
Uniqueness 144
Unmelted micrometeorites 87
Upwelling 205, 212
Uranium 256
Uranus 257

Vadose zone 215
Valence 255
Vibrational energy levels 40
Vibrational quantum number 48
Vibrational spectra 289
Virtual dipole moment 233
Viscosity measurements 251
Voyager 257

Waste disposal 213, 215, 216
Water fluxes 203
Watersheds 213, 217
Wave equation 40
Weak equivalence principle 9, 20
Weak interaction 69
Weathering 213, 177, 217
Weathering trends 87
Weathering, terrestrial 116
Wind stress 205, 212

X-ray diffraction 249, 254
Xe 268
Xenology 74
Xenon 77
Xenon isotopes 71

Zirconium pads 253
Zonal currents 206
Zone of aeration 215

Author Index

Abebe M
  see Walrafen G E 299
Adams D M 267, 268, 289, 299
Adelberger E G
  see Stubbs C W 18, 23, 24
Aeschliman D P
  see Mulac A J 272
Agnew G D
  see Moore G I 16, 17, 24
Ahrens T J
  see Rigden S M 302
Aitken M J 233
Akaogi M
  see Yagi T 251
Akella J 249, 254–257
  see Smith G S 254–257
Akimoto S
  see Shimomura O 251
  see Yagi T 251
Alfvén H 29, 30
Allan Cox 140
Allègre C J 148, 240, 242
  see Staudacher Th 72
  see Sarda P 148, 149, 150
Alley C O
  see Williams J G 13
Allin E J
  see Bhatnagar S S 290
Allman S L
  see Lehmann B E 73
Allsop A L
  see Aitken M J 233
Anastapoulos C
  see Besson J M 289
Ander M E 16
Anders E 38
  see Swart P K 49
  see Tang M 71
Anderson D L 242
Aoki K
  see Shimomura O 251
Appel P W U
  see Schidlowski M 189, 193
Appleby R
  see Adams D M 268, 269, 299
Arashi H 289
Arnold J R 61
  see Lal D 65
  see Nishizumi K 65
Aronson S H 4
  see Fischbach E 5, 6, 8, 16
Arrhenius G 195
  see Alfvén H 29
  see Brecher A 32
  see Mendis D A 29
Asell J F 273
Ashcroft N W 257, 290
  see Friedli C 290
Ashwood-Smith M J 197
Attolini M R 131
Awramik S M 192

Badrinathan C
  see Mathur D 77
Bahcall J N 73
Bains-Sahota S K 44, 46, 47
Baker F A 79
Balageas D
  see Bonny Ph 111, 113, 114
  see Darmon G 114
Balchan A S
  see Drickamer H G 250
Bally J 55
Banerjee S K 32, 229, 231
  see Jackson M 231
  see King J 231, 233, 234, 236
  see Levi S 231, 232, 236
  see Lu R 232
  see Lund S P 236
Barnes J 45
Barnett J D 268
  see Forman R A 251
Barnett J S
  see Piermarini G J 268
Barton M A
  see Moore G I 16, 17, 24
Basov N G 50
Bassett W A
  see Bird J M 154
Batchelor G K 141
Bates J B 273, 274, 277, 280
Baxter D 30
Becker B
  see Stuiver M 135
Becker T
  see Drake R E 72
Beer J 128
  see Cini Castagnoli G 131
Beg M A
  see Clarke W B 153
Begemann F
  see Ott U 72
Begun G M
  see Gilbert B 273
Belenov E M 48
  see Basov N G 50
Bell P M 258, 265, 298
  see Benjamin T M 256
  see Goettel K A 265
  see Hazen R M 300
  see Hemley R J 273, 290, 295
  see Mao H K 251, 255, 264–266, 268, 271, 295–297
  see Sharma S K 269, 271, 290–293, 296, 300
  see Xu J A 265
Bellusi G 256
Bender P L
  see Williams J G 13
Benedicks C
  see Lofquist H 155
Benedict U 255–257
  see Bellusi G 256

Benjamin T M 256
Bennet Wm R Jr 19, 23
Berder P L
  see Bertotti B 13
Bergmann R C
  see Rich J W 49
Bernatowicz T J 162, 164–167
Bernstein J 4
Bertaux J L
  see Langevin Y 101
Bertotti B 13
Besson J M 268, 289
  see Gauthier M 301
  see Polian A 258
Bhatnagar S S 290
Bhattacharya S K 46, 50, 53
  see Sharma P 230
  see Thiemens Mark H 35
Bibring J P 96
Bieri R H 157
Bird J M 154–156, 163, 169
  see Bochsler P 154, 155, 160–162, 164–166
Bizzeti P G 19
Bizzeti-Sona A M
  see Bizzeti P G 19
Block S
  see Barnett J D 268
  see Forman R A 251
  see Piermarini G J 251, 268
  see Walrafen G E 299
Bochsler P 154, 155, 160–162, 164–166
  see Stettler A 166
Bodenheimer P
  see Lin D N C 30
Boehler R
  see Young D A 302
Bohlke J K
  see Kirschbaum C 72
Bolt B A 137, 139, 145
  see Bullen K E 143, 145
Bonino G
  see Beer J 128
Bonny Ph 111, 113, 114
Bonte Ph 96, 102
Boone S
  see Cynn H C 301
Borg J 96
Botto R I 154, 162
Bourles D
  see Raisbeck G M 96
Bourot-Denise M
  see Christophe Michel-Levy M 101
Bowers W D
  see Francl T J 74
Boynton P E 19, 20, 23, 24
Bradley J P 88, 96, 101
Braginsky V B 20
Brasch J W 268

Author index

Braziunas F
  see Stuiver M 133, 135
Brecher A 32
Brenan J B
  see Watson E B 247
Bridgman P W 249, 250
Brister K E
  see Vohra Y K 254
Broecker W S 192, 202
  see Peng T-H 201
Bronson J C
  see Mills R L 268
Brown L 64
Brownlee D E 87, 88, 89
  see Bibring J P 96
  see Bradley J P 96
  see Maurette M 90
  see Schramm L S 105
Buchachenko A L 51
Buchsbaum S 298
  see Schiferl D 298
Bucker H 196
  see Horneck G 196
Buick R
  see Walter M R 190
Bullen K E 137
Bunch T E 109
Bundy F P 250
Bunting E N
  see Weir C E 251
Burne R V 190
Bussel G D
  see Aitken M J 233
Butler R F
  see Banerjee S K 231
Butterfield A W
  see Kelley S 72
Byahut S P
  see Sharma S K 289

Cabibbo N
  see Bernstein J 4
Cadogan P H 149
Callot G 91, 93, 116
Calvert L D
  see Mishima O 299
Carrigan R A 69
Carter W E
  see Williams J G 13
Cecchini S
  see Attolini M R 131
Chalam V
  see Gaur V K 24
Chan H A 14
Chang S 195
  see Bunch T E 109
  see Thiemens M H 44
Chaplot S L
  see Rao K R 289
Charney J G 205
Chassefiere E
  see Langevin Y 101
Chen C H 73
  see Lehmann B E 73

Chen L
  see Hemley R J 299, 300
Cheng H Y
  see Aronson S H 4
Chervin J C
  see Gauthier M 301
Chino T
  see Endo S 251
Choudhury N
  see Rao K R 289
Chow M F
  see Turro N J 52
Christenson J H 4
Christophe-Michel-Levy M 101, 102
Chu Y-H 56
Cicerone R J 54
Cini Castagnoli G 127–129, 131
  see Attolini M R 131
Ciufolini I 13
  see Bertotti B 13
Clanton U S 87
Clark S P
  see Turekian K K 151
Clarke W B 153
Clayton R N 35, 36
  see Matsuhisa Y 36, 42
  see Molini-Velsko C A 46
  see Onuma N 44
Clayton R W
  see Hager B H 242
Cohen S C 14
Cole G H A 290
Comisarow M B 72, 74
Connerney J E T
  see Northrop T G 30
Constable C G 234, 236
Copeland J
  see Ashwood-Smith M J 197
Counselman C C
  see Shapiro I I 13
Cowley R A 280
Cowsik R 1, 9, 13, 14, 19, 23
  see Gaur V K 24
Cox A 236, 237
Cox D E
  see Hemley R J 258
  see Mao H K 298
Craig H 148, 156–158, 161–163, 167, 168
  see Clarke W B 153
  see Krishnaswami S 65
  see Lal D 163
  see Lupton J E 156, 157
  see Marti W 155, 162, 168
  see Poreda R J 153, 161
  see Rison W 72
Crepon M 206
Cronin J W
  see Christenson J H 4
Crosby D
  see Boynton P E 19, 20, 23, 24
Cruishank D P 123
Currie D G
  see Williams J G 13

Curtis G H
  see Drake R E 72
Cynn H C 301

Daane A H
  see Spedding F H 255
Dabos S
  see Benedict U 255, 256
Dabos-Seignon S 256
Dalgarno A
  see Fox J L 36
Damon P E 230
Dansgaard W 167
Darby M I
  see Taylor K N R 255
Dardano C B
  see Clanton U S 87
Darmon G 114
Das P K 205, 206
Davies G F 242
Davis R Jr 62, 73
  see Bahcall J N 73
Dayne S J
  see Adams D M 268
De Angelis M 91, 93
DeLong S E 76
  see Spell T L 76
DePaolo D J 242
Deino A L
  see Drake R E 72
Delecluse P 206
Des Marais D J 151
  see Chang S 195
Devezeaux D
  see Bonny Ph 111, 112, 113
Dick H J B 154
Dicke R H 20, 68
  see Roll P G 20
  see Williams J G 13
Dixon F
  see Krishnaswami S 65
Doetsch G 209, 211
Dolino G 278
Dollfus A 123
Dose K
  see Horneck G 196
Downing R G 154, 155, 161, 164–167
Drake R E 72
Dran J C
  see Borg J 96
Drickamer H G 250
Driese S G
  see McSween H Y 106
Dube A P
  see Das P K 205, 206
Dubois A
  see Callot G 91, 93, 116
Dubrovskij B A 180
Duclos S J
  see Vohra Y K 254
Dufour C
  see Benedict U 255, 256

Dumont R
    see Levasseur-Regourd A C 123
Dunlop J S R 190
    see Walter M R 190
Dupree A K 29
Durrieu L
    see Borg J 96
Dymond J D
    see Bieri R H 157

Ebihara M
    see Anders E 38
Eckhardt D H 16, 24
    see Williams J G 13
Edmond J M
    see Measures C I 201, 203
Eichmann R
    see Schidlowski M 189, 192, 193
Ekstrom P
    see Boynton P E 19, 20, 23, 24
Elliot L D 301
Elmore D
    see Gove H E 61
Endo S 251
    see Mishima O 299
Eneyev T M 179
Epling G A 52
Epstein S
    see Ott U 72
    see Sheppard S M 167
    see Yang J 45

Fabre D 300
    see Thiery M M 300
Fahey A 38
Fairbanks W M 9
Falle J E 7, 0, 16, 24
    see Kayser P T 20
    see Niebauer T M 8
    see Williams J G 13
Faure G 189
    see Koeberl C 89
Fazzini T
    see Bizzeti P G 19
Fechtig H
    see Grün E 90, 97, 111, 112, 113
Fein A F
    see Maradudin A A 280
Fekete E
    see Eötvös B R von 5
Ferapontov N B
    see Basov N G 50
Ferraro J R 264, 273
    see Postmus C 268
Feshbach H
    see Morse P M 208
Finger L W
    see Hazen R M 273, 289, 298, 300
    see Hemley R J 258
    see Mao H K 298
Finkel R C
    see Beer J 128
Fireman E L 161
Fischbach E 4, 5, 6, 8, 16
    see Aronson S H 4

see Talmadge C 13
Fitch V L
    see Christenson J H 4, 19
Flasar F M
    see Hunten D M 299
Florio E
    see Epling G A 52
Flower W L
    see Mulac A J 272
Fong M Y
    see Nicol M 285
Forman R A 251
Fox J L 36
Francl T J 74
    see McIver R T Jr 74
Fraundorf P 113
    see Bradley J P 96
Friedli C 290
Fröhlich C 130
Fronabarger A K
    see McSween H Y 106
Frontov V N
    see Panov V I 24
Fujita E
    see Endo S 251
Fukunaga O
    see Shimomura O 251

Gabel E M
    see Clanton U S 87
Galimov E M 52, 177, 182
Galli M
    see Attolini M R 131
    see Beer J 128
Gaur V K 24
Gauthier M 301
Gerrity D P
    see Valentini J J 39
Ghose S 278
    see Rao K R 289
    see Sharma S K 286
Gilbert B 273
Gillete H
    see Dick H J B 154
Glassely W E
    see Kirschbaum C 72
Glenn R H
    see Grün E 90, 97, 111–113
Godwaland B K
    see Vohra Y K 256
Goertz C Z
    see Mendis D A 30
Goettel K A 265
    see Bell P M 265
    see Moss W C 253, 254, 258, 265
Goldanskii V I
    see Podoplelov A V 52
Goldman T 9, 10
    see Hughes R 14
Goldreich P 32
Goldsmith J R
    see Matsuhisa Y 36, 42
Gooding J L 89, 116
Goodrich C A 155
    see Bird J M 155

Goodwin B D
    see Moore G I 16, 17, 24
    see Stacey F D 6, 15, 16
Gordon L I
    see Craig H 158
Goswami J N
    see Fahey A 38
Gottel K A
    see Bernatowicz T J 162, 164–167
Gouhara K 278
Gove H E 61
Grace J D 300
Grady M
    see Tang M 71
Grady M M
    see Swart P K 49
Graham W B
    see Holmes N C 300
Grant B
    see Measures C I 201
Greenberg J M 195, 197
Grimsditch M
    see Polian A 258
Gross M L 72, 74
Grosshans W A 255, 256
    see Holzapfel W B 299
Grossman L
    see Clayton R N 36, 46
Groves D I
    see Dunlop J S R 190
Gruber W
    see Jackson M 231
Grün E 90, 97, 111–113
    see Mendis D A 30
Gschneider K A Jr 255, 256
Gundlach J H
    see Stubbs C W 18, 23, 24
Gurevich L E 179

Haberkorn R 52
Hagee V
    see Olson P 236
Hagen E
    see Koeberl C 89
Hager B H 242
Hall H T 250
Hallquist J O
    see Moss W C 253, 254, 258, 265
Halstead J B
    see Baker F A 79
Hamel G
    see Gauthier M 301
Hamilton D C
    see Nellis W J 257
Hammer C
    see Maurette M 87, 90, 91, 110, 111
Hanson R C
    see Young D A 302
Hargraves R B
    see Banerjee S K 32
Harmer D S
    see Davis R Jr 73
Harris F E
    see Ramaker D E 290

Author index

Harrison T M
  see DeLong S E 76
  see Spell T L 76
Hart R 149
Hart S R
  see Zindler A 150
Hawke R S 299
Haxton W
  see Aronson S H 4
Hayes J M 193, 194
  see Schidlowski M 193, 194
Hazen R M 273, 289, 298, 300
  see Mao H K 255, 298
Hebert P 300
Heckel B R
  see Stubbs C W 18, 23, 24
Heidenreich J E 38, 39, 51
  see Thiemens M H 38, 44, 55
Hemley R J 258, 265, 273, 274, 280, 281, 283, 284, 290, 295, 297–300, 302
  see Bell P M 258, 298
  see Hazen R M 297
  see Mao H K 271, 295–298
Hemminger J C
  see Sherman M G 75
Hennecke E W
  see Downing R G 154, 155, 161, 164–167
Herzberg G 39, 41, 44, 47
Hill H H
  see Kmetko E A 255
Hill J R
  see Mendis D A 30, 31
Hill R A
  see Mulac A J 272
Hirsch K R 264, 267, 299, 301
Hoaglin D C 106
Hodge P 87
Hoefs J
  see Veiser J 178
Hoffman D C 63
Hoffmann K C
  see Davis R Jr 73
Hofmann A W 242
Hogan L
  see Hart R 149
Hohenberg C M
  see Bernatowicz T J 162, 164–167
Holding S C
  see Stacey F D 6, 16
Holmes N C 300
  see Nellis W J 257
Holzapfel W B 256, 299
  see Bellusi G 256
  see Hawke R S 299
  see Hirsch K R 264, 267, 299, 301
Horneck G 196–198
  see Bucker H 196
Houpis H L F 31
  see Mendis D A 31
Hsui A T 16
  see Moore G I 16, 17, 24
  see Stacey F D 16
Hubbard W B 257, 290

Hughes R 14
Hughes R J
  see Goldman T 9, 10
Hulston J R 36, 47
Hummel F A
  see Sarver J F 275
Hunten D M 299
Hunter R L
  see Francl T J 74
  see McIver R T Jr 74
Hurst G S 72, 73
  see Chen C H 73
  see Lehmann B E 73
Hutcheon I D
  see Clayton R N 42
Hyodo M 231, 234, 235

Ikeda Y
  see Clayton R N 42
Imamura M
  see Yanagita S 36
Ip W–H
  see Mendis D A 30
Irvine W M 195
Irwin J J
  see Kirschbaum C 72
Isaila M V
  see Fitch V L 19, 23
Isakov V A
  see Basov N G 50
Ishibashi A
  see Endo S 251
Ishigame M
  see Arashi H 289
Ishii M
  see Yamanaka T 289
Itie J P
  see Benedict U 255, 256

Jackson M 231
Jackson T
  see Thiemens M H 44, 47, 55
Jacobson S B 242
Jamieson J C 251
Jayaraman A 251, 255, 264, 268, 273, 274, 290
Jeanmaire D L
  see Van Duyne R P 272
Jeffreys H 137
Jehanno C
  see Bonte Ph 96, 102
  see Maurette M 110, 111
Jekeli C
  see Eckhardt D H 16, 24
Jephcoat A P
  see Akella J 256
  see Hemley R J 258
  see Mao H K 298
Johansson B 255
Johnson E A 232
Johnson M L 299, 301
Johnson P G
  see Holzapfel W B 299
Johnson Q
  see Akella J 255, 257

  see Smith G S 255, 257
Johnston M 73
Jorgensen J D 279
Jouret C
  see Borg J 96
Junge C E
  see Schidlowski M 189, 192, 193
Jura M
  see Morris M 56

Kaplan I R
  see Hayes J M 193
  see Schidlowski M 193, 194
Kato N
  see Gouhara K 278
Kaula W M
  see Williams J G 13
Kaye J A 39, 41, 44, 54
Kayser P T 20
Keilis-Borok V 139
Kelley S 72
Kennedy B M 72
Kennedy G C
  see Grace J D 300
Kent D V 235, 236
Kessler J R 286
Khadem M
  see Measures C I 201
Kieffer S W 44, 289
King G A M 195
King J 231, 233, 234, 236
King R W
  see Shapiro I I 13
Kingsley J R
  see Sherman M G 75
Kinhead S
  see Young D A 302
Kirschbaum C 72
Kissel J
  see Langevin Y 101
Klein J
  see Nishiizumi K 65
  see Raisbeck G M 96
Klug D D 299
  see Wong P T T 267
Kmetko E A 255
Kobashi K
  see Fabre D 300
  see Thiery M M 300
Koeberl C 89
Koide M
  see Bieri R H 157
Kojima S
  see Nakamura T 269
Koumvakalis A
  see Cynn H C 301
Kozlov N N
  see Eneyev T M 179
Kraeutler B
  see Turro N J 51, 52
Kramer S D
  see Chen C H 73
  see Lehmann B E 73
Krishnan N
  see Cowsik R 5, 9, 14, 23

Krishnarao J S R 154
Krishnaswami S 65, 129
Krotkov R
 see Roll P G 20
Ku T L 201
 see Kusakabe M 201–203
 see Peng T-H 201
Kumar L
 see Ramaker D E 290
Kundu P K
 see McCreary J P 205
Kuroda K 7, 8
Kurz M D 155, 162
Kusakabe M 201–203
Kuwamura H
 see Shimomura O 251

Lagendijk A
 see Wijngaarden R J 295
Lago M T
 see Dupree A K 29
Lal D 61–65, 163, 230, 236, 237
 see Krishnaswami S 65, 129
 see Nishiizumi K 65
Lam P K
 see Sharma S K 289
Langer W D 56
 see Bally J 55
Langevin Y 101
Lawler R G 52
Lawrence F O
 see Hoffman D C 63
Lawson A W
 see Jamieson J C 251
Lazarewicz A R
 see Eckhardt D H 16, 24
Le Toulec R
 see Hebert P 300
 see Loubeyre P 298
LeSar R
 see McMahan A K 256, 257
Lean J 130
Lebedinskij A K
 see Gurevich L E 179
Ledford E B Jr
 see McIver R T Jr 74
Lee D S
 see Measures C I 201
Lee T D
 see Bernstein J 3, 4
Lehmann B E 73
Leshina T V
 see Podoplelov A V 52
Letokhov V 61
Levasseur-Regourd A C 123
Levi S 231, 232, 236
Levien L 279
Lewis R S
 see Swart P K 49
 see Tang M 71
Li Y H
 see Gouhara K 278
Libby W F 61
Liebenberg D H 268
 see Mills R L 268, 301

Lighthill M J 205, 206, 211
Lin D N C 30
Linthorne N P
 see Moore G I 16, 17, 24
Lippincott E R
 see Brasch J W 268
 see Weir C E 251
Litherland A E
 see Gove H E 61
Locke M J
 see Francl T J 74
 see McIver R T Jr 74
Lofquist H 155
Loosli H H
 see Lehmann B E 73
Loubeyre P 298
 see Hebert P 300
Lowe D R 190
 see Walsh M M 192
Lu R 232
Lund S P 230, 236, 237
Lupton J E 156, 157, 161
 see Craig H 148, 155, 157, 167, 168
Luther G G 6
Lynch M A
 see Kennedy B M 72

MacDonald G J F
 see Munk W H 139
MacDougall J D
 see Rubin K 64
MacFarlane J J
 see Hubbard W B 257
Mack R
 see Chang S 195
Mackinnon I D R 88
Magaritz M 168
Maines R G Sr
 see Jayaraman A 273, 274
Mamantov G
 see Gilbert B 273
Mammone J F 281–286
Mamyrin B A 72
Mandelbrot B B 142
Manuel O K
 see Downing R G 154, 155, 161, 164–167
Mao H F
 see Hemley R J 258
Mao H K 251, 255, 264, 265, 268, 271, 295–297, 298
 see Bell P M 258, 265, 298
 see Benjamin T M 256
 see Goettel K A 265
 see Hazen R M 298, 300
 see Hemley R J 265, 273, 290, 295, 297–300, 302
 see Sharma S K 269, 271, 290–293, 296, 300
 see Xu J A 265
Maradudin A A 280
Marckmann J P 299

Markin E P
 see Belenov E M 48
 see Basov N G 50
Markov M A 70
Marley M S
 see Hubbard W B 257
Maroni V A
 see Postmus C 268
Marshall A G 72, 74
Martens K D
 see Horneck G 196
Marti K
 see Craig H 155
 see Poreda R J 153
Marti W 155, 162, 168
Martin J M
 see Krishnaswami S 129
Martin K M
 see Adams D M 268
Martin S
 see Moss W C 253, 254, 258, 265
 see Reichlin R 258, 298
Marvin J
 see Jackson M 231
 see King J 231, 233, 234, 236
Mathur D 77
Matsuhisa Y 36, 42
Mattey D P
 see Ozima M 72
Mauer F A
 see Walrafen G E 299
Mauersberger K 39
 see Morton J 45
Maurette M 87, 90, 104, 110, 111
 see Bibring J P 96
 see Bonny Ph 111, 113, 114
 see Bonte Ph 96, 102
 see Borg J 96
 see Callot G 91, 93, 116
 see De Angelis M 91, 93
 see Raisbeck G 96
Mavin J
 see Lu R 232
Mayeda T K
 see Clayton R N 36, 42, 46
 see Molini-Velsko C A 46
 see Onuma N 44
Mcconville P 71
McCreary J P 205
McCrory M
 see Drake R E 72
McCrumb J L
 see Cicerone R J 54
McElhinny M W 233–236
 see Constable C G 236
McElroy M B 36
McHugh M P
 see Niebauer T M 8, 24
McIver R T Jr 72, 74
 see Francl T J 74
 see Sherman M G 75
McLagan K H
 see Kayser P T 20
McMahan A K 256, 257
 see Ross M 257
McSween H Y 105, 106, 109

Author index

Mckeegan K O
  see Fahey A 38
Measures C I 201, 203
  see Kusakabe M 201–203
  see Peng T-H 201
Mehringer P J
  see Verosub K L 236
Melveger A J
  see Brasch J W 268
Mendis D A 29, 30, 31
  see Houpis H L F 31
Menningmann H D
  see Horneck G 196
Meybeck M
  see Krishnaswami S 129
Miakisev G J 70
Michel K W
  see Haberkorn R 52
Michel-Beyerle M E
  see Haberkorn R 52
Michelson A A 67, 68
Middleton R
  see Nishiizumi K 65
  see Raisbeck G M 96
Miller S L 195
Miller S R
  see Chang S 195
Millman P M 89
Mills R L 268, 301
  see Buchsbaum S 298
  see Reichlin R 258, 298
  see Schiferl D 298
Milne V A
  see Dunlop J S R 190
Mio N
  see Kuroda K 8
Miono S 129
Mishima O 299
  see Klug D D 299
Misner C W 68
Mitchell A C
  see Nellis W J 257
Mitchell D 76
  see DeLong S E 76
  see Spell T L 76
Mitchell E W J 273
Mitra S S
  see Postmus C 268
Moffat J W 9
Molin Y N
  see Podoplelov A V 52
Molini-Velsko C A
  see Clayton R N 42, 46
Monberg E
  see Kessler J R 286
Moody M V
  see Chan H A 14
Moore D S
  see Schmidt S C 298
Moore G I 16, 17, 24
  see Stacey F D 16
Moore J G
  see Des Marais D J 151
Moore L S
  see Burne R V 190

Morfill G E
  see Wood J A 29
Moritz H 67
Morley E W 67, 68
Morosin B
  see Peercy P S 279
Morris M 56
Morrison G H
  see Botto R I 154, 162
Morse P M 208
Morton J 45
  see Barnes J 45
Moss W C 253, 254, 258, 265
Mosteller F
  see Hoaglin D C 106
Muir M D
  see Dunlop J S R 190
Mulac A J 272
Mulholland J D
  see Williams J G 13
Munro R
  see Walrafen G E 299
Murphy T
  see Johnson E A 232
Murthy M K 279
Musser J
  see Cowsik R 9, 14

Nace G A
  see Clanton U S 87
Nachtrieb N D
  see Jamieson J C 251
Nakamura T 269
Nanni T
  see Attolini M R 131
Narasimhan T N 213
Narayanaswamy P K 273, 277
Navon O 44, 54
Nedungadi T M K 277
  see Raman C V 273
Neih J C
  see Valentini J J 39
Nellis W
  see Ross M 257
Nellis W J 257
  see Holmes N C 300
Nelson D E
  see Kusakabe M 201–203
Newherter J L
  see Hoffman D C 63
Nicol M 285
  see Asell J F 273
  see Cynn H C 301
  see Johnson M L 299, 301
  see Kessler J R 286
  see Mammone J F 281–286
  see Nellis W J 257
  see Young D A 302
Nicol M F
  see Zinn A S 298
Niebauer T M 8, 24
Niell D R
  see Takahashi P K 269, 270
Nier A O
  see McElroy M B 36

Nieto M M
  see Goldman T 9, 10
  see Hughes R 14
Nishiizumi K 65
Nordvett K 13
Northrop T G 30
Nozaki Y
  see Kusakabe M 201–203

Oeschger H
  see Beer J 128
  see Lehmann B E 73
Ohta M
  see Miono S 129
Okabe H 51
Olsen E
  see Clayton R N 42
Olson P 236
  see Lund S P 230
Onuma N 44
  see Clayton R N 42
Oparin A I 195
Opdyke N D
  see Kent D V 235, 236
Oraevskii A N
  see Basov N G 50
  see Belenov E M 48
Ott U 72
Ozima M 72

Packer B M
  see Schopf J W 192
Paik H J
  see Chan H A 14
Palmer M A
  see Fitch V L 19, 23
Pan'kov V L
  see Dubrovskij B A 180
Paneth F 160
Panov V I 24
  see Braginsky V B 20
Parker E A 301
Passoja D E
  see Maurette M 104
Pauli W 69
Payne M G
  see Hurst G S 72
  see Lehmann B E 73
  see Thonnard N 72
Pederson A K 155
Peercy P S 279
  see Samara G A 285
Pekar D
  see Eötvös B R 5
Peng T-H 201
  see Broecker W S 202
Penzias A A 56
Perego A
  see Bizzeti P G 19
Perry E C 183
Phillips D L
  see Valentini D P 39
Phillips R C
  see Lehmann B E 73

Piermarini G J 251, 268
　see Barnett J D 268
　see Forman R A 251
　see Walrafen G E 299
　see Weinstein B A 269
Pillinger C T 47
　see Swart P K 49
　see Tang M 71
　see Ozima M 72
Pinceaux J P
　see Besson J M 268, 289
　see Loubeyre P 298
Pinnick D A
　see Young D A 302
Pistorius C W F T 299
Platzman G W 142
Plotkin H H
　see Williams J G 13
Plumlee R H 79
Podoplelov A V 52
Podosek F A
　see Bernatowicz T J 162, 164–167
Polian A 258
　see Hebert P 300
Poreda R J 153, 157, 161
　see Craig H 155, 161–163
　see Lal D 163
Porto S P S
　see Scott J F 274
Postma S 301
Postmus C 268
Pottier L
　see Callot G 91, 93, 116
Poultrey S K
　see Williams J G 13
Pound R V 6, 7
Pourchet M
　see De Angelis M 91, 93
　see Maurette M 87
Press F 144
Prewitt C T
　see Levien L 279
Price D L
　see Rao K R 289
Prospero J M 204
Provenzale A
　see Cini Castagnoli G 128
Pruzan Ph
　see Gauthier M 301
　see Mills R L 301

Quist A S
　see Bates J B 273, 274, 277, 280

Raab F J
　see Stubbs C W 18, 23, 24
Radousky H B
　see Nellis W J 257
Raisbeck G M 96, 97, 109, 111
Rama 173
Ramaker D E 290
Raman C V 273
Ramdas A K
　see Tekippe V J 273

Ramussen K 90, 91
Rao K R 289
Rapp R H 6, 14, 16
Raptis C
　see Mitchell E W J 273
Rebka G A
　see Pound R V 6, 7
Ree F H
　see Nellis W J 257
Ree R H
　see Ross M 297
Reeh N 91, 110
Reeh R
　see Maurette M 90, 91
Rees C E 47
　see Thode H G 47
Regnier S
　see Craig H 155
Rehm R G
　see Treanor C E 48
Reichlin R 258, 298
　see Akella J 255, 257
　see Moss W C 253, 254, 258, 265
　see Smith G S 257
Reid D M
　see Moore G I 16, 17, 24
Reitz G
　see Horneck G 196
Rempel D L
　see Gross M L 72, 74
Requardt H
　see Horneck G 196
Reynolds J H 79
　see Kennedy B M 72
　see McConville P 71
Rich J W 49
　see Treanor C E 48
Richez C
　see Crepon M 206
Rietmeijer F J M
　see Mackinnon I D R 88
Rigden J D 302
Ringwood A E 100, 242
Rison W 72, 161
　see Craig H 161
Roberts P H 141, 142
Robin E 97, 101, 110, 113, 118
　see Bonte Ph 96, 102
　see Maurette M 110, 111
Rodriguez S
　see Tekippe V J 273
Rogers W F
　see Stubbs C W 18, 23, 24
Roll P G 20
Rollet A P 301
Romaides A J
　see Eckhardt D H 16, 24
Romanenko V I
　see Belenov E M 48
　see Basov N G 50
Ronov A B 182, 184
Roof R B 255, 256, 257
Rosengren A
　see Johansson B 255
Ross M 257, 290, 297

Rourke F M
　see Hoffman D C 63
Roy R
　see Shafer E C 279
Rózsa M 19
Rubin K 64
Rujula de A 14, 22
Runcorn K 69
Ruoff A L
　see Vohra Y K 254

Safronov V S 30, 179
Sagdeev R Z
　see Podoplelov A V 52
Salam Adbus 69
Salje E 289
Samara G A 285
Samuelson R E
　see Hunten D M 299
Sands R W
　see Eckhardt D H 16, 24
Saraswat P
　see Cowsik R 5
Sarda P 148
　see Allègre C J 148, 149, 150
Sarver J F 275
Sato T 235, 236
Schaeffer O
　see Davis R Jr 62
Schidlowski M 189, 192, 193, 194
Schiferl D 298
　see Buchsbaum S 298
　see Reichlin R 258, 298
　see Young D A 302
　see Zinn A S 298
Schild A 68
Schliestedt M
　see Wilbrans J R 72
Schmidt L C
　see Mills R L 268
Schmidt S C 298
Schmidt V A
　see Banerjee S K 231
Schmitt H W
　see Thonnard N 72
Schneider K 73
Schneider T 290
Schock R N
　see Akella J 255, 257
　see Smith G S 255, 257
Schopf J W 189, 192
　see Awramik S M 192
Schott F 208
Schouten J A 298
Schramm L S 105
Schueler B
　see Morton J 45
Schwab M
　see Akella J 255, 257
　see Smith G S 255, 257
Schwager M
　see Bucker H 196
Schwake A
　see Johnson M L 301
Scott J F 274, 275

Author index

Sekanina Z
   see Southworth R B 113
Sélenyl
   see Rozsa M 19
Senanayake W E
   see McElhinny M W 233–236
Shafer E C 279
Shaner J
   see Mao H K 268
Shapiro I I 13
Sharma P 230
Sharma S K 263, 267, 269, 271, 273, 286, 289–293, 296, 300
   see Adams D M 267–269, 289, 299
   see Mammone J F 281–286
Shaw M S
   see Schmidt S C 298
Shelus P J
   see Williams J G 13
Shepherd T J
   see Kelley S 72
Sheppard S M 167
Sherman M G 75
   see Francl T J 74
Sherman W F 251
Shimomura O 251
   see Yagi T 251
   see Endo S 251
Shishkevish C
   see Ross M 257, 290
Shriver D F
   see Van Duyne R P 272
Sieh K E 141
Sikka S K
   see Vohra Y K 256
Silvera I F 268, 295
   see Wijngaarden R J 295
Silverberg E C
   see Williams J G 13
Slade M A
   see Williams J G 13
Slattery W L 257
Smith D E
   see Cohen S C 14
Smith G S 254–257
   see Akella J 255–257
Smith J L 155
Smith S P
   see Kennedy B M 72
Somayajulu B L K 65
   see Krishnaswami S 65
   see Sharma P 230
Sourthon J
   see Peng T-H 201
Southon J R
   see Kusakabe M 201–203
Southworth R B 113
Spain I L 251
Spedding F H 255
Spell T L
   see DeLong S E 76
Spirlet J C
   see Benedict U 255, 256
Stacey F D 6, 15, 16, 68

Stadtmuller A A
   see Sherman W F 251
Staudacher T
   see Allègre C J 148, 149, 150
Staudacher Th 72
Steinberg D
   see Mao H K 268
Stettler A 166
   see Bochsler P 154, 155, 160–162, 164–166
Stevenson D J 290, 298
   see Hunten D M 299
Stevenson R 300
Stoicheff B P 296
Stolper E M
   see Rigden S M 302
Stonecipher S A
   see Krishnaswami S 65
Strangway D W
   see Sugiura N 32
Strathearn G
   see Chang S 195
Strobel D F
   see Hunten D M 299
   see Kaye J A 44, 54
Stubbs C W 18, 23, 24
Stuiver M 133, 135
Sudarsky D
   see Fischbach E 5, 6, 8, 16
Suguira N 32
Sumiyama K
   see Endo S 251
Sunclair W S
   see Williams J G 13
Suzuki T
   see Yagi T 251
Swanson H E
   see Stubbs C W 18, 23, 24
Swart P K 49
Swisher C S
   see Drake R E 72
Syassen K
   see Hawke R S 299
Syfosse G
   see Gauthier M 301
Szafer A
   see Fischbach E 5, 6, 8, 16
Szumilo A
   see Boynton P E 19, 20, 23, 24

Tabor K D
   see Valentini J J 39
Taccetti N
   see Bizzeti P G 19
Takahashi E 240
Takahashi P K 269, 270
Takashige M
   see Nakamura T 269
Talmadge C
   see Fischbach E 5, 6, 8, 16
Tandon S N
   see Cowsik R 5, 9, 14, 23
Tang M 71

Tang T Y
   see Lawson A W 251
Tanguy L
   see Gauthier M 301
Tanner A B 173
Tarantola A 144
Tarle G
   see Cowsik R 9, 14
Tauxe L 234
   see Wu G 235
Taylor G R
   see Bucker H 196
Taylor H P
   see Magaritz M 168
Taylor K N R 255
Taylor M J 299
Tekippe V J 273
Thayer W
   see Akella J 255, 257
Thellier E 233
Thellier O
   see Thellier E 233
Thieberger P 19, 21, 23, 24
Thiemens M H 35–38, 41–49, 52, 55, 57
   see Bains-Sahota S K 44, 46, 47
   see Bhattacharya S K 46, 50, 53
   see Heidenreich J E 38, 39, 44, 51
Thiery M M 300
   see Fabre D 300
Thode H G 47
   see Hulston J R 36, 47
   see Rees C E 47
Thomsen H H
   see Maurette M 90, 91
Thomson W E
   see Baxter D 30
Thonnard N 72
Thonnard N
   see Lehmann B E 73
Thorne K S
   see Misner C W 68
Thouveny N 232, 235, 236
Tolstikhin L N
   see Mamyrin B A 72
Tomasko M G
   see Hunten D M 299
Tominaga Y
   see Nakamura T 269
Tomli Y
   see Endo S 251
Torgersen T 150
Torrenson O W
   see Johnson E A 232
Towler W R
   see Luther G G 6
Toyama N
   see Endo S 251
Trappeniers N J
   see Schouten J A 298
Treanor C E 48, 49
Treder H J 68
Troe J 53
Trower W P
   see Carrigan R A 69

Author index

Tsuji K
  see Shimomura O 251
Tuck G J
  see Stacey F D 6, 15, 16
  see Moore G I 16, 17, 24
Tucker P 232
Tukey J W
  see Hoaglin D C 106
Turekian K K 147, 148, 151
Turlay R
  see Christenson J H 4
Turner G 72, 73, 149
  see Kelley S 72
Turner S 69
Turrin B
  see Drake R E 72
Turro N J 51, 52

Udagawa M
  see Nakamura T 269
Uhrhammer R
  see Bolt B A 145
Unnikrishnan C S
  see Cowsik R 5, 9, 14, 23
Urey H C 35, 41, 42
Urmos J P
  see Sharma S K 286

Valentini J J 39, 45, 53
Valet J-P
  see Tauxe L 234
Van Duyne R P 272
Van Valenburg A 251
  see Weir C E 251
Van den Bergh L C
  see Schouten J A 298
Vanderborg C
  see Reichlin R 258, 298
Veiser J 178
Velde B
  see Besson J M 289
Verosub K L 236
Vigner E 290
Vijayakumar V
  see Vohra Y K 256
Viswanathan K
  see Salje E 289
Vitting J
  see Mao H K 255
Vogel C R 145
Vogel J S
  see Kusakabe M 201–203
Vohra Y K 254, 256
Von Eötvös B R 2–6, 8, 13, 16–20, 22
Vu H
  see Fabre D 300
Vuillard G
  see Rollet A P 301

Wacker J F
  see Lal D 163
Wahba G 145
Wakatsuki M
  see Shimomura O 251
Waldemeier M 132

Walrafen G E 299
Walsh M M 192
Walter M R 190
  see Awramik S M 192
Walton D 233
Ward W R
  see Goldreich P 32
Warren J L
  see Clanton U S 87
Wasserburg G J
  see DePaolo D J 242
  see Jacobson S B 242
  see Navon O 44, 54
Watanabe R
  see Stubbs C W 18, 23, 24
Waterstraat P
  see Verosub K L 236
Watson E B 247
Watson W D
  see Chu Y-H 56
Weathers M S
  see Bird J M 154, 155, 163, 169
  see Bochsler P 155, 160–162, 164
Weber P
  see Greenberg J M 197, 198
Wedeking K W
  see Hayes J M 193
Wedepohl K H 151
Weed H
  see Akella J 256
Weidner D J
  see Levien L 279
Weinberg S 4, 69
Weinstein B A 269
Weir C E 251
Weir S T
  see Vohra Y K 254
Welsch H L
  see Bhatnagar S S 290
Wernecke Ch
  see Salje E 289
Wetherill G W 162
Whaley E
  see Marckmann J P 299
  see Mishima O 299
  see Klug D D 299
  see Taylor M J 299
  see Wong P T T 299
Wheeler J A
  see Misner C W 68
Wheelock M M
  see Schramm L S 105
Whipple F 87, 91, 113
White W M
  see Hofmann A W 242
Whitehead A N 68
Wijngaarden R J 295
  see Silvera I F 268, 295
Wilbrans J R 72
Wilcockson J
  see Ashwood-Smith M J 197
Wilkinson D T
  see Williams J G 13
Will C M 9, 13, 68
  see Paik H J 14

Williams J G 13
Williamson D
  see Thouveny N 235, 236
Willis R D
  see Lehmann B E 73
Winter M B
  see Aitken M J 233
Wollenhaupt H
  see Bucker H 196
Wong P T T 267, 299
Wood D L
  see Jayaraman A 273, 274
Wood J A 29
Worden P W Jr 14
Wright I P
  see Tang M 71
Wright M C
  see Thonnard N 72
Wu G 235
Wyllie P J 239, 241

Xu J
  see Akella J 255
Xu J A 265
  see Sharma S K 271

Yagi T 251
  see Shimomura O 251
Yamakoshi Y
  see Raisbeck G 96
Yamanaka T 289
Yamoka S
  see Shimomura O 251
Yanagita S 36
Yang C N 3
Yang J 45
  see Ott U 72
Yen J
  see Young D A 302
Yiou F
  see Raisbeck G M 96, 97, 111
York D
  see Wilbrans J R 72
Young D A 302
  see Ross M 297
Yung Y L
  see McElroy M B 36

Zashu S
  see Ozima M 72
Zha C S
  see Hemley R J 258
  see Mao H K 298
  see Young D A 302
Zhan R
  see Stacey F D 6, 15, 16
Zindler A 150
Zinn A S 298
  see Young D A 302
Zinner E
  see Fahey A 38
Zook H A
  see Grün E 90, 97, 111, 112, 113
Zou G
  see Benjamin T M 256

Selected list of publications of Prof. D Lal on his work described in the text:

1. Production and interaction of mesons at very high energies, *Phys. Rev.*, **87**, 545, 1952 (with Y Pal, B Peters and M S Swami)
2. Cosmic ray produced beryllium isotopes in rain water, *Nucl. Phys.*, **1**, 196–201, 1956 (with P S Goel, S Jha, P Radhakrishna and Rama)
3. Cosmic ray produced Sr^{32} in nature, *Phys. Rev. Lett.*, **3**, 380, 1959 (with E D Goldberg and M Koide)
4. Radioisotopes P^{32}, Be^7 and S^{35} in the atmosphere, *J. Geophys. Res.*, **65**, 669–674, 1960 (with Rama and P K Zutshi)
5. Cosmic ray produced silicon-32 in nature, *Science*, **131**, 332–337, 1960 (with E D Goldberg and M Koide)
6. Cosmic ray production rates of Be^7 in oxygen and P^{32}, P^{33}, S^{35} in Argon at mountain altitudes, *Phys. Rev.*, **118**, 1626–1632, 1960 (with J R Arnold and M Honda)
7. The production rate of natural tritium, *Tellus*, XIII, 85–105, 1961 (with H Craig)
8. Record of cosmic ray intensity in the meteorites, *J. Geophys. Res.*, **66**, 3519–3531, 1961. In '*Researches on meteorites*' (ed.) C B Moore, John Wiley and Sons Inc., pp. 68–92, 1962 (with J R Arnold and M Honda)
9. On the investigations of geophysical processes using cosmic ray produced radioactivity, in *Earth Science and Meteoritics*, compiled by J Geiss and E D Goldberg, North-Holland Publishing Company, Amsterdam, pp. 115–142, 1963.
10. A method for the extraction of trace elements from sea water, *Geochim. Cosmochim. Acta*, **28**, 1111–1117, 1964 (with J R Arnold and B L K Somayajulu)
11. The measurement of radiocarbon activity and some determinations of ages of archaeological samples, *Curr. Sci.*, **34**, No. 13, 394–397, 1965 (with D P Agrawal and S Kusumgar)
12. Using natural silicon-32, *Nature* (*London*), **210**, 478–480, 1966 (with V N Nijampurkar, B S Amin and D P Kharkar)
13. Low-energy protons: average flux in interplanetary space during the last 100,000 years, *Science*, **151**, 1381–1384, 1966 (with V S Venkatavaradan)
14. Characteristics of global tropospheric mixing based on man-made C^{14}, H^3 and Sr^{90}, *J. Geophys. Res.*, **71**, 2865–2874, 1966 (with Rama)
15. Cosmic ray produced Mg^{28}, Si^{31}, S^{38}, Cl^{34m} and other short-lived radioisotopes in wet precipitation, *Tellus*, XVIII, 504–515, 1966 (with N Bhandari, S G Bhat, D P Kharkar, S Krishna Swami and A S Tamhane)
16. Cosmic ray produced radioactivity on the earth, *Handb. der Physik*, Springer-Verlag, Berlin, 46/2, 551–612, 1967 (with B Peters)
17. Activation of cosmic dust by cosmic ray particles, *Earth Planet. Sci. Lett.*, **3**, 299–310, 1967 (with V S Venkatavaradan)
18. Techniques for proper revelation and viewing of etch-tracks in meteoritic and terrestrial minerals, *Earth Planet. Sci. Lett.*, **5**, 111–119, 1968 (with A V Murali, R S Rajan, A S Tamhane, J C Lorin and P Pellas)
19. The radioactivity of the atmosphere and hydrosphere, *Ann. Rev. Nucl. Sci.*, **108**, 407–434, 1968 (with H E Suess)
20. On the energy spectrum of iron-group nuclei as deduced from fossil-track studies in meteoritic minerals, in *Meteorite Research* (eds) P M Millman, D Reidel Publishing Company, Dordrecht-Holland, 275–285, 1969 (with J C Loris, P Pellas, R S Rajan and A S Tamhane)
21. Chemical composition of nuclei of $Z > 22$ in cosmic rays using meteoritic minerals as detectors, *Nature* (*London*), **221**, 33–37, 1969 (with R S Rajan and A S Tamhane)
22. Recent advances in the study of fossil tracks in meteorites due to heavy nuclei of the cosmic radiation, *Space Sci. Rev.*, **9**, 623–650, 1969
23. Observations on space irradiation of individual crystals of gas rich meteorites, *Nature* (*London*), **223**, 269–271, 1969 (with R S Rajan)

Selected list of publications of D Lal

24. Collecting a sample of solar wind: an experimental study of its capture in metal films, *J. Appl. Phys.*, **40**, No. 8, 3257–3267, 1969 (with W F Libby, G Wetherill, J Leventhal and G D Alton)
25. Vertical structure of the troposphere as revealed by radioactive tracer studies, *J. Geophys. Res.*, **75**, 2974–2980, 1970 (with N Bhandari and Rama)
26. Man-made carbon-14 deep Pacific waters: transport by biological skeletal material, *Science*, **166**, 1397–1399, 1969 (with B L K Somayajulu and S Kusumgar)
27. Pattern of bombardment-produced radionuclides in Rock 10017 and in lunar soil, Proc. Apollo 11 Lunar Sci. Conf., *Geochim. Cosmochim. Acta*, **2**, 1503–1532, 1970 (with J P Shedlovsky, M Honda, R C Reedy, J C Evans, Jr., R M Lindstrom, A C Delany, J R Arnold, H H Loosli, J S Fruchter and R C Finkel)
28. Mixing of lunar regolith and cosmic ray spectra: new evidence from fossil particle-track studies, Proc. Apollo 11 Lunar Sci. Conf., *Geochim. Cosmochim. Acta*, **3**, 2295–2303, 1970 (with D Macdougall, L Wilkening and G Arrhenius)
29. Geochronology of lake sediments, *Earth Planet. Sci. Lett.*, **11**, 407–414, 1971 (with S Krishnaswami, J M Martin and M Meybeck)
30. Olivines: revelation of tracks of charged particles, *Science*, **174**, 287–291, 1971 (with S Krishnaswami, N Prabhu and A S Tamhane)
31. Fossil tracks in the meteorite, Angra dos Reis: a predominantly fission origin, *Nature (London)*, **234**, 540–543, 1971 (with N Bhandari, S Bhat, G Rajagopalan, A S Tamhane and V S Venkatavaradan)
32. Accretion processes leading to formation of meteorite parent bodies, Proc. of the Nobel Symposium 21, "From Plasma to Planet", Saltsjobaden, Sweden, September 6–10, 1971 (ed.) A Elvius, Almqvist and Wiksell, Stockholm (Wiley Interscience Division), 49–64, 1972
33. Hard rock cosmic ray archaeology, *Space Sci. Rev.*, **14**, 3–102, 1972
34. Silicon, radium, thorium and lead in sea water: *in situ* extraction by synthetic fiber, *Earth Planet. Sci. Lett.*, **16**, 84–90, 1972 (with S Krishnaswami, B L K Somayajulu, F S Dixon, S A Stonecipher and H Craig)
35. Dissolution and behaviour of particulate biogenic matter in the ocean: some theoretical considerations, *J. Geophys. Res.*, **78**, 7100–7111, 1973 (with A Lerman)
36. Cosmic ray effects induced in a rock exposed on the moon or in free space: contrast in patterns for "tracks" and "isotopes", *Moon*, **8**, 253–286, 1973 (with S K Bhattacharya, J N Goswami and S K Gupta)
37. Characteristics of tracks of ions of $14 < Z < 36$ in common rock silicates, *Earth Planet. Sci. Lett.*, **19**, 377–395, 1973 (with P B Price, A S Tamhane and V P Perelygin)
38. Silicon-32 profiles in the South Pacific, *Earth Planet. Sci. Lett.*, **18**, 181–188, 1973 (with B L K Somayajulu and H Craig)
39. Chronology of marine sediments using the ^{10}Be method: intercomparison with other methods, *Geochim. Cosmochim. Acta*, **39**, 1187–1192, 1975 (with B S Amin and B L K Somayajulu)
40. Size spectra of biogenic particles in ocean water and sediments, *J. Geophys. Res.*, **80**, 423–430, 1975 (with A Lerman)
41. Silicon-32 specific activities in coastal waters of the world oceans, *Limnol. Oceanogr.*, **21**, 285–293, 1976 (with V N Nijampurkar, B L K Somayajulu, M Koide and E D Goldberg)
42. Large volume *in situ* filtration of deep Pacific waters: mineralogical and radioisotope studies, *Earth Planet. Sci. Lett.*, **32**, 420–429, 1976 (with S Krishnaswami, B L K Somayajulu, R F Weiss and H Craig)
43. Irradiation and accretion of solids in space based on observations of lunar rocks and grains, *Phil. Trans. R. Soc. London A.*, **285**, 69–95, 1977
44. Regeneration rates in the ocean, *Am. J. Sci.*, **277**, 238–258, 1977 (with A Lerman)
45. The oceanic microcosm of particles, *Science*, **198**, 997–1009, 1977
46. Observations on the spatial distribution of Dhajala meteorite fragments in the strewnfield, *Proc. Indian Acad. Sci.*, **86A**, 393–407, 1977 (with J R Trivedi)
47. On estimation of mass ablation of meteorites based on studies of cosmic ray tracks, in *Nuclear Track Detection*, **2**, 37–49, 1978 (with S K Gupta)
48. Depositional history of Luna 24 drill core soil samples, *Earth Planet. Sci. Lett.*, **44**, 325–334, 1979 (with J N Goswami, M N Rao and T R Venkatesan)
49. Formation of the parent bodies of the carbonaceous chondrites, *Icarus*, **40**, 510–521, 1979 (with J N Goswami)
50. Comments on some aspects of particulate transport in the oceans, *Earth Planet. Sci. Lett.*, **49**, 520–527, 1980

Selected list of publications of D Lal

51. Solar modulation effects in terrestrial production of carbon 14, Proc. of the 10th International Radiocarbon Conf., *Radiocarbon*, **22**, 133–158, 1980 (with G Castagnoli)
52. An approach to determining pathways and residence time of groundwaters: Dual radiotracer dating, *J. Geophys. Res.*, **86(C6)**, 5299–5300, 1981 (with S K Gupta and P Sharma)
53. Some comments on exchange on CO_2 across the air-sea interface, *J. Geophys. Res.*, **88**, 3643–3646, 1983 (with H Suess)
54. Cosmic ray record in solar system matter, *Ann. Rev. Nucl. Sci.*, **33**, 505–537, 1983 (with R C Reddy and J R Arnold)
55. Gas-rich meteorites: Probes for particle environment and dynamical processes in the solar system, *Space Sci. Rev.*, **37**, 111–159, 1984 (with J N Goswami and L L Wilkening)
56. Atmospheric P_{CO_2} changes recorded in lake sediments, *Nature (London)*, **308**, 344–346, 1984 (with R Revelle)
57. Production of ^{10}Be and ^{26}Al by cosmic rays in terrestrial quartz *in situ* and implications for erosion rates, *Nature (London)*, **319**, 134–136, 1986 (with K Nishiizumi, J Klein, R Middleton and J R Arnold)
58. Ionization states of cosmic rays: ANURADHA (IONS) Experiment in Spacelab-3, *Pramana–J. Phys.*, **27**, 89–104, 1986 (with S Biswas, R Chakraborti, J N Goswami, H S Mazumdar and M K Padmanabhan)
59. GEOSECS ATLANTIC ^{32}Si profiles, *Earth Planet. Sci. Lett.*, **85**, 329–342, 1987 (with B L K Somayajulu, R Rengarajan, R F Weiss and H Craig)
60. Cosmogenic ^{10}Be in Zaire alluvial diamonds: implications for ^3He contents of diamonds, *Nature (London)*, **328**, 139–141, 1987 (with K Nishiizumi, J Klein, R Middleton and H Craig)
61. *In situ* produced cosmogenic isotopes in terrestrial rocks, *Ann. Rev. Earth Planet. Sci. Lett.*, **16**, 355–388, 1988
62. Cosmogenic ^{32}P and ^{33}P used as tracers to study phosphorus recycling in the upper ocean, *Nature (London)*, **333**, 752–754, 1988 (with T Lee)

Figure 1. Pandit Nehru visiting the Tata Institute of Fundamental Research. Dr H. J. Bhabha faces away from the camera.

Figure 2. Dr Lal in his Radiocarbon laboratory in the Tata Institute of Fundamental Research.

Figure 3. Dr Lal with Prof. C. V. Raman.

Figure 4. Dr Lal receiving the Apollo moon samples from the American Ambassador.

Figure 5. Dr Lal with his young co-workers at the Tata Institute of Fundamental Research.

Figure 6. Mrs. Indira Gandhi inaugurating the Silver Jubilee Celebrations at the Physical Research Laboratory.

Figure 7. (left to right) Profs. Yash Pal, Lal, Peters and Sreekantan in Lal's office at the PRL.

Figure 8. Prof. and Mrs. Lal with Prof. V. L. Ginzburg during the latter's visit to the PRL.